The
POWER
OF LIFE

The

POWER
OF LIFE

*The Invention of Biology
and the Revolutionary Science
of Jean-Baptiste Lamarck*

JESSICA RISKIN

Riverhead Books • New York
2026

RIVERHEAD BOOKS
An imprint of Penguin Random House LLC
1745 Broadway, New York, NY 10019
penguinrandomhouse.com

Image credits may be found on pages 391–394.

Book design by Christina Nguyen

LIBRARY OF CONGRESS CATALOGING-IN-PUBLICATION DATA
Names: Riskin, Jessica author
Title: The Power of Life : The Invention of Biology and the Revolutionary
Science of Jean-Baptiste Lamarck / Jessica Riskin.
Description: New York : Riverhead Books, 2026. | Includes bibliographical references and index.
Identifiers: LCCN 2025049918 (print) | LCCN 2025049919 (ebook) |
ISBN 9780593852576 (hardcover) | ISBN 9780593852583 (ebook)
Subjects: LCSH: Lamarck, Jean Baptiste Pierre Antoine de Monet de, 1744–1829 |
Naturalists—France—Biography | Evolution (Biology)—History—19th century |
Evolution (Biology)—History—18th century | LCGFT: Biographies
Classification: LCC QH31.L2 R57 2026 (print) | LCC QH31.L2 (ebook)
LC record available at https://lccn.loc.gov/2025049918
LC ebook record available at https://lccn.loc.gov/2025049919

Printed in the United States of America
1st Printing

The authorized representative in the EU for product safety and compliance is
Penguin Random House Ireland, Morrison Chambers, 32 Nassau Street,
Dublin D02 YH68, Ireland, https://eu-contact.penguin.ie.

FOR C.L.K.

with love and gratitude
as abundant
as all the insects and all the worms

Portrait of Lamarck by Charles Thévenin, ca. 1802

There . . . exists in nature a peculiar, powerful, and ever active cause, which has the faculty of forming combinations, of increasing and varying them. . . . Now this powerful cause . . . resides in the organic activity of living bodies. . . .

[Living] organization, whether animal or vegetable, has, by means of the power of life, gradually composed and complicated itself, from the greatest simplicity to the greatest complexity.

• J.-B. Lamarck, *Zoological Philosophy*, 1809

CONTENTS

J ean-Baptiste Lamarck created his radical science of biology in the midst of political revolution, and he continued to develop it through reactionary reversals and new revolutions, under revolutionary committees and representative assemblies, then under dictators and more monarchs. Throughout it all, he defended not only a particular set of scientific ideas but a model of what science should be, against the most powerful members of the scientific and political establishment who promoted a starkly different notion of science and of its relation to authority. He persisted through a lengthy illness and the deaths of many of those closest to him. Today Lamarck's ideas have long outlived him, though largely in exile from mainstream science. His central idea was that to live is to act, transforming oneself and the world around. Creative agency was not consolidated in any divine or earthly power but distributed throughout living nature. It followed that science was not the description or manipulation of the world's design but the study of nature's continual self-making and remaking.

I began writing Lamarck's story and the story of his science during a global pandemic, and I am finishing in the midst of a war on science, scholarship, education, the environment, human rights, and representative self-government. Throughout this succession of dark moments, I have found a source of optimism in Lamarck's model of science and the principles at its core: that life at its essence is creative agency; that living beings, especially the smallest and humblest of them, continually remake the world; that knowledge of nature comes not from standing apart and imposing an imperial order upon it but from a sensitive and participatory engagement from within, using every kind of understanding we can muster—humanistic, scientific, aesthetic, all of it. The implication I propose to draw from Lamarck's mode of science is that neither natural knowledge nor transformative power can ultimately be controlled or monopolized.

Jessica Riskin, Norwich, Vermont, September 5, 2025

The Garden Where
It All Began

(NO, NOT THAT GARDEN)

In which the Project of the present Book is laid out; we set the Scene for the Story to come by taking a Walk in a Garden and witnessing a Creative Commotion of Botanists, Musicians, Novelists, Librarians, Painters, and Zoologists in making certain important Discoveries, namely, that Plants have Sex, that Elephants have Passions and Musical Tastes, and that Giraffes have Moral Habits. These various Beings, Discoveries, and Approaches to Nature provide the Setting in which our Hero will Invent the Science of Life. His Radical Ideas will be Cast Aside for almost Two Centuries, but they will then Return to transform the Field he founded in areas such as Epigenetic Inheritance.

Among the living beings to whom my daughter Madeleine formed ardent attachments as a little girl is a *Prunus serrulata*, or Japanese cherry tree—Sato-zakura "Shirotae" (snow white)—living in the Garden of Plants in Paris. At about sixty years old, it is by far the youngest of the five officially designated "Remarkable Trees" in the garden. The tree sprawls out laterally, forming a giant, lumpy tent over the ground, which it grazes with the tips of its outermost branches. In March and April, garnishing itself all over in snowy blossoms, it draws hordes of enchanted iPhone-wielding admirers, whom Madeleine considers fair-weather fans. She finds the tree diversely lovely at every stage of its annual cycle. She first began visiting it at age eight with her grandmother, who walked her there for botanical pilgrimages followed by hot chocolate in the garden's restaurant. During the terrible winter and spring of 2021, while on a pandemic-induced leave of absence from college, she passed the tree each day on her way to and from work. Sometimes, I met her there, and from behind our masks we took comfort in its indifference to the travails of the human world.

After she continued on her way to work, I lingered, at first in frustration. This was Paris during the pandemic lockdown, so the libraries and archives, along with everything else except grocery stores, were closed. I was writing a book—this one, in fact—about the tumultuous life and revolutionary science of Jean-Baptiste Lamarck, who proposed a science of living things to be called "Biology" (*Biologie*) and developed, at its foundation, the first modern theory of evolution. (He didn't call it "evolution," however, but simply referred to "change.")

Lamarck lived and worked in the Garden of Plants, and I would have liked to be inside the library of the National Museum of Natural History, situated at the garden's west entrance, which houses a major collection relating to the early history of evolution including Lamarck's papers. But at least the parks and gardens were open.

If you walk into the garden through the opposite entrance on the Seine, there he is, chin in hand, gazing contemplatively down at you from a bench atop a high pedestal, his long cloak thrown dashingly back over his shoulders. The bronze statue by Léon Fagel went up in 1909, but it seems fairly true to the few likenesses done during Lamarck's lifetime. A relatively small man, slender and delicate, Lamarck appears in contemporary portraits with a broad forehead, prominent cheekbones, a long, Gallic nose and chin, and large, dark, pensive eyes.

As I dawdled among the trees and along the paths, I began to think about how the garden itself sets the scene for the story I was trying to tell, the one you are now reading.

It was here in this garden that the first ideas of evolution emerged, between roughly the middle of the eighteenth century and the early nineteenth. This was a peculiar moment: a brief interval during which people, particularly in France, sought natural causes for everything in the world, including living beings. Before this, a supernatural god had played the starring role in theories of the origins of living things. Afterward, evolutionary biologists—not Charles Darwin himself but his most influential interpreters, people such as the German biologist August Weismann—described evolving organisms as passive in the process of evolutionary transformation, thereby making room for an implicit divine Engineer directing that process from a distance. But in between, naturalists working in and around the Garden of Plants did an extraordinary thing: They described living beings as actively producing and transforming *themselves.*

In so doing, these early evolutionary theorists made an assumption of "causal pluralism," to borrow a phrase from the late evolutionary biologist, paleontologist, and historian Stephen Jay Gould, with whom I once spent an afternoon exploring the Garden of Plants and the National Museum of Natural History, talking about the work that went on there during their early years. By "causal pluralism," I mean that rather than assuming a divine presence as the single, ultimate cause of all things, people supposed the world must be shaping itself in myriad ways, through many kinds of causes, and so they sought to grasp it by drawing upon every form of understanding at their disposal. Their science included history, literature, every variety of philosophy, and all the arts.

This holistic, pluralist mixture of ideas and approaches constituted the distinctive intellectual world of the early Garden of Plants and of the first authors of evolutionary theory. It was the intellectual world in which Lamarck developed his momentous theory that living things are in a continual state of self-transformation. The giraffe became the emblem of Lamarck's theory; although he devoted only a couple paragraphs to it in his voluminous writings, it perfectly encapsulated his idea that animals have the power to transform themselves heritably. Lamarck proposed that giraffes had acquired their distinctive form by stretching to reach the leaves on high-up branches, lengthening their necks and forelegs by tiny yet heritable amounts that had then accumulated over many generations.

Charles Darwin followed Lamarck in assuming that living beings could transform themselves heritably by their own behaviors; Darwin called it "the inherited effects of use and disuse." But soon after Darwin's death, at the end of the nineteenth century, there came the reactionary overthrow I've mentioned, when an influential generation of evolutionary theorists moved to eradicate Lamarck and his self-transforming organisms from biology. These new theorists, Weismann and others, described living beings as purely passive in the process of evolutionary transformation, thereby eliminating the threat that they might usurp God's monopoly on creation. Coinciding with this passive image of living things came a new and different model of evolutionary biology, and of science in general. If the world was a mechanical device full of engineered artifacts, it required an engineering sort of science to grasp it: a dry, dispassionate documentation of parts and their constrained movements. This was a mode of science that defined itself by its strict contrast with historical, literary, and artistic ways of understanding.

Whereas the older model of science had been open, holistic, and causally pluralist, the new one was isolated, reductive, and causally monist. This new model eliminated the many agencies at work in the older one—such as the desires, behaviors, and willful actions of living beings—leaving only the mechanical fitness of form to function. Evolutionary theorists and popularizers writing in the second half of the nineteenth century including Weismann, the English philosopher and social theorist Herbert Spencer,[1] and the English biologist Thomas Henry Huxley announced the separateness of science from all other modes of understanding. They insisted on the distinction between the "scientific" and the "unscientific," favoring catchphrases such as "scientific method," the "scientific spirit," "scientific men," and "scientific minds," and they remade evolutionary theory to exemplify this new, exclusionary model of science. In their hands, science became a thing apart, exceptional, enthroned, existing in splendid isolation.

In this way, the history of evolutionary theory—the history of its development and popularization—is also a history of the changing nature and public image of science. The concerted reactions against Lamarckism represented the major political forces and interests—imperial, industrial, commercial, racial, religious, socioeconomic—that reshaped science over the course of the nineteenth century. Contemplating Lamarck's science and its fall from favor, I've become interested in the larger development it represented: the redefinition of science, setting it apart from all other forms of knowledge, which took place within fifty years around the turn of the twentieth century. To seek a whole, integral, interpretive understanding of the world as Lamarck had done became, in many areas and in popular understanding, a thing of the past.

This book is about Lamarck, through whose life flowed all the Shakespearean currents of tragedy, comedy, and history: hardship, struggle, and frustration, resolution and intense creativity, love and farce and solitude. It is about the science that Lamarck named "Biology," with the first evolutionary theory at its core: how that science began and what it became. And it's about how science came to be, as it is today, in a position of supreme intellectual authority yet also extreme intellectual isolation.

As I walked the paths of the Garden of Plants, I began searching for remnants of the lost intellectual world that generated the theory of evolution: its geography, its landscape, and its inhabitants, some of whom are still alive today. Come and meet them on a prefatory stroll through the lanes and centuries of the garden.

SEXY TREES

At the opposite end of the chronological spectrum from Madeleine's Japanese cherry, the oldest of the garden's Remarkable Trees is a pistachio (*Pistacia vera* L.). Less eye-catching than the cherry tree in full bloom, the venerable pistachio is nevertheless impressive, a striking combination of age and youth. Its bent, gnarled trunk is supported by the iron pole that holds its plaque, but its twisted branches bear an abundance of glossy dark green leaves. This impressive tree enjoys the additional distinction of having demonstrated to the human world the sexuality of plants. It was planted in 1702, but I'll begin its story sixty-seven years earlier, in 1635, when Louis XIII acquired a château and estate on the Left Bank in central Paris and, under the influence of his doctor, Guy de La Brosse, made a garden there for doctors and apothecaries.

This garden, the Jardin du Roi (King's Garden), which would later become the republican Garden of Plants when the Revolution removed the *roi*, opened to the public in 1640 as a place to stroll or to attend free public lessons in anatomy, chemistry, and botany, offered accessibly in

Pistacia vera by Pierre-Joseph Redouté, ca. 1801–1819

French rather than, as at the university, in Latin. In 1700, Louis XIII's son Louis XIV dispatched his chief botanist, Joseph Pitton de Tournefort, on an expedition to the Levant to find out everything he could about the plants, metals, minerals, illnesses, and remedies of that region. Tournefort returned to Paris two years later bringing marketable intelligence and goods including the seeds from which the Remarkable Pistachio took root.

As the tree grew, it produced flowers, but these yielded no fruit, only pollen, which in Tournefort's opinion was the tree's excrement. However, Tournefort had an upstart colleague and former student named Sébastien Vaillant, who got it into his head that pollen was not tree excrement but tree semen. In fact, the idea of plant sex was abroad in the land, although Tournefort rejected it; Vaillant was drawing on suggestions that several naturalists had begun to make over the previous decades. In 1682, the English doctor Nehemiah Grew had asserted that plants were hermaphroditic, "or Male and Female," that a plant's "Attire [stamens] doth serve, as the Male, for the generation of seed," and that "the Seed-Case is the Womb."

A dozen years later, the German doctor Rudolf Jakob Camerarius, professor of medicine and director of the botanical garden at Tübingen, wrote a long letter to a friend and colleague describing a series of experiments he had performed on plant sexuality. Unlike Grew, Camerarius believed that while some plants were hermaphroditic, others were dioecious, meaning that individuals were either male or female. Dioecious plants, Camerarius thought, were best suited to experiments designed to test the hypothesis of sexual reproduction in plants. His experiments suggested that for female plants to bear fruit, they needed to come into contact with pollen from the anthers of male plants.

Vaillant, eager to test this hypothesis for himself with regard to the garden's barren pistachio tree, thought of another pistachio just fifteen minutes' walk away in the Garden of the Apothecaries that also produced flowers, though slightly different ones, and no fruit. Taking a flowering branch from the King's Garden pistachio, he carried it carefully over to the Garden of the Apothecaries, where he shook it over the flowers of the pistachio there. Several weeks later, Vaillant's experiment bore fruit when baby pistachio nuts appeared on the female tree in the Garden of the Apothecaries.[2]

Soon afterward, at the decidedly unsexy hour of 6:00 on the morning of June 10, 1717, Vaillant stood triumphantly before the audience assembled for the first lecture of the annual course on botany in the King's Garden and announced that plants had sex. Vaillant, a lowly underdemonstrator of plants, was filling in for his boss, the regular professor of the course and director of the King's Garden, Antoine de Jussieu. While the unsuspecting Jussieu traveled in Spain, Vaillant exposed his students' minds to the salacious idea that flowers were sex organs. Stamens, he said, were testicles, and pistils ovaries. Petals surrounded these organs to cover and protect them, and to preserve their vegetal version of modesty. During reproduction in certain flowering plants, Vaillant recounted with Rabelaisian relish, "the tension or swelling of the male organs occurs so rapidly that the lips of the bud, giving way to such impetuous energy, open with surprising speed. In that moment, these fiery organs, which seem to think only about satisfying their violent desires, abruptly discharge in all directions, creating a tornado of dust which expands, carrying fecundity everywhere; and by a strange catastrophe they now find themselves so exhausted that at the very moment of giving life they bring upon themselves a sudden death." Vaillant's listeners, if somnolent at the start of the dawn lesson, were surely wide awake by its end. Indeed, after Jussieu's return from Spain, his students clamored to have Vaillant continue lecturing to them, which Jussieu reluctantly allowed.

Tournefort did not live to witness this irreverence. Nine years earlier, on a spring day in 1708, rushing with a big bundle of plants across a street bordering the King's Garden, he'd been struck by a cart and crushed against a wall, and died from his injuries several months later. But he had followers in the Academy of Sciences, botanists who had hitched their careers to his views. These Tournefortians rose up against Vaillant's betrayal of the master and his bawdy approach to botany. Jussieu himself was working on an edition of Tournefort's works; once back from Spain, he presented his own discourse decisively omitting the idea of sexual reproduction in plants. When Vaillant died five years later, the academy did its best to relegate him to professional oblivion, declining to publish the usual eulogy. Jussieu did value Vaillant's personal herbarium suffi-ciently to preempt its sale to an English botanist, claiming it instead for the garden, where it still remains.

In the end, Vaillant's conviction that plants have sex prevailed. A few decades later, the Swedish botanist Carl Linnaeus invented a standard-ized biological nomenclature that is still mostly in use, beginning with a "sexual system" of plant taxonomy inspired crucially by Vaillant. "This sun affords such joy to all living things that words cannot express it," rhapsodized Linnaeus in his university thesis. "Why, all animals feel the sexual urge. Love even seizes the very plants," whose "nuptials" he then goes on to describe. A lover of classical poetry, especially Virgil and Ovid, throughout his career Linnaeus pursued his pet theme of love and sexual desire as universal among living things, including plants.

Vaillant was a causal pluralist. He saw causes in the pistachios' pol-len, stamens, and pistils, and also in the desires of all organic beings. Un-derstanding a world made by more than one kind of cause requires an approach of more than one dimension. When Vaillant contemplated the barren pistachio in the King's Garden and had a momentous idea, poetry and sympathetic identification were as fundamental to his discovery as experimentation and careful observation. The same goes for Linnaeus. When he encountered Vaillant's idea and built a new system of taxon-omy around it, sentiment and literary expression were again essential to the process. It's not that these people were writers as well as scientists, or that they wrote poetically about their science. No, this is what their sci-ence *was*: literary, philosophical, empirical, experimental. We owe our understanding of plant reproduction and our system of biological no-menclature to poetry as much as to what we would now call biology.

In addition to co-opting Vaillant's herbarium, Jussieu appropriated the now-vacant position of under-demonstrator of plants, which he gave to his younger brother Bernard. It was thanks to Bernard de Jussieu that the King's Garden came to house the most famous of its Remarkable Trees: the majestic cedar of Lebanon (*Cedrus libani*) that presides over the entrance to the twisting, climbing path called the Labyrinth. Just how Bernard de Jussieu came to plant the cedar in the garden has been a matter of imaginative speculation. Versions of the story proliferate, and curiously, all of them involve a hat. According to the Christian version, Bernard carried the tiny seedling all the way from the Holy Land, cradled lovingly in his hat, depriving himself of water to keep the baby tree moist. A more prosaic version has Bernard bringing the seedling home from England in a pot that broke during the crossing, whereupon he resourcefully transferred the baby cedar to, once again, his hat.[3]

Jussieu's hat does seem to have played a role, in fact. It's well established that during a trip to London in 1734, he received a pot containing two cedar of Lebanon seedlings, that he placed the pot in his hat to protect it during the return journey—so he often told his nephew Laurent—and that he planted one of the cedar seedlings in the King's Garden. A commemorative drawing by Charles Monnet from 1798 shows Bernard extending his already famous hat containing the tiny cedar sprouts toward a gigantic hole that has evidently just been dug: The digger stands in rolled shirtsleeves leaning on his pick, while assembled onlookers gesticulate excitedly. In the background, a camel peers down its long nose at the unfolding scene. *A camel in central Paris?* you are perhaps exclaiming. This brings us to the menagerie.

ELEPHANT EROTICA

The northwest quadrant of the Garden of Plants houses the world's second-oldest zoo, after the Tiergarten Schönbrunn in Vienna, though

Bernard de Jussieu planting the cedar of
Lebanon in the Garden of Plants, by
Charles Monnet, 1798

Monnet's addition of a camel to the scene shows artistic license, since
when Bernard planted the Lebanese cedar, the zoo didn't yet exist. It
came to exist soon thereafter, first as a gleam in the eye of Georges Buffon, who served for half a century as the King's Garden's director and
mastermind.

Buffon managed to capture the intendance of the King's Garden in
1739, as an aspiring young man of thirty-one, and he kept it until his
death in 1788. During the intervening decades, his ambition and political skill transformed the garden into a major center of research in natural
history. He doubled the size of the garden, extending it northeast all the
way to the Seine; acquired an adjacent mansion for himself to live in as

intendant; expanded the King's Cabinet, the royal natural history collection, acquiring another adjacent mansion to house it; and commissioned the architect Edme Verniquet to design an elegant, neoclassical amphitheater for lectures.

It was in this amphitheater that Buffon's protégé Lamarck would later present to the world his momentous theory of the continual self-transformation of living beings. Buffon himself proposed the beginnings of an idea of species change in 1753. From his vantage point in the King's Garden, he had begun work on a monumental tableau of the entire natural world that ultimately grew to thirty-six volumes. In the fourth volume, which embarks upon the nature of animals, Buffon observed that the family resemblances among different kinds of animals might just tempt a naturalist to consider the possibility that all forms of animal life could have originated over time from a single animal. Little wonder that he wanted a zoo and tried to have the royal menagerie at Versailles transferred to the King's Garden, where it could become a part of the research collections, but unusually, he failed.[4]

Thanks in part to this uncharacteristic failure, the menagerie in the Garden of Plants is the world's oldest *national* zoo, having originated in a revolutionary, republican act. On 3 pluviôse year 2 of the one and indivisible French Republic (a.k.a. January 22, 1794), the National Convention's Committee on Public Instruction drew up a decree abolishing the royal menagerie at Versailles and transferring its occupants to a new menagerie to be built in what was no longer called the King's Garden, but now the Garden of Plants, in Paris.

Several months earlier, the National Convention had reconstituted the King's Garden and King's Cabinet as a single national research and teaching establishment, with a staff of twelve professors, to be called the Museum of Natural History. Lamarck, recently attached to the King's

Cabinet as the botanist in charge of herbaria and eager to shield the royal botanical garden and natural history collection, including his own position, from the gathering political storm, had proposed this plan of reorganization during the first stages of the Revolution. Diplomatically referring to what had hitherto been the King's Garden as the Garden of Plants, Lamarck accomplished a bloodless king removal by nomenclature. The Convention, having finished the job by guillotine, folded the Versailles menagerie into the new national research and teaching facility, creating the world's first public zoo (its above-mentioned competitor, the Tiergarten Schönbrunn, originated in 1745, but as the imperial menagerie of the Holy Roman emperor Franz I and empress Maria Theresia).

In fact, there were few remaining occupants in the Versailles menagerie since it had been pillaged during the summer of 1792; many of the animals, including "a beautiful dromedary," several quadrupeds, and a great many birds, had been either eaten or sent to the knacker. The director of estates at Versailles, Louis-Charles Couturier, under instructions from the revolutionary government, had written to the man who was serving as intendant of the King's Garden after Buffon's death, the botanist and Romantic novelist Bernardin de Saint-Pierre, offering him the survivors for the natural history cabinet, presumably to be killed and displayed either stuffed or as skeletons. After receiving repeated entreaties, Bernardin finally made a visit to Versailles in January 1793 to inspect the formerly royal, apparently doomed creatures.

When Bernardin and his colleagues arrived at Versailles, they found only five survivors: a quagga, a now-extinct subspecies of southern African zebra; a bubal, a now-extinct kind of North African hartebeest; a crested pigeon from the Banda Islands; an Indian rhinoceros; and a "beautiful lion" from Senegal. The lion had been raised together with a puppy, and the two presented a scene to delight the author of *Paul et*

Virginie. "Their friendship," Bernardin observed, "is one of the most touching spectacles that nature could offer a philosopher for his speculations." The dog, a pointer, seeing the inspection party arriving at his shared enclosure, came rushing over, tail wagging, while the lion paced gravely, rubbing his massive head against the bars. When the visitors spoke to the animals, the dog threw himself playfully at the lion's mane and bit him on the ears. According to their keepers, the lion often invited the dog to play by lying on his back with his paws in the air and hugging the dog in his arms.

Later, after the pair had moved to the new menagerie at the Garden of Plants, the museum's librarian, Georges Toscan, wrote an essay, "Story of the Lion and His Dog," describing how the majestic and ferocious lion's character was "softened by the joys of friendship." The "natural gaiety" of the dog, and his "frank and open air," Toscan reported, "tempered the grave and serious humor of the king of animals." The lion grew tender, as when "with one paw he softly pressed his friend to his breast, while with his tongue he licked under his belly." When the dog became ill and died soon after arriving at the menagerie, Toscan recounts, the poor lion was inconsolable, his grief a monument to friendship.

Following his visit to Versailles, Bernardin made an earnest plea for including living animals, not just stuffed ones, among the collections at the Garden of Plants. The garden, he wrote, contained an active soil with growing plants, but "no animal that feels, loves, thinks." How could the study of natural history proceed without these? "What use will it be," Bernardin demanded, "to know [animals] only dead if we are never to see them living?" He urged that a living menagerie was essential to many areas of learning, such as the arts, allowing artists to sketch, paint, and sculpt the animals from life. Best of all, a menagerie would present opportunities for moral and social instruction. The lion and his dog friend

LE LION *du Muséum d'Histoire Naturelle avec* Son Chien.
Dessiné d'après nature).

The Lion of the Museum of Natural History with His Dog
by Nicolas Maréchal, an 2 (1794/5)

demonstrated that the most savage of beasts could become sociable. As another example of interspecies sociability, Bernardin told the story of a cockatoo he'd encountered on the Île de Bourbon (now Île de la Réunion) that "took such a great affection for a spaniel" that whenever it saw the dog, it flew up to him and followed him around "uttering cries of joy." When the dog went indoors to sleep, the bird sat for hours with its head between its talons.

Even the ferocious and "apparently stupid" rhinoceros at Versailles was "sensitive to friendship": Bernardin had witnessed this sensitivity in 1770 during the rhino's passage through the Île de France (now Mauritius). He loathed pigs and crushed them with his horn, but he developed a warm affection for a certain goat, which he allowed to eat its hay between his legs. These relationships were suggestive. Perhaps humans' famous friendship with dogs could help us befriend other animals. Dogs and other domestic animals might act as ambassadors, bringing wild animals into closer communion with humans. Bernardin also proposed that it might be possible to foster not just friendship but interspecies love, producing new creatures altogether. According to Aristotle, the Indians bred dogs with tigers to get "dog-tigers"; why not "dog-lions"?

Wild beasts began showing up at all hours at the gates of the garden. On 14 brumaire year 2 (November 4, 1793), the naturalist Étienne Geoffroy Saint-Hilaire, then holder of the principal chair in zoology at the Museum of Natural History, was going about his business "in the calm of his cabinet" when he received "a very unexpected piece of news": A panther, a civet, a monkey, and a polar bear were waiting outside in the street. This was the immediate result of an order of the municipal police banning the display of "dangerous animals such as lions, leopards and others" in fairs or on public squares; the creatures must be either slaughtered or delivered to the Garden of Plants, where the owners would receive indemnities. The first four surprise arrivals had been displayed on the Place de la Révolution (now the Place de la Concorde), where Louis XVI and Marie Antoinette had been guillotined a few months earlier. Close on their heels came a "tiger-cat," or serval, which is a wildcat from sub-Saharan Africa, another polar bear, two mandrills, and three eagles. Geoffroy hastily arranged temporary cages under the windows in the museum and began the process of applying to the National Convention for support to build a menagerie, which request was granted the following spring.

At last, on 7 floréal year 2 (April 26, 1794), four of the five Versailles beasts arrived at the Garden of Plants; the rhinoceros had died meanwhile. Then, one day in May, Geoffroy Saint-Hilaire and Lamarck made a trip to Raincy—the former estate of a cousin of Louis XVI, the duc d'Orléans, who had been guillotined several months earlier—to commandeer some large quadrupeds. A few months later, the professors succeeded in acquiring two dromedaries that had belonged to the aristo-crat and military officer Charles-Joseph Lamoral, seventh prince of Ligne, who had fled the Revolution and taken up residence in Vienna.

Lamarck, having devoted his career principally to botany, was now, at age fifty, reinventing himself as a zoologist in order to take up the lowli-est of what would be three zoology chairs in the new Museum of Natural History, the professorship of insects and worms. He himself had pro-posed the creation of this position, but he had had others in mind to fill it, since, being a botanist, he had hoped to take charge of the botanical collections himself. However, the other botanists associated with the King's Garden had resented Lamarck's presence ever since his arrival in 1789. "Professor of insects and worms" was a sufficiently lowly sounding position that others were content to leave it to him. But on the trip to Raincy, Lamarck was after bigger prey. The estate was positively thrum-ming with animals. Lamarck and Geoffroy claimed some goats, some geese, eight sheep, eleven deer, four swans, a bull, and a water buffalo. There was also a camel, which had already been bought by a revolution-ary leader named Merlin de Thionville, but he cheerfully donated it to the new national menagerie. Geoffroy must have told his apparently very credulous son, Isidore, that he and Lamarck had personally driven this motley crowd of creatures back through the streets of Paris to the garden. That's the story Isidore told in his biography of his father. But in fact, the two professors merely brought back a list of animals, and a few weeks later their Raincy caretaker delivered them.

Animals continued to arrive at the new menagerie, most of them tro-phies of the revolutionary army. Among the most spectacular were two elephants whose entrance along the Seine on the evening of March 24, 1798, was greeted by throngs of admirers. Hans and Parkie (renamed Marguerite upon her arrival in France) had begun life in Sri Lanka, where they were captured in 1784 and became gifts of the Dutch East India Company to William V, Prince of Orange. The elephants had been living in the menagerie at William's estate of Oude Loo, near Apeldoorn, until moving to Paris following the French occupation of Holland. When separated from each other for the journey, the elephants showed signs of extreme distress, and when they were reunited in Paris, the poet Louis-François Jauffret recounted that "they caressed one another with their trunks and made the air resound with cries of joy. Their eyes . . . were wet with tears. It was the most touching scene."

The elephants were the subject of one of the first experiments at the Garden of Plants menagerie. Two months after their arrival, on May 29, fourteen musicians came over by arrangement from the Conservatory of Music to play a concert for the pachyderms, to see how they might respond to different kinds of music. Toscan, the librarian, once again told the story in a two-part article for *La Décade philosophique*. The painter,

The Elephants Represented at the Moment of Their First Caresses
by J.-P. Houel, an 12 (1803)

draftsman, and engraver Jean-Pierre Houel also came to observe the event, which he later included in his illustrated account of the elephants. According to these contemporary reports, the elephants' English keeper, one Thompson, to safeguard the element of surprise, distracted them with food while the musicians set up out of sight in a gallery above the enclosure, connecting to it by a trapdoor. When the musicians began to play, the elephants stopped eating and ran toward the sound, exhibiting "curiosity, astonishment and disquiet." They circled beneath the trapdoor, directing their trunks at the opening. Gradually they calmed down and began to respond to the music.

Unexpectedly, what follows reads like a contemporary erotic novel. Imagine Laclos or the marquis de Sade, but with elephants. The musicians began with Gluck's Dance in B minor from his opera *Iphigénie en Tauride*, arousing the elephants to move rhythmically to the music and utter piercing cries. Next came the song "O ma tendre musette," played in C minor on solo bassoon, which Toscan reported put the elephants into "a sort of enchantment." Gently waving their trunks, they "seemed to inhale the loving emanations." Marguerite appeared more moved than Hans, caressing both him and her own teats with her trunk, which she next stuck into her mouth and then into his ear in an effort to rouse him. When the orchestra moved on to the popular revolutionary tune "Ça ira," in D major, which it repeated several times during the concert, both elephants grew impassioned, and Marguerite's passion developed into a delirious frenzy, as she galloped about, uttered various cries, flapped her ears furiously, and tried to get Hans to respond to her caresses. Finally, according to Houel, "the soft harmony of two human voices" descended like refreshing rain from the upper gallery, and Marguerite grew calm.

After a brief intermission, the orchestra resumed playing, this time in full view of the elephants, from right next to their enclosure. The pachyderms showed no interest in Haydn, but Hans was visibly aroused by the sound of a clarinet playing from the overture to *Nina, o sia La pazza per amore* by Giovanni Paisiello. His mood died abruptly when the orchestra returned to "Ça ira" for the fourth time. Although Thompson and Houel both reported later witnessing the elephants in various suggestive positions, Marguerite's lust remained unfulfilled. Houel thought Hans must still be too immature for a consummation.

A couple of weeks later, also according to Houel, a group of student trumpeters came to test whether the elephants would respond to a chaotic cacophony of trumpets, oboes, and clarinets playing all at once differently than to music. Though unmoved by the noise, which the humans in attendance found earsplitting, the elephants responded as before when the horn players switched to playing harmonious music, with tender caresses.

The musicians, artists, writers, and naturalists who participated in the elephant concert experiments wanted to understand the elephants' experiences and emotions using music, measuring the elephants' responses against their own. These experimenters took a variety of causes into consideration—sound, music, vision, emotion, physiology—and used various means of observation and representation, including lyrical descriptions and drawings. The same plurality of causes and methods characterized Toscan's and Bernardin's accounts of interspecies friendships and romances as well as their musings on the creative possibilities of animal and human sociability, which they thought might even produce new forms of animals altogether. Storytelling and social theory were fundamental to their description of a dynamic interconnectedness among living organisms, which was in turn essential to the earliest ideas of evolution.

Walking along the wall enclosing the menagerie, I see the fuzzy arms of a red panda (which is not actually a panda but more like a raccoon) appear at the top of the climbing structure, then its tail, then its whole head as the creature peers over the wall. I feel a pang of compassion for this Himalayan animal, evolved to roam over an open mile of high mountain, now locked in an enclosure in central Paris.

The menagerie is a poignant contradiction. Bernardin insisted that to observe animals' manners and habits, zoologists must study them not only alive but healthy, happy, and thriving. Hunting animals to kill and taxidermy them, you would "never see them except fleeing and trembling" or dead and stuffed. Likewise, Toscan, describing the elephant concert, praised "those Artists armed, not with scalpels and instruments of torture, but with oboes, flutes, and violins." He thought it was both more humane and more logical "to study the springs and functions of life in life itself, rather than seeking these out in death, or in the convulsions of an expiring animal."

Causal pluralism has a humane implication. To take the emotions and behaviors of living organisms as natural causes suggests that one cannot understand how animals work without having a measure of regard for their feelings. The zoologist Bernard-Germain de Lacépède, professor of reptiles and fish at the Museum of Natural History, promised that the new menagerie would eliminate the small, squalid enclosures of old and would "erect, so to speak, on the ruins of these prisons," an establishment worthy of a free and enlightened citizenry. The new zoo would resemble "a varied and cheerful countryside, where the different species of animals will enjoy all the freedom it is possible to allow them without danger" and visitors would witness them in their natural state.

At the same time, in both its royal and its republican guises, the Garden of Plants, including its menagerie, was an expression not just of curiosity and research but also of power: of conquest and political might. Furthermore, despite Bernardin's and Lacépède's idealism, the animals continued to live in more or less terrible circumstances for want of space and funds. Many animals died because of their poor living conditions. Often, visitors harassed the animals, throwing stones at them or letting their dogs chase them to make them move about. Conditions did improve a little during the first decades of the nineteenth century when an influx of government money, perhaps thanks to the arrival and enormous success of a giraffe (more on that in a moment), permitted several new buildings and other renovations. But during the four-month Siege of Paris in 1870–71, which culminated in the French defeat ending the Franco-Prussian War, many of the zoo animals were eaten.

A plaque at the entrance to the Garden of Plants explains that "during the Siege of Paris, the animals served as food for the Parisians," and the historical timeline on the garden's website includes the following entry: "The darkest hours: during the siege of Paris by Prussian forces, many animals were killed by the bombardments, and others were slaughtered to feed the population." This official history is misleading. Many people were indeed starving, but they weren't the ones who ate the zoo animals. Rather, the administrators of the menagerie, including Isidore Geoffroy Saint-Hilaire (who had inherited his father Étienne's chair in zoology) sold some animals—including two elephants named Castor and Pollux—to Coutier and DeBoos, elite butchers on the prosperous Boulevard Haussmann. To relieve the tedium of the siege and dramatize the extraordinary lengths to which brave Parisians would go in refusing to surrender, a wealthy clientele feasted on elephant soup and roast camel.[5]

The world's oldest national zoo remains, according to some animal rights activists, inhumane. They cite for instance the caging for the last half century of an orangutan known as Nénette.[6] Trapped in Borneo in 1972, Nénette has borne four sons in captivity and entertained generations of zoo visitors. Another enclosure houses a pair of snow leopards, whose wilderness habitat ranges from five to fifteen miles. Orangutans, snow leopards, and red pandas are all endangered, and the museum participates in research and breeding programs for the sake of conservation. But whether these efforts necessitate or justify the confinement of the animals is debatable.

Many of these caged animals' predecessors arrived as military trophies and symbols of political power. Yet they also helped to shape a holistic and pluralist science of life founded in the principle that living beings actively create themselves and the world around them. The authors of this new science of life plainly saw a contradiction in their science of animal agency emerging from a practice of subjugating animals. If Bernardin, Toscan, and Lacépède were right, the moral failing was also a scientific mistake.

GIRAFFE PERSONHOOD

The republic gave way to an empire, which gave way to a restored Bourbon monarchy. *Plus ça change, plus c'est la même chose.* In 1826, Muhammad Ali, pasha of Egypt, offered a young Nubian giraffe to Charles X as a diplomatic gift, and the following spring Étienne Geoffroy Saint-Hilaire went to pick her up and bring her to the Garden of Plants. Lamarck was by this time very frail and blind; ironically, despite posterity's firm association of him with giraffes, he never got to see a living one, even though for the last several years he had had one for a neighbor. She was the first living giraffe in France.

The journey was long and complex. Bernardino Drovetti, the French consul general of Egypt, first arranged for the giraffe to sail across the Mediterranean from Alexandria to Marseille aboard the Sardinian brig

I due Fratelli. A retinue of handlers looked after her, including Hassan El Berberi from Sennar, Drovetti's head groom, who would return to Egypt from Marseille, and Atir from Darfur, who had been enslaved to Drovetti and would remain in Paris as the giraffe's caretaker. The very young giraffe required twenty to twenty-five liters of milk each day, so she traveled with three cows. She was already too tall to fit below deck, so her transporters removed a panel at the base of the main mast and, by lining the opening with straw, allowed her to pass her neck through in safety and comfort. They also stretched an oiled canvas above the opening attached to four posts, creating a canopy to shield her head from sun and rain.

In this way, the giraffe crossed the Mediterranean and arrived in Marseille in good health on October 23, 1826. After a three-week quarantine, the prefect Count Christophe de Villeneuve-Bargemon put her up in his garden. There she wintered, safe from the Parisian cold, resting and maturing. The prefect himself took her for long afternoon strolls in the countryside. She had such success in Marseille that Geoffroy Saint-Hilaire, arriving the following April to pick her up, found it difficult to persuade the prefect and his wife to part with her.

Finally, on May 20, 1827, the giraffe and her retinue set out on foot for Paris. The three milk cows went first, followed by the giraffe and her handlers, then a cart with luggage, food, and various other caged animals bound for the menagerie. The giraffe wore a blue oilskin raincoat decorated on one side with golden fleurs-de-lys, representing the French throne, and, on the other, the coat of arms of the viceroy of Egypt. Geoffroy Saint-Hilaire often walked with her; at other times he rode in a coach, or else traveled ahead to make arrangements. Each night, the giraffe required special accommodations, which frequently involved remodeling stables on the fly. The cortege walked through Avignon, through Lyon.

Everywhere, great crowds of admirers greeted the giraffe, and Geoffroy Saint-Hilaire delivered public lectures. In Joigny, a Madame Jeanniot placed a haystack in her second-floor window so the giraffe could graze on it in passing.

The giraffe arrived in Paris on June 30 and was greeted by royalty, by the military, by essentially everyone in the city. Arriving at the Garden of Plants, she took up residence in the Rotunda, which had been built under Napoleon to house large carnivores but had proved unequal to the task and instead held big herbivores (starting with Marguerite the elephant, who had died a decade earlier). The giraffe received 600,000 visitors during the first summer. Giraffe merchandise became a craze. People rushed to produce, buy, and sell giraffe portraits and giraffe-themed books, clothing, dishes, toys, sculptures, and cartoons.

Geoffroy published his observations of the giraffe, in which he espe-

Study of the Giraffe Given to Charles X by the Viceroy of Egypt by Nicolas Huet, 1827. The person shown seated with his back partially turned to the viewer is Atir from Darfur, who had been enslaved to the French consul in Egypt before making the voyage to Paris. He lived with the giraffe in the Rotunda and gave a daily performance in which he groomed her using a currycomb on a pole.

cially dwelled upon her character, with its striking combination of mildness and courage. She was so "good-natured," so "gentle with us . . . so easy to handle, so docile, so good a person" that she even allowed a baby mouflon (a sort of wild sheep from the Caspian region of central Asia), born during the trip, to cavort all over her in "his childish games." And yet this sweet creature would stand up to a lion and very possibly win. If the lion didn't take her by surprise, Geoffroy explained, she would dispatch it with the "first blow of her hoof, the accelerated and violent throw of her front legs." This defensive action was so instinctive that when the giraffe's human keepers inadvertently startled her, she would begin to assume a kicking posture but would immediately repress her own response "by means of her extreme goodness or her tame habits."

Giraffe habits, by the time Geoffroy wrote these words, had become notorious as one of his friend and colleague Lamarck's examples of how animals could alter their own forms over time in ways that their offspring could inherit. Lamarck's influence is clear in Geoffroy's keen interest in the customs and behaviors of animals. Geoffroy believed, like Lamarck, that one could explain the forms of animal bodies as resulting at least partly from the animals' own actions, whereas his devout Lutheran colleague-turned-rival Georges Cuvier believed that the only way to explain anatomical structures was in terms of their functions, assuming that a divine designer had suited structure to function. Geoffroy and Cuvier aired their differences the year after Lamarck's death in the winter and spring of 1830 in an official debate at the Academy of Sciences. During the debate, Cuvier presented his view that function determined animal structures, implicitly assuming a rational god who was the only and ultimate cause in nature.

Generations of historians have represented Cuvier as a champion of empirical modern science against dogmatic philosophical tradition, yet it was he who clung to the ancient idea that animal forms remained fixed from generation to generation, and to the principle that the purposes of animal parts explained their structures (again, implicitly assuming a purposeful designer). Geoffroy rejected that approach, arguing as Lamarck had done that function must follow structure, not precede it. A structure must exist before it can function in a certain way. Therefore, function could not be considered a cause of structure. Rather, according to Geoffroy, multiple causes—including animal habits—might bring about variations in structure, which then resulted in variations in function.

. . . . .

Lamarck lectured in the Garden of Plants for almost three decades, beginning at the height of the Revolution in 1795 and ending during the Second Restoration in 1819. He and his family lived in rooms on the second floor of the mansion that Buffon had acquired to serve as the intendant's residence, near the entrance to the garden. This housing benefit was essential, since Lamarck would never escape financial struggles and came to have eight children. From his residence, it was a short walk across the garden to deliver his lectures at the amphitheater. There, Lamarck drew students from all over the world. His course register lists people from twenty-seven countries including Brazil, Jamaica, and Turkey. His students also came from every walk of life: artists, doctors, pharmacists, chemists, geologists, botanists, zoologists, military officers, engineers, lawyers, merchants, and writers including Balzac and Sainte-Beuve.

To this diverse crowd, Lamarck announced a new science for which he coined a name: "biology." This new science would take in everything to do with living bodies, and it would study the way living forms tended to move, develop, grow, and transform, producing new organs and structures. Lamarck's lectures were so popular that even fictional characters attended them. In his autobiographical novel *Volupté: The Sensual Man*, Sainte-Beuve sent his lightly fictionalized doppelgänger, the lovelorn young Amaury, to Lamarck's course. In Amaury's voice, Sainte-Beuve offered an evocative description of what it was like to sit in the garden's amphitheater and listen as Lamarck spoke. He described the "passionate and almost painful tone" in which Lamarck raised "serious primordial questions." Lamarck's worldview, he said, "had much simplicity, starkness, and much sadness. . . . According to him, things made themselves by themselves, all alone in continuity." Sainte-Beuve/Amaury loved contemplating "this mysterious power of life, as small and as elementary as possible. . . . I loved these questions of origin and end. . . . My reason suspended, and as though inclining toward these limits, enjoyed its own bewilderment."

Sainte-Beuve's Romantic description gives us the feel of Lamarck's lectures and his mode of science, which mixed emotionalism, poetry, and philosophy with material causation and rational investigation. Later, in writing his famously hostile "eulogy" of Lamarck, Cuvier would dismiss his theory of life as poetry rather than science, condemning not only the theory that organisms create and transform themselves but the expansive and lyrical mode of science that gave rise to it. In this sense, Cuvier did indeed represent the emerging vision of science as the detached, reductive account of passive mechanisms.

The garden's amphitheater, where Sainte-Beuve's fictive young Amaury encounters Lamarck's starkly beautiful, tragic, mysterious, vital view of nature, is just steps away from Vaillant's sexy pistachio. In fact, in addition to the neighborhood and landscape of the Garden of Plants, Lamarck and Vaillant had some other things in common: a conviction about the dynamic interconnectedness of all living things; a willingness

to insist on their ideas in the face of prevailing authority; and, in consequence, the dangerous disapproval of powerful people. Lamarck vindi-
cated his fellow sufferer, praising Vaillant's knowledge and wisdom and
singling out his discourse on the structure and function of flowers as
having established the sex of plants. He declared Vaillant's refutations of
Tournefort nothing less than "excellent."

On our prefatory stroll, we've traveled from the Japanese cherry to
the pistachio to the menagerie to the intendant's mansion to the museum to the amphitheater. It was in this landscape of tree erotica, interspecies romances between lions and dogs, elephant concerts, and giraffes
with kindly habits that the theory of evolution first emerged. We'll end
our walk where we began, at the *Prunus*, in whose celebrity, and the ardent affection it inspires, we can see a continuation of the centuries-
long interspecies engagement—sensitive, intellectual, emotional—that

The Garden of Plants Cedar by Jean-Baptiste Hilaire, 1794

informed an organizing idea of modern biology, the idea that living things continually create and re-create themselves, one another, and the world around them. The *Prunus*'s splendid effusion seems to display Lamarck's fragile yet inexorable power of life, and the insufficiency of any reductive approach to grasping it.

Now we'll begin the story at the beginning. Part 1 follows our protagonist's original generation and first creative self-transformations from unwilling seminarian to adolescent soldier to aspiring musician to unbearably bored banker to, at last, enraptured botanist. Here we go, and be ready for plot twists: hidden perils, unexpected strokes of luck, reversals of fortune, and sudden opportunities, a bit like the history of life itself, at least according to Lamarck.

PART ONE

Flowers and Trees

In which our Hero is born, grows up, rejects the Priesthood, runs away to fight in the Seven Years' War, and becomes an impromptu Military Officer. Garrisoned in Provence, he grows Fascinated with Flowers, leaves the Army, moves to Paris, and begins to study in the Garden of Plants. He meets Jean-Jacques Rousseau out Botanizing and becomes a Botanist. He decides that Plants compose the World from primitive Elements; that Plants are the original Creators of all Composed Things; and that Botany is therefore the most fundamental Science. He meets Rosalie de la Porte and lives with her for Fifteen Years; she dies soon after the birth of their sixth child.

ONE

The Little Priest

He didn't want to be a priest! Nevertheless, Jean-Baptiste Lamarck, aged eleven, found himself summarily packed off to study for the priesthood at the Jesuit school at Amiens, an ancient city in Picardy in northern France, about a day's travel by horse from the village where he'd been born and spent his first decade. The plan of an ecclesiastical career was the obvious solution to his parents' problem of what to do with their eleventh and youngest child and fourth son. "Unfortunately," as his own son Auguste later put it, "he felt not the least vocation" for the priesthood. He envied his older brothers, who were soldiers. They returned from their campaigns in splendid uniforms and were greeted with grand parties attended by all the local nobility.

To be sure, the career of one of these brothers, Jean-Antoine-Bernard, had ended almost before it began: He'd died at age seventeen, in 1747, after being taken prisoner during the War of Austrian Succession at the Siege of Bergen op Zoom. Jean-Baptiste, the baby of the family, was three years old at the time. But his eldest brother, Louis Philippe, was a page to King Louis XV in the Petite Écurie du Roi (the royal stables) at Versailles before joining the army, where he became a captain. The next surviving brother, Philippe François, with whom Lamarck had the closest bond, was the chevalier de Bazentin. With a recommendation from Benjamin Franklin, he would later go to fight in the American War of Independence, be taken prisoner by the English and held in Florida in dire conditions, then be liberated, again at Franklin's instigation, in a prisoner exchange. But military careers were expensive—a commission could cost thousands of livres—and launching these older sons had drained the family's resources.[1] At Amiens, Lamarck could lodge in the seminary and study for free. Adding insult to injury, his family took to calling him "the little priest."

Much later, when he observed that living beings continually respond to their circumstances in transformative ways, Lamarck was no doubt thinking of his own life. If you're a giraffe living on the savanna and the only food is high up in the trees, you have to stretch your neck and forelegs to reach it, and they grow longer. If you're the eleventh child of a family of impoverished military gentility, you have to leave home at age eleven to attend the free Jesuit school about a day's ride away, studying for a career in the Church, even though that's the last kind of career you want. But you learn Latin and Greek, grow intimate with the books in the school library, and gain access to the well-stocked natural sciences cabinet with its state-of-the-art instruments and charts.

Books and scientific instruments were surely not a prominent feature of Lamarck's earlier environment. On both sides, he descended from the *noblesse d'épée* (nobility of the sword). His mother came from a grand military family, the Fontaines of Picardy; his father was an army lieutenant as well as lord of the unremarkable village of Bazentin. The village was completely obliterated in July 1916 during the Battle of the Somme, then rebuilt after the armistice. Apart from that battle, Lamarck himself is apparently the biggest thing to have happened in Bazentin. There's a bust of him on the site where the Lamarck house stood and a statue of a giraffe situated—triumphantly, one can't help but feel—right next to the church. Otherwise, the main attraction in Bazentin is the military cemetery, which contains 182 graves from World War I.

Of course, when Lamarck was a child, neither his career nor the Battle of the Somme had happened yet, so there must have been very little indeed to distinguish Bazentin from many other such villages in *la France profonde*, effectively as far from Paris, the capital of the Enlightenment, as from Beijing. Indeed, Bazentin had the same primary means of connection to both these distant cities and the larger world: the Jesuits, with their books, reports, collections, correspondence networks, libraries, and schools.

The house where the reluctant little priest was born was made of brick, its corners and windows of white limestone, about fifty feet by twenty-five, with two stories and a high, curved roof. We know this because Alpheus Spring Packard, an American entomologist and evolutionary biologist at Brown University who became a devoted follower and the first English-language biographer of Lamarck, made a pilgrimage to Bazentin in 1901. He found the Lamarck house uninhabited, "almost a ruin," being used by the peasants next door as a storeroom for wood and rubbish. Packard's grainy photographs and a set of sketches he commissioned from the naturalist artist Louis Joutel, along with some nineteenth-century postcards, are all that remain of the house. The village schoolteacher obligingly showed Packard inside, where he found a narrow hall opening into spacious rooms on either side. The drawing room was ornate with low white wainscoting, an imposing cornice, and a great fireplace over which there hung a bucolic painting on a wooden panel depicting a shepherdess and her lover "engaged in other occupations than the care of the flock of sheep visible in the distance."

This was the rustic, provincial setting beyond whose confines, while grudgingly studying for the priesthood, Lamarck began to discover a larger world. But when his father died in 1759, Lamarck's first words were "Well, then! I won't be a priest!" and he flung off the hated clerical collar.

TWO

The Ersatz Chevalier

T he Seven Years' War was under way: a great struggle initially between France and Great Britain for land in North America, which had drawn in contenders for territory in central Europe, principally Prussia, Austria, and Saxony. Lamarck, aged seventeen, declared to his mother that he was going to fight, with such energy and conviction that he quite overwhelmed her resistance.

Next, he went to appeal for help from the biggest local notable, the Countess of Lameth, a well-connected lady who lived about two and a half hours' ride from Bazentin in a grand seventeenth-century château known as the little Versailles of Picardy. (It was the same obliging countess who later secured Benjamin Franklin's support for Lamarck's brother the chevalier de Bazentin.) The countess's brother was the military commander Victor-François, duc de Broglie, then marshal of France, who was in Hesse, a principality of the Holy Roman Empire (now in central Germany), commanding the vanguard army of General Charles de Rohan, prince de Soubise. The countess wrote Lamarck a letter of introduction to François IV de Lastic, the colonel of an infantry regiment fighting under her brother in Westphalia.[1]

Taking along a small suitcase, a modest packhorse, and the young farmhand who looked after his mother's turkeys to serve him as valet and companion, Lamarck made the five-hundred-kilometer trek through France and Germany. With impeccable timing, he arrived at the regiment in Westphalia on July 14, 1761, on the eve of the Battle of Vellinghausen, in which the French were (spoiler alert) utterly routed. Lamarck presented himself grandiosely to the distracted colonel as the "chevalier de Saint-Martin," improvisationally borrowing the title from his paternal line. Colonel de Lastic, who was in the midst of preparing for battle, flew into a fit of pique against the countess for saddling him at such a moment with a short, puny, and totally inexperienced teenager. Amid the blizzard of orders and counterorders, and with the decision to attack at dawn, the colonel had no time to figure out what to do with his commander's sister's protégé but nevertheless offered him the hospitality of his tent and dinner table before being called away to military headquarters.[2]

Imagine the colonel's surprise when, galloping back to his regiment at sunrise, he glimpsed young Lamarck marching into battle at the head of his first company of grenadiers. "What are you doing there?" the colonel shouted. "That's not your place! Get back, my friend, and follow the troops!" "Colonel," Lamarck responded (according to his son Auguste), "I'm here to serve. Don't refuse me permission to march with these brave men, I hope my presence won't dishonor them." Charmed, or perhaps merely out of patience, the colonel allowed Lamarck to remain in position.

Situated behind a hedge, Lamarck's company of grenadiers was sheltered from direct attack but not from enemy fire. Among the first victims was the captain of the company; when his head was blown off, his brains splattered Lamarck. The lieutenant was killed soon thereafter. The day proceeded disastrously. Marshal de Broglie and the prince de Soubise had agreed that the former would attack from the north and the latter from the south, but they fumbled their timing, with de Broglie too far in advance of Soubise. Soon the battle was decided: The Germans had won; the French army was in retreat.

There now remained of the grenadiers only fourteen men and no officers: no one, that is, to order them to withdraw. They turned hopefully to the infant aristocrat who had arisen in their midst. "Young man," they said, "it's up to you to command us now. What are we doing here? Our side is retreating, we should follow them." "Comrades," Lamarck replied, "I thank you for the honor you do me, but we've been placed here. We can't leave our post without being relieved. If you're afraid of being captured, leave: as for me, I'm staying." Feeling obliged to obey the young ninny (the Revolution was still almost three decades off), the men lay down in the grass to wait.

Meanwhile, the French army, arriving at a safe distance from the battlefield, turned around to pay its respects to the victorious, advancing German army. Colonel de Lastic now saw that he was missing an entire company and realized that the grenadiers had been given no order to retreat. Summoning an adjutant, he ordered the poor man to run back toward the front line, at risk of being captured, and bring the company back even if only a single man remained. Probably he had his commander's sister's troublesome protégé in mind.

The adjutant left, slipping as silently as he could through wooded areas, behind hedges and hillocks, until he spied the remains of the company of grenadiers waiting in the grass. Not daring to shout, as the plain was full of enemy soldiers, he vehemently waved his handkerchief at them until some finally noticed and pointed out the signal to their ersatz captain. "Now we can leave," declared the upstart, "here are our orders." He then led the remains of the company on the dangerous retreat through what was now enemy territory to what remained of Colonel de Lastic's regiment. That evening, the colonel—enchanted by Lamarck's bravery, according to Auguste, or perhaps simply doing the obvious—brought Lamarck to his commander, the Countess of Lameth's brother, Marshal de Broglie, who declared that he was making the embryo chevalier an officer on the spot. Two weeks later, Lamarck was officially nominated to the rank of ensign in the regiment; later he was promoted to sublieutenant. He remained with the regiment for seven years.[3]

You may be wondering if there's an excuse for taxing your patience with this shaggy-dog story of Lamarck's youthful escapades. Does it hold lessons regarding his later invention of the science of biology, apart from the sheer incongruity of such an outcome? Well, it does perhaps hold one significant lesson: Keep in mind, if you would, that the author of biology was a born Romantic.

Also, it was while Lamarck was garrisoned with his regiment in Provence that he discovered a love of botany. The plants of that Mediterranean, rocky terrain, so different from in northern Picardy, caught his eye. Indeed, even for a brash young adventurer—and again, our hero was a Romantic—it would be hard to miss the flowers that brightly blanket the Provençal landscape in every season: the copious purples of wisteria, irises, thyme, lavender, rosemary; the sundry yellows of sunflowers, mimosas, and broom; the pink and white peonies; the brilliantly red poppies; and everywhere, too, flowering trees, almond, plum, apricot, apple, peach, cherry. He later reminisced to his daughter Rosalie that during a brief leave at home in Bazentin, he had traded some precious music notebooks (he played the viola da gamba and had an intense, lifelong love of music) to his brother, the chevalier de Bazentin, in exchange for a copy of a botanical handbook, Chomel's *Summary of the History of Common Plants*. Back in Provence, Lamarck took the handbook with him on rambles, collecting specimens for an herbarium and confirming their names with the help of a local pharmacist.

THREE

The Violist, No, Banker, No . . .

L amarck left the army when he was twenty-four, in 1768. We have the story from Cuvier, who wrote the first official biographical essay about Lamarck in the form of a remarkably non-eulogistic eulogy. The Lutheran Cuvier hated Lamarck and his radical idea that living beings were their own creators, leaving not much room for divine Providence. Cuvier said Lamarck retired from military life because of a mishap that occurred while the regiment was garrisoned in Provence: One of his ingenious companions playfully attempted to lift him by the head, causing "a serious derangement in the glands of the neck." If Lamarck's life were a novel, its unsubtle author would obviously have intended this neck-stretching episode to foreshadow Lamarck's giraffe-themed posterity. Perhaps indeed Cuvier used some poetic license, because according to Auguste the neck problem was in fact scrofula, an infection and swelling in the lymph nodes of the neck.

In search of a cure, Lamarck went to Paris and consulted with the celebrated surgeon and hospital reformer Jacques-René Tenon. Cuvier says the condition required an intricate surgery from which Lamarck had to convalesce for a year. But Auguste told a different story: Tenon immediately recognized the problem to be an abscess below the ear and effected a radical cure with a few swift and strategic strokes of the lancet.

"And that was how he made a total change in his life," continued Auguste, writing to his own son, Eugène, decades later. "Or no, I'm wrong, there was a time back at the family home, with his mother. This was a time of idleness that one would like to be able to erase from his life. But finally, his mother too died." Disencumbered of both parents, Lamarck returned to Paris. (Auguste was apparently wrong again, though, since Lamarck's mother died in 1775, at which point he was already in Paris.) The world his oyster, or so it seemed, he briefly pondered a career as a musician. But his brother Bazentin—the same who had relieved him of his music notebooks, perhaps indeed with an ulterior motive—had apparently assumed the role of parental encumbrance upon the flighty youngest sibling and adamantly opposed a musical career.[1]

Merde, alors! Banking, then. For a change, Lamarck tried a practical, unromantic endeavor. He got a job keeping the books for a banker in central Paris, in what was then the rue Thévenot (before Napoleon III and the Baron Haussmann, during their imperial grandification of the city, swallowed it along with many other little streets into the rue Réaumur, which blazes straight across the center of the city like a conquering army). Displaying a moderate degree of stoicism, Lamarck stuck out the job for a year before succumbing to the inevitable: Banking he couldn't bear.

Perhaps craving a change of scene, around this time Lamarck and Bazentin withdrew for a year to live together in a peaceful village on the outskirts of Paris. There, they devoted themselves to the study of history and science, relieving their studies with long rambles in the countryside. One day, out on a botanizing jaunt, they encountered none other than the celebrated philosophe Jean-Jacques Rousseau, who was also out collecting botanical specimens.[2] Rousseau's writings—radical reflections on the foundations of human individual and social life in nature—had made him so famous that he was known generally by his first name, like Elvis or Beyoncé. Jean-Jacques was then about sixty, living in Paris, and in the thick of his botanical phase, the preoccupation of his late years.

Rousseau was writing a series of letters on botany to his friend Madeleine-Catherine Delessert, intended to help initiate her young daughter Madelon into a love of plants. Later, when these were published, they helped popularize botany, which had been an esoteric, scholarly pursuit conducted in Latin, and inspire a great botanizing craze across Europe. Rousseau often made field trips to the outskirts of the city. A friend of Lamarck's later recounted that Rousseau, who was notoriously temperamental, condescended to let Lamarck co-botanize with him only on the condition that Lamarck pretend to ignore him altogether.

Having discovered a love of botany, as we know, while garrisoned in Provence, Lamarck had begun pursuing the subject in the King's Garden, attending lectures by the doctor and botanist Louis-Guillaume Le Monnier, holder of the chair in botany there. He also studied with Bernard de Jussieu, he of the baby cedar tree in his hat, who had assumed Vaillant's position as under-demonstrator of plants in the garden and was the great mentor of all aspiring botanists. Lamarck's encounter with the exalted Jean-Jacques perhaps decided the matter, as how could it not? Botany it was: Lamarck's passion, his chosen calling.

FOUR

The Romantic Botanist

If you're an herbivore, constantly in danger of being devoured by a lion, you must do a lot of running, and you become slender and swift. If an idiot friend has attempted to lift you by the head, causing a career-changing neck injury (unless it was scrofula), and then your overbearing older brother has refused to let you become a musician, and then you've unsurprisingly turned out to detest banking, and then that same overbearing older brother has borne you off to a bucolic retreat for a year, where you've fortuitously gotten to go botanizing with Jean-Jacques himself, but your family still wants you to adopt a practical profession, such as medicine, you decide to try a bet-hedging sort of combined approach: medicine, okay, but *also* botany.

At age twenty-eight, Lamarck began training in both subjects concurrently. For several years he followed courses in anatomy, physiology, and botany at the King's Garden and the School of Medicine in Paris, including those taught by the young comparative anatomist Félix Vicq d'Azyr, four years Lamarck's junior. Vicq d'Azyr performed dramatic and rather creepy demonstrations that, even decades later, remained in Lamarck's mind, such as when he showed that a severed frog's heart would respond to stimulus as much as twenty-four hours after its removal from the frog, thereby demonstrating the great irritability of tissues in simple animals.

To these formal curricula, Lamarck added a third subject in the form of an independent study. From the high, narrow window of his Parisian garret, he could see nothing but clouds; contemplating these, he became a student of meteorology. The formation and dispersal of clouds was the subject of Lamarck's first remarks before the Royal Academy of Sciences. Around the same time, he also began a collection of shells that grew to become one of the biggest and most beautiful in the city; this was a fashionable pursuit and an expensive one, but he probably struck a deal with some Parisian shell merchants, trading his growing expertise for occasional rare specimens.

In this way, by self-exertion and adaptive response to environmental pressures and possibilities, Lamarck by around age thirty had undergone a veritable transformation of species from a brash, adventure-seeking teenager to a contemplative naturalist immersed in the many forms of living things and their constantly changing milieu.

In his early thirties, while engrossed in these studies, Lamarck also became romantically involved with a woman named Rosalie de la Porte. The two moved in together to an apartment right at the geometric center of Paris in the rue des Deux Ponts, which vertically bisects the Île Saint Louis, where it sits in the middle of the Seine separating the Left and Right Banks. They would remain together until her death fifteen years later, and they would have six children together.[1]

Rosalie de la Porte was apparently good for Lamarck's productivity. Soon after they got together, in 1778, Lamarck produced his two first works—*French Flora* in three volumes and a baby girl named for her mother, Rosalie Joséphine, who would remain her father's closest companion for the rest of his life. Much later, when Lamarck was old and blind, this Rosalie would be his amanuensis, reading to him, conducting research for him, and even writing for him. Soon after baby Rosalie's birth, the family moved southward onto the Left Bank following the direction of Lamarck's botanical ambitions, leaving the rue des Deux Ponts for the rue Copeau (now the rue Lacépède), a street that runs down the Montagne Sainte-Geneviève directly into the northwest corner of the King's Garden.

And it was there in the garden, around that same time, according to Auguste, that a casual bet with some friends led Lamarck to write *French Flora*. One day, strolling with some fellow botany students, he remarked that he could enable the first passerby, whoever it might be, to identify any plant in the garden. All he'd need would be to first explain to the person the basic distinguishing features of plants. When the friends took him up on the bet, he requested a little preparation time. Soon afterward, they gathered back in the garden, accosted the first passerby, and hey, presto! Before their very eyes, Lamarck transformed the person into an amateur botanist, clinching the metamorphosis with a successful plant identification. Today, the idea of turning any random passerby into an amateur botanist may not seem extraordinary. But at the time, it was a radical and downright Rousseauian bet. It meant that botany was not the preserve of scholars writing in Latin. On the contrary, botany was a walk in the park, a pastime for the people.

The method that Lamarck developed to win his bet was the same one he then presented to great acclaim in *French Flora*. He called it a "method of dissection"; today it's called a "dichotomous key." It works like this: You take an unknown specimen and answer a branching series of "this or that" questions; for example, does it have immediately apparent stamens and pistils or not? The answer to each question determines the next; for instance, if your specimen has immediately apparent stamens and pistils, are the florets grouped in calyxes, or are they independent? If the florets are grouped in calyxes, are the florets all the same, or are some cone shaped and others tongue shaped? If they're all the same, are they cone shaped or tongue shaped? If they're cone shaped, *eh ben voilà!* Your specimen is *Carduus marianus*: a milk thistle.[2]

Apart from achieving popular success by bringing botany to the people, Lamarck's major purpose in *French Flora* was to exploit what seemed to be a little opening in the field of botany in the form of a rift among botanists regarding nomenclature. The reigning system of botanical nomenclature in Europe—the system that in fact still serves as the basis for botanical and zoological classification today—was the creation of the Swedish naturalist Carl Linnaeus, whose rhapsodies on the sex lives of plants we briefly encountered during our prefatory stroll through the Garden of Plants. By the 1770s, Linnaeus had pretty much conquered Europe—all, that is, except for the King's Garden in Paris, where Georges Buffon, the garden's director, adamantly rejected his system as arbitrary and artificial.

Plant sex was once again at the heart of the dispute. Having learned from Vaillant to regard flowers as the sex organs of plants, Linnaeus based his system of taxonomy and nomenclature on these organs—the stamens and pistils—reasoning that since all plants had them, they would make a convenient and consistent basis for classification. But Buffon protested that a system of tidy categories based upon a single criterion belied the lush and jumbled profusion of nature. He instead championed as more holistic and therefore more natural the classification scheme of a fellow Frenchman, in fact a fellow denizen of the King's Garden: Tournefort, the same who had brought the Remarkable Pistachio to the garden, who had maintained that its pollen was tree "excrement," and whom a passing wagon had spared from living to witness his former student Vaillant's announcement that the pollen was in fact tree semen.

Buffon's opposition to Linnaeus was passionate, famous, and mutual: Linnaeus got lasting revenge by giving the name *Buffonia* to a marsh plant that serves to shelter toads, punning on *bufo*, Latin for "toad," the extra *f* a nudge-wink that has propagated itself down through centuries of botanical textbooks. In this conflict, Lamarck perceived his opportunity. He would develop his own system, drawing upon both Tournefort and Linnaeus while supplanting both. Of Tournefort's approach, with its "thousands of species" breeding confusion and tedium, Lamarck exclaimed, "What chaos!" Of Linnaeus's, he objected, "Why neglect the multiple resources that Nature gives us to help us know her?" Lamarck promised a happy compromise between Tournefort's chaos and Linnaeus's

Portrait of Georges Buffon
(1707–1788) by François-Hubert
Drouais, 1761

reductive simplicity: a method that would be, like Little Bear's porridge, just right.

This was one of the rare moments in Lamarck's life when he pulled off this kind of savvy maneuver. Usually, as we'll see, he landed on the wrong side of political machinations, or just clinging by his fingernails to professional survival. But this first time, with beginner's luck, it worked: Buffon warmly embraced Lamarck's project of superseding Linnaeus and arranged for *French Flora* to be printed by the Imprimerie Royale (the royal printer) at the government's expense. He also appointed himself Lamarck's patron and protector. This was a great advantage, since Buffon was a towering figure in natural history, with extensive resources and influence. With Buffon's advocacy, Lamarck immediately gained admission as an adjunct member to the botanical section of the Royal Academy of Sciences.

Buffon did all this, moreover, even though Lamarck not only used primarily Linnaean names in *French Flora* but also endorsed Linnaeus's principle of the primacy of plant sex: the very thing, Lamarck observed, that distinguished plants from minerals as organic beings. Like other living things, plants existed in sentient and even sensuous relation to one another. The self-centered human observer, roaming admiringly amid the brilliant "greenery enameled with a thousand colors," might well feel it was all intended just for him, these lovely flowers with their "sweet perfumes" and "innocent welcome" to their "laughing and lively" presence. But in fact, the flowers were not meant to beguile humans. These "most sensitive and most universal organs" were meant to fulfill the plants' own sex lives.

Lamarck surely labored over these Romantic effusions. He was not someone to whom words came easily; on the contrary, he was generally a stiff, dry, awkward writer. "Lamarck's pen," as one contemporary expressed it, was "neither elegant nor correct." It is a remarkable thing that Buffon adopted him anyway: Buffon, who was all graceful eloquence and wit, so facile with both pen and tongue that the mathematician and philosopher Jean d'Alembert disparagingly dubbed him the "King of Sentences"; Buffon, who famously announced to the Académie Française, on the day of his election to that august body of guardians of the French language, that *style is the man himself*," providing a slogan to generations of litterateurs and sophisticates. Perhaps he was partly mocking the academicians; if so, it was also self-mockery, and it was only partly. In fact, Buffon used his brilliance as a stylist to say something that was simultaneously both lightly ironic and also perfectly earnest. Literary style was the essence not only of the man but of the science itself. He worried about Lamarck's lack of style and asked his friend and coauthor Louis Jean-Marie Daubenton to look over the "Preliminary Discourse" of *French Flora*.[3]

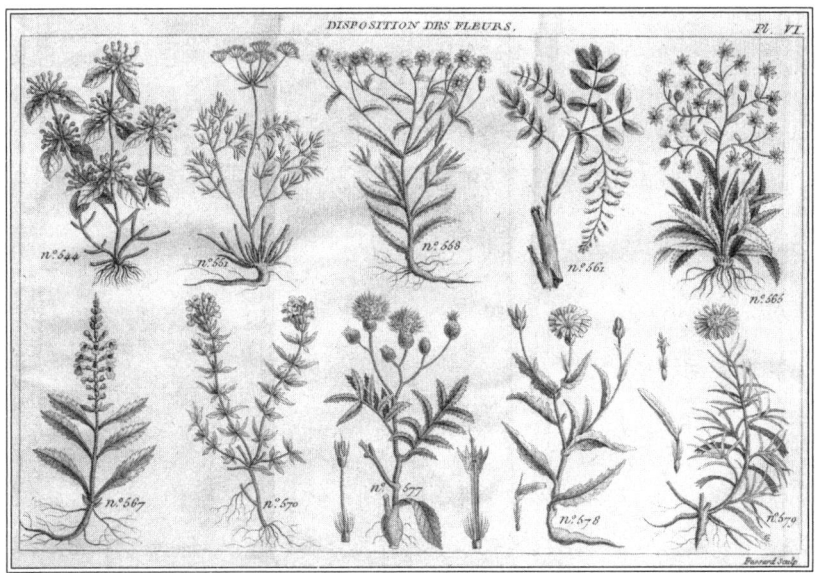

Flowers by Étienne Fessard, from Lamarck, *French Flora*

Daubenton redelegated the job to his own student, the budding mineralogist René-Just Haüy, who was also a professor of humanities at the University of Paris. Haüy essentially rewrote the "Preliminary Discourse" before it went to press. Lamarck, who in later life would often express his feelings of grievance against the injustices his colleagues inflicted on him, also had a quick and keen sense of obligation. Throughout his life, both in conversation and in print, he warmly acknowledged the debt to Haüy. "I confess," he wrote, "that the style part [of the 'Preliminary Discourse'] is entirely his." The body of *French Flora*, however, with its brilliantly enameled greenery, its laughing and lively festival of nature, and its sensitive floral organs, was Lamarck's own.

FIVE

The Traveling Naturalist

F*rench Flora* brought Lamarck not only Buffon's patronage and a foot—or at least a big toe—in the professional door as an adjunct member of the Academy of Sciences but also, for once, after it was published in 1778, a bit of extra money. Using these funds, he organized an expedition in search of rare plants. He included two friends of about his own age, in their early thirties, both named André: André Thouin, who had inherited his father's role as chief gardener in the King's Garden when he was just seventeen; and André Michaux, a fellow student from Le Monnier's botanical course who would go on to a career as botanist and explorer. Lamarck and the two Andrés traveled to the Auvergne region in south-central France, where they explored the volcanic Massif Central highlands around the Puy de Dôme, a mountain of lava; they investigated the Mont-Dore area, where the thermal springs provided an ancient Gallo-Roman spa town; and they perused the lush plateaus and river valleys of the Cantal range.

It was during this trip that Lamarck began to form a scandalous idea. It came to him after he found some vegetable matter buried in the earth that appeared to be "half-transformed into flaky, schistose clay," which he later brought back with him to Paris. He speculated that in this bit of stuff he had actually caught living matter in the act of metamorphosing itself into mineral matter, and that in fact living things themselves *created* the inanimate world in this way, composing it from elements. Plants, after all, could produce themselves from just sunlight, air, and water. Lamarck therefore surmised that all minerals—inanimate compound matter such as chalk, clay, marble, flint, quartz—arose from the actions of living beings. He reasoned that plants must play the primary role in this process by producing their own substance, which then became the primary matter for animals and natural forces, such as heat and gravity, to act upon, transforming it into further animal and mineral matter. Plants therefore supplied the first stage of compound matter for making minerals.

Lamarck proposed an easy experiment to demonstrate plants' ability to do this: Suspend a hyacinth or narcissus bulb in a carafe with nothing but distilled water and you'll see that it won't die but, quite the contrary, will grow an entire flowering plant. This, when weighed, will be much heavier than the original bulb, proving that the bulb has actually created matter just from "the pure water, pumped up by its roots, from the air it has absorbed, and from the expanding fire" of the sun. Lamarck also described a similar experiment using a turnip: Carve a small hollow in its side and suspend it in the air by its root, Houdini-style, nothing up its sleeves! Keep the hollow filled with water and again you'll see it grow and create flowers literally out of thin air.

When he described these experiments, after returning from his trip with the two Andrés, Lamarck was in his mid-thirties, living with Rosalie de la Porte in their small apartment in the rue Copeau near the King's Garden. Their second child, André (named after his godfather, André Thouin), had recently been born, when Rosalie Joséphine and *French Flora* were both about two and a half years old, in the spring of 1781. It's easy to imagine Lamarck himself performing some of these experiments on the creative capacity of plants in his own apartment. In fact, as further evidence of the unique ability of plants to create composed matter from elements, he reported that "some people" he knew liked to raise peas in earthenware dishes with just pure water and cotton wool to support the roots, lending greenery to their apartments during wintertime.[1] Perhaps he himself, with his young family, liked to brighten the dark, gray Parisian winter by watching decorative pea plants unfurl themselves from earthenware dishes, conjuring their delicate pale green tendrils from thin air.

. . . .

While André was still a newborn, Lamarck left Paris again on a journey of several months, this time as the escort of Buffon's hapless teenage son, Georges, nicknamed Buffonet. Poor Buffonet seems to have counterbalanced all his father's good fortune and success. Already as a child he had a reputation for mediocrity. The writer Antoine Rivarol referred to him as "the poorest chapter of his father's Natural History." Another story has it that the Prussian emperor Frederick II presented Buffonet to the ladies at his court saying, "Here, mesdames, is the son of the illustrious Buffon; but it's not his greatest work."

Much later, soon after Buffonet's own marriage and in the thick of the Revolution, his wife would leave him for the colonel of his regiment, the most dashing of aristocratic revolutionaries, Louis Philippe II, duc d'Orléans, then styling himself by the radical moniker Philippe Egalité. Finally, just when Buffonet had happily remarried and become a colonel in his own right, he would fall victim to the Reign of Terror. He was arrested and guillotined at age thirty, just a few short days before Maximilien Robespierre, leader of the National Convention's Committee of

Public Safety that was presiding over the executions, fell from power. As Buffonet mounted the scaffold, he called out in proud defiance, "Citizens, my name is Buffon!" His first, last, and greatest distinction was his name.[2]

Buffonet was infinitely precious to his father. Buffon's much younger and intensely beloved wife, Marie-Françoise de Saint-Bélin-Mâlain, had died after falling from a horse when Buffonet was just five years old. Their one other child, a daughter, had died a decade earlier when she was about a year and a half. The death of Buffon's wife plunged him into deep despair; he wrote to a friend the following month, "At first it was a cruel wound, which is now degenerating into an illness which I regard as incurable and which I must accustom myself to bearing as a necessary evil. My health has deteriorated, and I have abandoned, at least for a time, all my occupations."

When he at last emerged from this stage of deep bereavement, Buffon concentrated all his love and concern on his son. Even when Buffonet was an adult and a military officer, Buffon doted on him. For instance, on one occasion when Buffonet had suffered a serious illness and his father had arranged a convalescence leave for him, Buffon wrote in an ecstasy of concern, "Come, my dear son; but, if you'll listen to me, do not come alone. . . . I cannot be comfortable if you are not accompanied by some responsible person. What would become of you alone if you fell ill on the way? Do not be in a hurry to leave; do not hurry on the road; eat but sparingly, and only wholesome things and little meat; even less liquor or overly strong wine. I am only expecting you on the 7th; it is the day of my birth, and it will be that of my happiness, if I can kiss you in good health of body and mind."

Having spent a significant portion of his own fortune on the King's Garden, which he referred to as his "eldest son," Buffon wanted to make it a sort of inheritance for its little brother, Buffonet, whom he hoped would succeed him as the garden's director. This scheme ultimately fell through, a further example of Buffonet's ill fortune, but meanwhile Buffon invested much time and care in his son's education as a naturalist, including sending him at age seventeen on a tour of central Europe with Lamarck.

For seven months, Lamarck escorted Buffon's cherished, ill-starred seventeen-year-old son around Europe, and as a result became a corresponding member of the King's Garden and King's Cabinet. Buffon clearly intended this arrangement to benefit both of the people he was most keen to help, his son and his new protégé. He charged Lamarck, in his new, official capacity, with visiting the botanical gardens of central Europe and putting them in communication with the King's Garden as a sort of botanical ambassador.

On the eve of Lamarck and Buffonet's departure, Buffon wrote to a close friend, Philippe Guéneau de Montbeillard, "My son leaves tomorrow for his great voyage, with M. le chevalier de Lamarck, of the Academy of Sciences. I was very happy to find him such a companion." Several weeks later, writing to the same friend, he reported that Buffonet was "currently in Amsterdam, and it seems to me that his companion is pleased with him; it's all that I could wish, since he's a wise man."

Lamarck and Buffonet traveled to Holland, then Germany, then Hungary, visiting gardens, universities, cabinets, and museums. They brought along letters of introduction from Buffon to monarchs, nobles, industrialists, and naturalists. In Berlin, they met the Botanical Garden's director, Johann Gottlieb Gleditsch, another investigator of plant sex who carried out a version of Vaillant's pistachio experiment with dwarf palms. Gleditsch also argued that insects played a crucial role in pollination and devised experiments to show that beetles could reason.[3]

The trip wasn't all strolling in gardens and admiring flowers, however. In the Harz Mountains, Hanover, Freiberg, and Saxony, Lamarck and Buffonet toured mines. In Chemnitz, they visited the limestone quarries. In Freiberg, they met the mineralogist Abraham Werner, professor at the school of mines, who established methods for deriving chronological succession from rocks. In Vienna, the emperor Joseph II received them graciously and invited them to descend by a mechanical lift into the silver mines at Kremnitz (now Kremnica, Slovakia). They remained in the depths of the vast, subterranean operation for several hours, and Lamarck later sent a detailed account of the visit to Buffon.[4] Exploring the mines, Lamarck gathered evidence to support his radical new idea that all minerals arose from the agency of living beings, from their remains and the other debris they produced.

"In one of the mines of Freyberg in Saxony, where I went down, I found manifest proof," Lamarck recalled triumphantly. As he descended through the layers, he could see that with the passage of time and under the heat and pressure of the earth the organic material at the surface changed into ever harder and less organic mineral compounds, from clay to schist to mica to quartz. The soft, crumbly, clayey bluish-gray schist of the surface became harder and denser until, "from the second galleries . . . about 140 fathoms or 840 feet deep," the clay had been replaced by layers that were "almost entirely quartzose," that sparkled when struck with flint. He had made similar observations "at Claustahl in Hartz, at Schemnitz & at Cremnitz in Hungary." In fact, he reported, "I have constantly noticed [this] in all the mines where I have been down."

Lamarck carried back to Paris some more transitional specimens that appeared to be in the process of transforming from organic to mineral matter. For instance, he collected a bit of pectin, a starchy plant fiber, that was at one end in "the most obvious clayey state" and at the other in "a completely glassy state."

As he continued his voyage with Buffonet, Lamarck collected not only plant specimens and seeds but also minerals to add to the collections of the King's Garden and King's Cabinet, as well as examples of partially mineralized organic matter. Eventually, however, troubles arose. One day, when Buffonet wanted to go out alone without his highly conscientious escort, he deliberately splattered ink all over Lamarck's clothes to prevent his chaperone from leaving their lodgings. Lamarck still resented this childish act of rebellion almost forty years later when he recounted the story to a visitor. Buffon, reading between the lines of the letters he received, gathered that the situation had deteriorated and recalled the travelers to Paris, where Lamarck could contemplate his new ideas about the origins of natural substances in peace.

Back in Paris, he continued to ponder the remarkable ability of plants to merge elements into compounds. This ability was unique to plants, he pointed out, since no animal could subsist on just air, sunlight, and water. By now, Lamarck was confident that plants were the first link in the cosmic chain, bringing the primitive elements together into elaborate, compound forms. Botany was therefore the most fundamental of the sciences, because it studied plants, the original creators of all things on Earth, whether living or inanimate. But then animals took up the baton from plants, by eating and digesting vegetation (and other animals), in turn producing more complex compounds: animal bodies, with all their parts and products. Together, animals and plants shared the defining ability of living beings: to build, compose, and create. Working in concert, they built the world itself, usurping the traditional job of a divine creator.

The Stealth Geologist

W hile pursuing his new idea about the composition of matter, Lamarck also devoted himself to the mammoth task of editing the botanical volumes of the revised *Encyclopedia*, a gargantuan publishing project in an age of gargantuan publishing projects. First among these had been Denis Diderot's and Jean d'Alembert's twenty-eight-volume, almost seventy-two-thousand-article compilation of all knowledge about everything in the world, arranged in alphabetical order, the *Encyclopedia, or, Rational Dictionary of Sciences, Arts, and Crafts*, the last volume of which had appeared in 1772. Now, ten years later, in 1782, the writer and editor Charles-Joseph Panckoucke was undertaking an expanded and reorganized reedition of the *Encyclopedia*, replacing its alphabetical order, which could drive a person crazy rushing hither and yon among the massive folio volumes in pursuit of a given topic, with a more user-friendly system that subordinated the alphabetical order to an arrangement of volumes by subject matter. To edit the volumes on botanical subjects, Panckoucke recruited Lamarck, which was a great opportunity since Lamarck—although he secured a promotion at the Academy of Sciences in 1783 from adjunct to associate—still had no paid position and had very limited funds. The costs of the revised *Encyclopedia* were supported by subscription.

Hidden in plain view in one of Lamarck's articles for the *Encyclopedia* is his first presentation of the momentous idea that plants and animals compose the material world. He snuck it into his article "Classes of Plants," where he includes some discussion of minerals, which might seem incongruous in an article on botanical classification, but we know the reason: Lamarck had decided that minerals were deeply akin to plants, as their original creations. In his article on botanical classification, he announced that minerals were never the result of "direct formation," but instead they arose over time from the debris and detritus of living organisms. He also insinuated the idea into the cumbersome title of a chart: "Inorganic Entities, Without Life, Produced by the Successive Alterations of Composed Substances That Have Been Part of Living Beings."

In this unassumingly revolutionary chart, Lamarck divided the original, organic sources for inorganic compound matter into four different kinds of "topsoil": animal topsoil made by crustaceans; animal topsoil found in cemeteries and garbage dumps; plant topsoil produced in marshes; and plant topsoil developing in fields and forests. Lamarck's chart shows these four kinds of organic products giving rise to inanimate composite matter including—from softer to harder—chalks, sulfurs, bitumen, clays, calcified stone, niter, alum, steatite, marble, borax, gypsum, schist, alabaster, alkalis, vitriols, talc, calcareous spars, pyrite, fluorspar, native metals, gritstone, schorl, pebbles, gem crystals, flint, feldspar, petro-silex (felsite), agate, jasper, and, at the very hardest end of the spectrum, converging upon quartz and rock crystal.

Lamarck's claim that only living organisms could produce compound matter was certainly radical in his day, and it might sound far-fetched even today. Yet all the inanimate substances he lists in his chart can, according to current geology, result from materials or processes involving organisms, and many of them indeed positively require living beings. Almost two and a half centuries after Lamarck created his chart, the role of organisms in the formation of minerals has inspired renewed interest among geologists. According to current science, the first organisms, more than 3.5 billion years ago, changed the chemistry of the atmosphere and the ocean, precipitating large-scale mineral deposits, and ever since living beings have shaped Earth's surface mineralogy. Geologists studying "mineral evolution" speculate, for instance, that biological activity might even have helped to stabilize the continents by increasing the rate of granite production.

These geologists have suggested that without life there would be much less free oxygen in the atmosphere, or perhaps none at all. Oxygenation associated with biological activity led to the Great Oxidation Event, a sharp increase in atmospheric oxygen that took place a little over two billion years ago, when emerging microorganisms seeded the air with oxygen. This was probably the crucial turning point in the proliferation of diverse minerals on Earth. The free oxygen interacted with existing minerals to form thousands of new ones.

The Great Oxidation Event also brought multicellular life and the production of skeletal biominerals that, in their turn, transformed the surface mineralogy of the planet. During the early Cambrian period, about 538 million years ago, a rapid rise in skeletal minerals—calcite, aragonite, magnesian calcite, apatite, opal—greatly increased the volume of rock on Earth's surface. Microbes also foster clay mineral production; some geologists describe a "clay mineral factory" that took place during the Neoproterozoic Era between half a billion and a billion years ago. In addition to producing new minerals, the clay mineral factory would have further increased the availability of free oxygen in the atmosphere since clay absorbs and sequesters carbon, preventing it from bonding with the oxygen produced by photosynthesis.

Overall, people studying mineral evolution suggest that biochemical processes are responsible for most of Earth's mineral species. Although Lamarck didn't describe these biochemical processes specifically, he did claim that for as long as it has existed, life has produced and shaped the inanimate world. Geologists working today share this view. "The origin of life depends on minerals," as one mineral physicist has put it, "but the origin of minerals depends on life."[1] Living beings have built Earth's surface landscape on the grandest scale, from granite for continents to limestone for mountains. If you're not a geologist, you probably didn't know that there's such a thing as "biogenic" quartz, for instance—quartz that forms from silica-rich plankton. But you might have known that the White Cliffs of Dover are made of chalk from pulverized shells of single-celled algae called coccolithophores. Or that the summit of Mount Everest is made of limestone, also a primary component of the Rockies, sections of the Alps, and the mountains of Guilin in China, and that limestone is made from calcium carbonate from corals and seashells.

The idea that living beings could make mountains was quite new in Lamarck's lifetime. For centuries, when people found marine fossils far from any ocean, such as on the tops of mountains, they'd taken these as evidence of the biblical Great Flood of Noah, in which water covered the whole earth. But around the time Lamarck began taking an interest in botany and natural history, certain people started suggesting that the rock of mountains such as the Jura range, along the border between France and Switzerland, didn't just have seashells on or in it but was in fact made of the remains of living things. "All calcareous rock is produced from the animal kingdom," wrote Linnaeus, and for once Buffon agreed. "All calcareous earth," he concurred, "can be regarded as animal earth because it is produced only from the detritus of shells."

Lamarck contributed a subtle yet crucial change of emphasis: Not only were these mountains made *from* the remains of marine organisms, he said, but they—along with the rest of the inanimate world—were made *by* living things. The original source of the mountains was not dead organisms but the incessant activity of living ones.

An inconspicuous chart in an encyclopedia article about botanical classification might seem an unlikely place to present a subversive idea. But for anyone who understood the full significance of Lamarck's "inorganic entities" chart, it was world changing. It meant that when you look around the earth, what you see is a dynamic consequence of the lively, unceasing, creative agency of organisms.

SEVEN

The Cross-Dressing, World-Traveling Botanist and Her Lover

(LAMARCK WAS NEITHER, BUT HE ENDED UP WITH THEIR COLLECTION)

To give the botanical volumes of the *Encyclopedia* global coverage, Lamarck made use of all that came to the King's Garden from around the world. The garden in its position at the epicenter of the French empire received a steady influx of botanical, mineralogical, and zoological plunder. The botanical specimens that Lamarck used included materials collected by Jeanne Barret, the cross-dressing, world-traveling botanist referred to in the title above, and her lover Philibert Commerson, both of whom traveled with the admiral and explorer Louis-Antoine de Bougainville on his circumnavigation of the world between 1766 and 1769.

Their voyage provided important support for Lamarck's idea of a world made by living beings, so it deserves a detour in our story. Let's begin with Commerson, who was a doctor, explorer, and naturalist from Bourg-en-Bresse, a town in eastern France, not far from Lyon, at the foot of the Jura Mountains. His close friend the astronomer Jérôme Lalande, also from Bourg-en-Bresse, described Commerson as short and slight, about five feet three, with big black eyes and an aquiline nose, a delicate complexion, extremely lively, "eating only by necessity and often without noticing." Botany was his great passion, and he regarded social occasions and entertainments to be a waste of time.

Like Lamarck, Commerson "would have liked to turn everyone into a Botanist," as Lalande remarked. Here was another champion of the radical Rousseauian principle that botany should be for the people. Commerson himself boasted to his friend and fellow botanist Louis Gérard, "I infect all who approach me with my botanomania." His conversation, Lalande said, was "a torrent of fire . . . his expressions of the greatest energy." He spoke too freely, sometimes making enemies, but none outdid him in generosity. He was "ardent, impetuous, violent and extreme in everything; at games, in love, in his animosities as in his friendships."

Among those infected with Commerson's botanomania was his wife, Antoinette-Vivante Beau, who brought to their union not only loveliness and wealth—as Commerson triumphantly told Gérard—but, above all, sympathy with his botanical fervor. They met while Commerson was out searching for rare plants in the area of Charolais, as he recounted: "I found in that country a sensitive one that I am about to introduce not into my herbarium but into the nuptial bed. . . . She is indeed a girl philosopher. . . . I'm sure I'll share all my tastes with her. I have already inspired in her a decided one for natural history and our walks have become true herborizations." When, soon after their marriage, Antoinette-Vivante Beau died in childbirth, Commerson was "crushed by sorrow," writing to a friend, "I exist today only for the memory of having belonged to her," and signing off, "the most desolate of men."

To look after the baby and the household, Commerson hired a young peasant woman from Autun in the Loire valley: Jeanne Barret. Apparently, he found consolation in his bereavement. The two were soon lovers, and Barret became pregnant. Lalande had already been urging Commerson to move to Paris to work with the naturalists at the King's Garden, especially Bernard de Jussieu. Now that Barret was pregnant, following Lalande's advice would have the further advantage of avoiding a scandal. Commerson and Barret made the move, leaving Commerson's two-year-old son, Archambault, to be raised by his maternal uncle, a somewhat censorious yet ultimately accommodating priest named François Beau. They took a third-floor apartment in the rue des Boulangers, a crooked little street near the King's Garden right next to the rue Copeau, where Lamarck would later live (though at this point in 1764, Lamarck was still garrisoned with his regiment in Provence). Their own child was born soon thereafter but died in the care of a wet nurse.

In Paris, Commerson and Barret threw themselves into botanizing, working closely with Bernard de Jussieu. Like Lamarck a few years later, they probably accompanied Jussieu on some of his weekly herborizations. These were rigorous botanical expeditions that departed from the King's Garden each Wednesday morning at 4:00. Sometimes the botanists would stay overnight in a village where they would feast and dance the night away, lighting fireworks and making a joyful din, though careful not to rouse Jussieu, whose sleep was sacrosanct. The party brought along lots of supplies: firearms, changes of clothing and shoes, hunting and pruning knives, boxes to hold seeds and cartons to contain drying samples, a writing case, books, thermometers, barometers, compasses, and—importantly—volatile alkali, Jussieu's signature cure for snakebites.[1]

By the autumn of 1766, Commerson had succeeded in establishing himself in Parisian botanical circles so effectively that Bougainville invited him to serve as the naturalist on his expedition to circumnavigate the world. Three years earlier, France had lost all of its North American colonies by the Treaty of Paris that ended the Seven Years' War. Louis XV and his ministers hoped that Bougainville's voyage would restore some French prestige and perhaps identify new possible colonies to compensate for the loss of imperial territory.

If Commerson accepted Bougainville's invitation, he would have to embark in just fifteen days, without having returned home to see the now four-year-old Archambault. "I find myself," Commerson wrote to his brother-in-law, François Beau, "in the slipperiest circumstance that could be," with just three days to make his decision. "I've thrown myself into the first café . . . to write you this letter in haste." Commerson said he was driven by "a lively passion . . . for great and difficult things" and a "fury to see" new phenomena; when confined to "common circumstances," he lapsed into a terrible "apathy." On the voyage with Bougainville, he hoped to witness "men—they say—8 or 9 feet tall, good, human, not yet at all perverted by society." The circumnavigation would "mark an epoch in the political and literary world." Commerson begged Beau to write and send his blessing: "I will be in despair not to receive letters from you before departing. . . . I embrace you and my son a thousand and a thousand times."

Barret and Commerson planned that she would go along disguised as Commerson's male valet, since women were banned by a royal ordinance from the king's fleet. This decision proved crucial to the expedition's scientific importance since Barret ultimately did most of the work, while Commerson developed an illness from which he never fully recovered, becoming increasingly weak and fragile. They boarded the storeship *Étoile* at La Rochelle on February 1, 1767, with so enormous a quantity of equipment, including more than two hundred books, that the obliging captain gave up his own quarters to them.

After three months' sailing, they reached the east coast of South America, disembarking here and there as they continued southward to Tierra del Fuego, with a detour to the Malouines Islands (now the Falklands); then they passed through the Strait of Magellan into the Pacific Ocean and explored the west coast of Patagonia. They stopped at various South Pacific islands on their way to Tahiti, landed in Papua New Guinea, and continued into the Indian Ocean, making more island landings before arriving at Île de France (now Mauritius).

Here, Barret and Commerson left Bougainville's expedition. In Tahiti, Barret's secret had been discovered. According to Lalande, a group of Tahitians had "recognized her, perhaps by smell." Other accounts suggest it was the Tahitian prince Ahutoro who first discovered that Barret was a woman; he joined the expedition and, according to the ship's doctor, François Vivès, became good friends with her. At any rate, Bougainville's sailors had already begun to suspect Barret, and she had told them she was a eunuch; she also always carried two pistols, which she displayed whenever she felt threatened. It seems likely that Bougainville had been in on the secret for at least a while. But once the truth was fully out, Barret and Commerson couldn't remain on the expedition. Barret chivalrously represented the pretense as hers alone, telling Bougainville tearfully that she had had no other livelihood and that the chance of making a tour of the world had "piqued her curiosity."

Barret and Commerson stayed in Île de France, where the colonial intendant was a good friend of Bernard de Jussieu's, a man named Pierre Poivre whom Commerson had met earlier in Paris. Poivre invited Commerson to remain and join him in studying the flora of the Mascarene

Islands and Madagascar. He offered resources including the services of his cousin and godson, Pierre Sonnerat, as a draftsman. Sonnerat, then a young man of about twenty, was a skilled artist who would go on to a career as a naturalist and naval and colonial commissioner. Commerson traveled to Madagascar, contributed to Poivre's botanical garden, the Jardin de Pamplemousse (Grapefruit Garden), and added to his and Barret's enormous collections, until he died in 1773. The contemporary accounts agree that Barret was with him to the end and closed his eyes.

During his final illness, Commerson diagnosed himself with dysentery and rheumatic gout, but even before that, and throughout the journey, his health had been poor. He suffered from recurring leg ulcers, which might have resulted from a rabid dog bite, as well as pleurisy and nephritis. While, like all European naturalists, Commerson relied on the information and help of local people, he also made a dramatic point of collecting specimens for himself, going out into the most treacherous of terrains. But it was Barret who supplied the physical strength and stamina for the pair. On their botanical explorations, as Bougainville recalled, she tirelessly climbed up and down mountains carrying everything—weapons, instruments, cases, notebooks, specimens—even in the snow and "on the frozen slopes of the strait of Magellan." Bougainville greatly admired Barret's industry and expertise, describing her as "a very experienced botanist," and later had a hand in inducing the French Ministry of the Marine to award her an annual pension of 200 livres.

Barret impressed Vivès, the ship's doctor, too. He reported that carrying all she did would "normally take eight or ten hands" and marveled, "One had to have seen her . . . exhausted from the rigors of the cold, in the water for shells, or in the foamy mud and snow for entire days seeking plants. . . . One can say in praise that it is impossible to conceive the work she did." Lalande commended her "indefatigable courage." A young gentleman passenger on the *Boudeuse*, one of Bougainville's ships, a nobleman named Charles-Nicholas Othon, prince de Nassau-Siegen, also expressed great esteem for Barret. The prince himself stumbled around the beaches and mountains in his customary aristocratic attire of high heels and velvet coats, wearing his wig of long, silky, flowing curls, apparently leading some of the indigenous people to suspect that *he* was a woman. Of Barret, he commented that rather than assume Commerson had induced her to come on the voyage, "I prefer to accord her alone all the honor of so bold an undertaking." Barret even received a favorable mention from Diderot in his philosophical dialogue about Bougainville's voyage, where he observed that "these delicate machines [that is, women] sometimes hide very strong souls."[2]

As for Commerson, he described Barret like a mythological heroine or even a goddess, reporting that she crossed "with agility the highest mountains of the Strait of Magellan and the deepest forests of the austral islands." He continued:

> Armed with a bow, like Diana, armed with intelligence and gravity, like Minerva, helpful and virtuous, inspired by some propitious god, she eluded the traps of beasts and men, not without countless times risking her life and her honor. She will be the first woman to have completely circled the terrestrial globe, having traveled more than fifteen thousand leagues.[3]

After Commerson's death, Barret opened a bar in Port Louis, the capital of Île de France, where she scandalously allowed people to play billiards and drink spirits even during Sunday Mass. Soon afterward, she married a French drum major stationed there, and in 1775 she sailed back to France with him.

Once back in France, Barret apparently set about trying to rescue her and Commerson's collection, which later estimates placed at about thirty thousand specimens, including five thousand new species, as well as fifteen hundred drawings. A doctor in Île de France, after Commerson's death, had shipped a small fraction of it to the King's Garden, where the naturalists were busy appropriating it for their own purposes, removing and dispersing the specimens. Meanwhile, the rest remained unattended in Île de France. Barret seems to have alerted the new king Louis XVI's government to the situation when she arrived in France, because the ministers took a sudden interest in the languishing collection and ordered Sonnerat, then in the Indian Ocean, to recover it. He sent back some more of it in 1775, all mixed in with his own collection.

At this point Lamarck was still living in his Parisian garret apartment, watching the clouds form and disperse from his small window, and studying medicine at the School of Medicine and botany at the King's Garden. But in just a few years, working on the botanical volumes of Panckoucke's *Encyclopedia*, he would find himself up to his elbows in the herbarium of the King's Cabinet with, as Lamarck later described it, "plants collected in almost all parts of the world," especially "the famous collection that Commerson formed in the Island of Java in Madagascar, in the Îles de France and Bourbon [now Réunion], in Brazil, in [the Strait of] Magellan, etc."

Around the same time, Lamarck also inherited Sonnerat's collection, including whatever it might have contained of Barret and Commerson's specimens. Whenever any traveler arrived in Paris with plants to show, Lamarck made a practice of being the very first to pay them a visit. When Sonnerat returned in 1781 from several years' tenure in Pondicherry (now Puducherry), the capital of French India, Lamarck was not just the first but the only one to call on him, and a disappointed Sonnerat bestowed the extensive herbarium he had brought back from India upon his sole visitor.

Lamarck was the first to publish an illustration of one of Barret and Commerson's most surprising and enchanting discoveries, which they encountered in the countryside near Rio: a plant that continues to lend its loveliness to gardens all over the world. Commerson named it "Bougainvillea" after the expedition's leader. The plant is remarkable for the fact that its beauty derives not so much from its flower, which is small and unremarkable, as from a kind of leaf, the brilliantly colored bracts surrounding the flowers. The botanist Antoine-Laurent de Jussieu (Bernard's nephew) noted that the whole family to which the bougainvillea belonged consisted of plants foreign to Europe and was full of unexpected features for European botanists, such as these bright bracts. He named the family Nyctaginaea from the Greek root *nukt* for "night," because its flowers bloom in the evening.[4]

The bougainvillea appeared to European readers for the first time in one of Lamarck's botanical volumes for Panckoucke's *Encyclopedia*. The copperplate engraving, based on drawings by Louis-Denis Fossier, a scientific illustrator connected with the Academy of Sciences, shows a stem of the plant with leaves, bracts, and flowers, as well as close-up views detailing the stamens, the pistils, and the insertion of the flowers into the bracts.

Another example of a surprising plant that Barret and Commerson collected came from a genus native to Madagascar. Like the Nyctaginaceae family (as it is now called), this genus defied traditional old-world classifications. For instance, it had many different sorts of leaves, of various sizes and shapes, and it was a new experience for French botanists to see these appearing together on a single plant. Moreover, these tricky leaves appeared to shield and disguise the flowers . . . and we remember

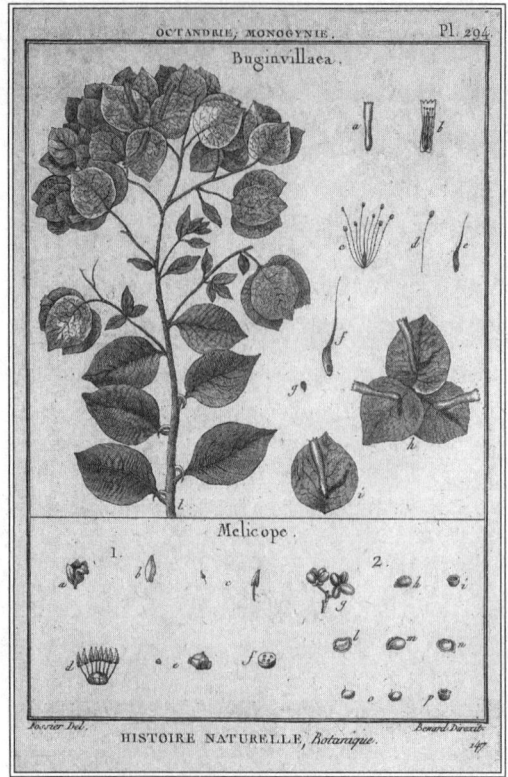

Bougainvillea by Louis-Denis Fossier, from Lamarck, *Tableau encyclopédique*

from Vaillant, Linnaeus, and Lamarck just what *flowers* are. "I'm crazy about it," Commerson rhapsodized to his botanical colleague in Port Louis, Joseph-François Charpentier de Cossigny, regarding the new plant: "It has its stamens [the male fertilizing organ] carried without filament at the edge of a little cup inside the flower fold that we call *nectarium*." In addition to this suggestive placement of the stamen nestled in a fold of the flower, Commerson admired the striking distribution of leaves, with the fullest at the top and the most sinuous on the lower branches. Also, "the fruit, sir, the fruit, don't forget it I beseech you," he exclaimed.[5]

The sexy plant with its hidden parts, surprising features, and enticing fruit reminded Commerson of his partner, and he took to calling it "bonafidia" in reference to "Bonnefoi," a nickname Barret sometimes used. Later, Commerson made this allusion to Barret even more explicit by naming the plant *Baretia* and dedicating it to her. "This plant with its deceptive finery or foliage," he wrote, "is dedicated to the valiant young woman who, taking on the dress and temperament of a man, had the curiosity and the audacity to travel the whole world, by land and by sea." He acknowledged how much he owed to Barret's indefatigable help: "We are indebted to her heroism for so many plants never before collected, for so many herbaria constituted with care, for so many collections of insects and shells, that it would be an injustice on my part . . . not to give her the deepest tribute by dedicating this flower to her."

Immersed in Barret's specimens and the other plants that had arrived at the King's Garden from all over the world, Lamarck drew a bold conclusion: Everywhere on the planet Earth, living beings were composing the inanimate world around them. "Although the animals and plants found in Madagascar, India, the Moluccas, Peru, the West Indies, etc., are very different from those who inhabit Europe," he wrote, "their remains nevertheless everywhere produce clayey, marly, or limestone soils; everywhere, we find marbles, gypsums, schists, magnesium stones, gneiss, quartz rocks, rock crystal, etc." Life in all of its infinite diversity of forms retained its defining capacity to create compound matter. Moreover, Lamarck concluded that nowhere in the world was there any other creative force at work. He considered the global collection in the King's Garden to be "very convincing proof" that all "raw and inorganic materials . . . come directly or indirectly from the remains of beings who have been endowed with life."

EIGHT

Revolution in the Garden

A n acute attack of bladder stones in the spring of 1788 carried off Buffon at age eighty: a painful death, to be sure, but in another way perhaps a further example of his charmed existence, since he just managed to avoid the travails of the Revolution. Buffon died during the very last moments of a world in which he had been born to a lofty position, and where he had masterfully presided throughout his life. This world on the eve of its total dissolution remained deceptively intact, so Buffon never knew it was all about to come apart. To him, the order of things was comfortably eternal.

The intendancy of the garden passed not to Buffonet, despite Buffon's fondest hopes, but to the marquis de La Billarderie, who had an indirect family connection with Lamarck. When Lamarck appealed to him for professional help, the marquis honored their connection in traditional fashion by creating a position for Lamarck as keeper of herbaria (the dried plant collections) of the King's Cabinet. This Old Regime act of patronage took place in June 1789, just weeks before the storming of the Bastille on July 14. For the first time, Lamarck had a salaried position, of 1,000 livres annually. In May 1790, he would also at last become a paid member of the Academy of Sciences, with an annual allowance of

1,200 livres. But his good fortune proved largely illusory, coming as it did with the collapse of the Old Regime. Within three years, the revolutionary government would abolish the Academy of Sciences in its purge of royal, elite, and expert institutions.

The keeper of herbaria position would prove similarly short-lived. It was a clumsy, impolitic idea anyway, since it greatly annoyed the two occupants of botanical positions in the garden: René Desfontaines, who had recently succeeded Le Monnier in the chair of botany, and Antoine-Laurent de Jussieu, Bernard's nephew, who now held the position of botanical demonstrator. Both resented the encroachment on their territory.

Lamarck wanted to completely reorganize the herbaria; he had a palpable hankering to impose order on chaos, and "this beautiful Collection, in the state it is in, cannot be of the slightest use to the progress of Botany." He submitted a comprehensive proposal—also intended to justify his precarious position as keeper of herbaria—to Bernardin de Saint-Pierre, the botanist and Romantic novelist whom we encountered in the garden during our prefatory stroll, who succeeded La Billarderie as intendant in 1791. First, each plant must be "fixed with strips on a half-sheet of white paper, and this half-sheet be enclosed in a folder," so that "those who come to consult the collection will not break the plants." Next, each plant must have a label with a number at the top; beneath that, the name according to Linnaeus or any "more modern author"; next, Tournefort's designation; then the plant's native region; the herbarium or garden it had come from; the name of the person who had given it to the cabinet; and the year it was given.

As for the categories into which Lamarck proposed to organize the specimens, he diplomatically chose the system of his disgruntled colleague Antoine-Laurent de Jussieu as the one that best illuminated the natural relationships among the plants and also the one that corresponded to the organization of the King's Garden (since Jussieu's job as botanical demonstrator involved organizing the garden). Once he had affixed, tagged, and arranged the specimens, Lamarck proposed to publish a catalog of the whole collection accompanied by a table of all the plants arranged according to "Linnaeus's sexual system," for the use of those botanists—virtually all of them—who followed Linnaeus.

Once the general herbarium was established with a single example of each plant in the collection, Lamarck meant to use the duplicate specimens to form four more specialized herbaria. He reckoned it was useless to keep the specimens grouped according to the travelers who had collected them. Instead, he proposed grouping them by place of origin: Europe, Asia, Africa, and the Americas. Each of these would be further subdivided by region, and in Lamarck's list we can easily identify a map of France's global imperial presence and ambitions: North America, South America, the Antilles, the Cape of Good Hope, Ethiopia and Senegal, the Barbary Coast, India, China, Japan, the Molucca and Philippine Islands, and the islands of France, Bourbon, and Madagascar.

Any plants left over after all this work was done would go into a reserve to be exchanged with botanists beyond the garden for specimens that would increase the collection. Lamarck also proposed to expand the herbaria by planting new plants in the garden and drying new specimens from these.

Soon after assuming his new position as intendant of the King's Garden, Bernardin noticed that he never seemed to see the keeper of herbaria around the place and wrote to Lamarck to ask him about it. Lamarck came immediately to find Bernardin and explained that he'd like nothing better than to do his job of keeping the herbaria, but he'd been altogether unable to gain access to them. Bernardin then checked with Desfontaines and Jussieu, who told him that Lamarck's position was altogether "useless," that La Billarderie had created it only as a special favor, and that they themselves were perfectly capable of looking after the herbaria.

Bernardin reported to the new governing authority—the revolutionary National Assembly, made up of representatives from the three "estates" (clergy, nobility, and commoners)—that he suspected Lamarck's plan to reclassify the herbaria, subordinating Tournefort's nomenclature to Linnaeus's, had been the main provocation driving Desfontaines and Jussieu to go so far as to bar Lamarck from the collection. "I already knew," Bernardin commented wryly, "that it was more difficult to arrange botanists than plants." In order to keep Lamarck on salary but out of his colleagues' way, Bernardin proposed sending him back out botanizing around the kingdom with the aim of achieving a more complete edition of *French Flora*. "That way," he told the members of the National Assembly with resolute optimism, "everyone will be employed and content."

But general contented employment was very much not the order of the day, and Bernardin's solution didn't end up being enacted; instead, as we'll see, Lamarck would be removed from botany altogether. No sooner had he gained an official, salaried position in the garden than everything was thrown into uncertainty. Following the summer of 1789, the king and his court no longer had jurisdiction over what was no longer the King's Garden. Lamarck wrote two desperate pamphlets pleading to maintain his job as keeper of herbaria, which the National Assembly's Finance Committee threatened to eliminate. Indeed, the garden itself—along with all academic and royal institutions—was in peril from the beginning of the Revolution. In the summer of 1790, the National Assembly requested that the officers of the King's Garden draft a plan of reforms to reduce costs. In response, Lamarck wrote a *Memoir on*

Cabinets of Natural History and Particularly on That of the Garden of Plants, deftly renaming the garden to purge it of its now-dangerous royal connection. Lamarck proposed the creation of several new conservators, including one of "insects and worms." Little did he suspect who would end up in that position! For himself, he requested the position of conservator of botanical collections, further infuriating Desfontaines and Jussieu.[1]

Over the summer of 1790, the officers of the garden drew up their plan for reform, in which they adopted many aspects of Lamarck's proposal. They called for the establishment of a Museum of Natural History encompassing the garden and cabinet, with twelve professorships, equal in rank and pay, each with a salary of 2,500 livres. There would be no intendant; rather, the professors would run the garden collaboratively, choosing a director from among themselves on a rotating basis by vote. The professorships included chairs in chemistry and the chemical arts, gardening, human and animal anatomy, mineralogy, and naturalist painting. For the two chairs in botany, the plan named Desfontaines and Jussieu. In zoology, there were to be a professorship in oviparous quadrupeds, cetaceans, and snakes for Lacépède; and one in quadrupeds and birds for Geoffroy Saint-Hilaire. As for Lamarck, the plan proposed that he leave botany behind to occupy the third zoological professorship in "the Natural History of Insects and Worms."

. . . .

Lamarck and Rosalie de la Porte now had four children. Their third and fourth had been born in 1786 and 1787 in the rue Copeau: Antoine, hard of hearing, would become a painter, never marry, but live a modest life into old age and the middle of the Second Empire; and René would die in adolescence in 1805, soon after Napoleon crowned himself emperor of the French. In the midst of the revolutionary upheaval, in January 1791, their fifth child, Auguste, was born. Auguste would be the most successful and established of Lamarck's children, the source of the family's posterity up to the present day, and also the source of much of the personal information about his father's life that made it into the historical record.

The autumn following Auguste's birth, the National Assembly ceded power to a new body, the Legislative Assembly, a step closer to representative government since the election of the assembly was by male census suffrage: one household, one vote. That summer, with his new (though short-lived) modicum of prosperity from the keeper of herbaria position and the Academy of Sciences pension, and also with the help of a substantial loan, Lamarck bought a bigger house with a little garden, and he and Rosalie moved their family of five children around the corner from the rue Copeau to No. 4, rue du chemin de Gentilly (now the rue Geoffroy Saint-Hilaire), which ran partly along the Garden's southwestern border.

A month later, in August 1792, an insurrection within the Revolution brought the end of the monarchy, the arrest of the king, and the replacement of the Legislative Assembly by the National Convention, France's first republican government. Cornélie, Lamarck and Rosalie de la Porte's sixth and last child, was born around this time: Her birth certificate is lost, but her name reflects the revolutionary moment, evoking the virtuous Roman matron and mother Cornelia, who educated her sons to be great republican politicians.

Like her older sister, Rosalie, Cornélie remained single throughout her life and stayed with her father until his death. She appears with him on the bronze bas-relief decorating the plinth of the statue of Lamarck by the sculptor Léon Fagel that now presides over the north entrance to the Garden. The French National Institute installed this statue in 1909 to commemorate the hundredth anniversary of the year of publication of Lamarck's *Zoological Philosophy*, which also happened to be the hundredth anniversary of the year of Darwin's birth, thereby competing with the centenary Darwin festivities at Cambridge University. The bas-relief bears the defiant inscription "Posterity will admire you. Posterity will avenge you, my father."[2]

Rosalie de la Porte lived in the new house at No. 4, rue du chemin de Gentilly for only a few short months; she died there around the time of Cornélie's birth, in October 1792. Lamarck became a widower and single father of Rosalie Joséphine (fourteen), André (eleven), Antoine (six), René (five), Auguste (one and a half), and the baby Cornélie. However, young André, the second child, didn't remain long in the new house either, if he lived there at all. From earliest childhood, he was passionately interested in the sea, in ocean explorations and voyages, in maritime adventure. André "spent whole days in his father's library," as his younger brother Auguste later recalled, "devouring the stories of Captain Cook and other famous navigators." At seven, he ran away from school with a piece of bread in his pocket and a map of the world in his hand, to follow the bank of the Seine "like Christopher Columbus, in search of some unknown land." Happily, unlike Columbus he was forestalled from brutal acts of enslavement and colonization. When evening fell and his bread was finished, "he perceived that he hadn't foreseen all the difficulties of his undertaking" and accepted a ride home from a charitable coachman.

But at eleven, around the time of the family's move, his baby sister's birth, and his mother's death, André achieved a closer approach to his Columbian ambitions by shipping as a cabin boy aboard the frigate of the slave trader and pirate Jean-François Landolphe, who had been extending French imperial interests and establishing slave-trading posts in Benin. When Landolphe's frigate returned to France, André was no longer on it, and no one knew where he'd gone; for the next seven years, the family would receive no news of him.

. . . .

In January 1793, the National Convention convicted Louis XVI of high treason and executed him by guillotine on the Place de la Revolution (soon afterward renamed the Place de la Concorde). The following summer, the "Montagnards"—the most radical members of the National Convention, so called because they occupied the highest benches at the riding hall of the royal equestrian academy where the Convention met— seized control of the Convention and began making liberal use of the guillotine to consolidate their power. These radicals were also known as Jacobins since they were members of the dominant political club, the Jacobin Club, which in turn got its name from the Couvent des Jacobins (a Dominican monastery), where the club met.[3]

Just as this increasingly violent phase of the Revolution began, the National Convention's Committee on Public Instruction at last took up the question of the administrative reform of the Garden of Plants and on June 10, 1793, issued its decree establishing the National Museum of Natural History, which essentially follows the plan that the officers of the garden had written. Lamarck was appointed professor of the natural history of insects and worms; he was also to take charge of the museum's library.[4]

If you're a bird who doesn't much like swimming but who needs to seek food by wading into the water, your feet constantly sinking into the muddy bottom, your determined efforts to keep your body dry will result in elongated stilt-like legs. If, during the tumult of the Revolution, the only way to retain a job in the botanical garden and natural history museum is to take a position no one else wants, a "professorship in insects and worms," you will turn from a career in botany and, in middle age, remake yourself as a zoologist specializing in these lowly creatures, for whom you invent a new taxonomic category, "invertebrates," on the basis of which you will forge a momentous new theory of the history of life.[5]

PART TWO

Insects and Worms

In which Lamarck turns reluctantly from Botany to the Zoology of lowly Creatures, which prove to be unexpectedly Inspiring, leading him to the Principle of Animality, namely, Generation. He shows that the meanest of Creatures are the Creators of Mountains and Continents. He marries young Charlotte Reverdy and buys a House in her Native Countryside, where he observes some Swallows cooperating to Build a Nest and reflects upon the Collaborative, Creative Agency of Living Beings; he offers this Observation to a Friend as an Example of Animal Cooperation. Charlotte Reverdy dies at Twenty-Four, leaving Lamarck with two more Children and big Debts. He marries his Brother's Wife's Sister, Julie Mallet. He angers his Colleagues at the National Institute by Rejecting their Theory of the Composition of Matter: Only Living Things, he says, can Compose the World. He presents to his Students, for the First Time, his Theory that Animals can Transform Themselves by means of their Habits and Behaviors, that is, by Acts of Will.

NINE

The Reluctant Zoologist

H e didn't want to be a professor of insects and worms! Nevertheless, Jean-Baptiste Lamarck, aged forty-eight, found himself declared one. On June 10, 1793, as we've seen, he arrived
at this curious pass: Upon going to bed the night before, he'd been a
botanist, but that day the revolutionary committee in charge of reorga-
nizing the King's Garden, with a stroke of its pen, turned him into—not
a frog, exactly—but a zoologist. Robespierre was maneuvering into position at the helm of the Committee of Public Safety, accelerating the use
of the guillotine to separate the heads from the necks of many aristocrats
and anyone else the committee deemed counterrevolutionary. Lamarck,
a middle-aged widower with six children, had little choice. He kept his
head but had to remake himself overnight in middle age, leaving botany
to take up the professorship in insects and worms, creatures about which
he knew, as yet, very little.

Professors in the new museum were entitled to lodgings in the Garden of Plants. So, Lamarck completed his residential trajectory toward
the garden by moving *into* it. Baby Cornélie and her siblings Rosalie Joséphine (fifteen), Antoine (seven), René (six), and Auguste (two) moved
with their widowed father, leaving their house on the outskirts of the

garden at No. 4, rue du chemin de Gentilly (André, then aged twelve, was still missing after shipping out to sea). They took up residence in the grand old fifteenth-century mansion Buffon had acquired at the south-west corner of the garden, which had served as the intendant's residence until the Revolution. Initially they lived in rooms on the third floor but were soon able to move down to the large apartment on the second floor.[1] In exchange for these comfortable surroundings and grand vistas, La-marck agreed to spend his days contemplating the tiniest and humblest of creatures.

By the end of the summer, a rallying cry had sounded in the National Convention and was resonating throughout the country: "Let terror be the order of the day."[2] And yet life in the garden carried precariously on. The family's new, more expansive living quarters included two cellars—one for wine and cider, the other for firewood—a kitchen, and two din-ing rooms. The apartment also had a living room, a study, various smaller cabinets and antechambers, and four bedrooms. A thorough inventory taken some years later invites us on a mind's-eye tour of the apartment; we'll begin in the fancier dining room. Here it is, overlooking the street bordering the garden. The room seems cheerful, with blue and white curtains and two paintings of seascapes in gilt frames, perhaps remind-ing those at home of the seafaring André.

Leaving the dining room, we continue along the street side of the house to Rosalie's room. She's twenty-two at the time of the inventory and has the second-largest bedroom in the apartment with an adjoining *cabinet de toilette* for washing and dressing. We find a wicker chair and pine table where Rosalie might sit and work on her father's projects, a bookshelf, an oak cupboard, and an earthenware stove with fireplace tools. She also has a clock in a copper box with glass lid, two sliver-plated candlestick holders, and a print in a gilt frame. Her sleeping alcove seems cozy, protected by red calico curtains, with a woolen mattress covered in checkered linen, a ticking feather bolster, and a white woolen blanket.[3] Nosily, we rummage through her clothing and note that she sleeps in a muslin-trimmed nightshirt and nightcap. Getting dressed, she has a choice among muslin or taffeta petticoats and dresses, or else a dark red felt dress, and cotton or silk stockings. Once dressed, she ties a pair of canvas pockets around her waist to hold small items: coin purse, comb, watch, pincushion, pencil, small book, or notebook. She picks a handkerchief, one of her three pairs of shoes, and perhaps one of her two straw hats. If the weather is cool, she can take a lawn scarf or white muslin shawl; on a warm day, she might pick up her fan on her way out.

Following Rosalie out of her room, we turn a corner to the master bedroom, with its two windows overlooking the garden, their canvas curtains trimmed in muslin. A pair of mirrors between the windows and a third on the mantel calmly decline to show us the scenes they've reflected, but another seascape in a gilt frame reveals a possible lack of imagination in interior decoration. Wandering over to the mantelpiece, we admire a watch mechanism like Rosalie's.[4] Lamarck evidently engages in a great diversity of activities in his bedroom: eating, drinking, sitting around and talking, toileting, washing, sleeping, producing ever more children. The big, built-in cupboards on either side of the fireplace contain two Japanese porcelain teapots, an assortment of porcelain cups, plates, and bowls, a cracked sugar bowl, and a set of six cordial glasses on mahogany saucers. The room has a wealth of places to sit and lounge: two horsehair armchairs and four side chairs all upholstered in a yellow-patterned Utrecht velvet; a marble pedestal table garnished with copper; and a rosewood Regency chest with a marble top.

The privacy of the sleeping alcove is protected by woolen curtains in yellow, red, and white stripes, but we're invisible on this imaginary tour, so we pull them aside and peek within. We find two matching backrests in the double sleeping berth. The berth looks comfortable with its three woolen mattresses covered in checkered canvas, a bolster filled with feathers and another with horsehair, a woolen blanket, and a yellow taffeta quilt. A wardrobe behind the sleeping alcove holds a chamber pot and faience washing bowl, each on its own stand. The capacious room also has a spot for a third person to sleep. Inside a cabinet next to the sleeping alcove, we discover a bunk with a pallet, a walnut bedside table, and an old cane chair. Brazenly, we rifle through Lamarck's wardrobe and discover two pairs of cashmere breeches, one yellow, the other gray, and three suits: a gray silk, a black wool, and a dark red felt one like Rosalie's warmest dress. Lamarck has a choice of cotton or silk stockings and muslin cravats, two hats, and two pairs of shoes. Depending on the season and the occasion, he might wear his frock coat, a waistcoat, or perhaps his fur-trimmed greatcoat.

After the master bedroom, still on the garden side of the house, we find another bedroom—perhaps for Antoine, the oldest boy at home—which contains a backgammon set with boxwood and ebony pieces, the only game or toy in the inventory. The younger children evidently sleep in the fourth bedroom, on the street side, which has two sling beds and an old sunken sleeping berth.

Arriving at the corner of the building, we come upon Lamarck's study. From the ornate, double-sided oak desk he can look out in one direction to the garden's entrance and, in the other, across the garden to the Seine. Here he is now, contemplating the garden from his desk. We turn our gaze around the room and glimpse a rosewood secretary desk with a hinged lid; a small walnut writing table where, perhaps, Rosalie is now writing letters; and a mahogany cupboard with locking shutters to hold sensitive papers and documents. The room offers several inviting places to sit: two wicker chairs, two polka-dotted horsehair cabriolet chairs, and a shabby silk armchair. When visitors come to the garden to see Lamarck's herbarium or collection of shells, he seats them comfortably in one of these, and during breaks from work he surely sits in one himself to play his viola da gamba, which he keeps near at hand.

We wander over to inspect Lamarck's library of more than a thousand volumes, housed in four big gray wooden cupboards. There's an oak step stool with five steps so we can climb up to see the books on the highest shelves. We recognize plenty of old friends (and enemies): works by Linnaeus, Tournefort, Benjamin Franklin, and Georges Cuvier. Housed in its own, special cabinet apart from the other works is Buffon's *Natural History* in 38 volumes. And we can hardly miss Panckoucke's *Encyclopedia*, all 241 volumes of it; Lamarck himself edited and oversaw the botanical volumes. There are works on botany, zoology, and geology: Duhamel du Monceau's *Physics of Trees*, Elizabeth Blackwell's *Curious Herbal*, Brisson's *Ornithology*, and Bourguet's *Treatise on Petrifications*.

We notice, of course, plenty of works on insects and worms: d'Argenville's *Conchology*, Geoffroy's *Abridged History of Insects Found in the Vicinity of Paris*, and Bazin's *Natural History of Bees*. The colonial projects of France and the other imperial powers figure prominently here in Lamarck's library as in his herbarium: Patrick Browne and Hans Sloane on the natural history of Jamaica, Poiret's *Voyage en Barbarie*, and Charles Plumier's *Treatise on the Ferns of America*. We pick up Jean-Joseph Sue's *Essay on the Physiognomy of Living Bodies Considered from Man to Plant* and leaf through it. The author was a professor of anatomy at the Royal College of Painting and Sculpture, and his *Essay* straddles anatomy and art. He writes that all living things, including plants, vividly express their sensibilities through their "physiognomies." Lamarck said something similar in *French Flora*, when he described the "innocent welcome" of flowers and their "laughing and lively" presence.

In certain respects—its beautiful and convenient location, its size, some of the nicer furnishings—the apartment seems opulent. But the only "jewel" mentioned in the inventory is a watch in a silver case on a silk cord; the cutlery is characterized as "plain" and "mismatched." There's no record of the family having servants, though a servant or two might have slept in the extra berths in the master bedroom or smaller street-side bedroom; Lamarck's wife and daughters likely did much of the housework. By the standards of most French lives around 1800, Lamarck's was surely one of good fortune and ease, yet he had reason to be perpetually worried about money, his children's futures, and his own professional standing and security. These worries were slightly mitigated when, the autumn after being declared a zoologist by the National Convention, in October 1793, he renewed his life in another way by marrying a very young woman named Charlotte Reverdy. At barely twenty, she was three decades his junior and just five years older than his daughter Rosalie. The young bride brought a small dowry. In marrying Lamarck, she also of course inherited all the children, ranging in age from the then-fifteen-year-old Rosalie to the one-year-old Cornélie.

The following spring of 1794, Lamarck inaugurated his course on insects and worms, which he would teach annually under the terms of his professorship in the garden. For the following quarter century, through the rest of the revolutionary years, then the Napoleonic period, then the Bourbon Restoration, he walked across the garden from his residence to the amphitheater, built on the eve of the Revolution at the very end of Buffon's intendancy, to give his lectures.[5] It was in this course that Lamarck would first present his radical theory of life to the world.

Lamarck continued teaching his first course throughout the summer of 1794, which brought "Thermidor," the great upheaval in the midst of the Revolution nicknamed for the date in the revolutionary calendar—9 thermidor (July 27) year 2—when comparative moderates rebelled against Robespierre's Reign of Terror, as they now named it. These anti-Jacobins seized control of the National Convention and arrested Robespierre, then demonstrated their superior judiciousness by executing him and his closest associates the next day without a trial. In the autumn, Lamarck became the secretary of the group of professors at the museum; for the remainder of year 3 (until autumn 1795), the minutes are in his wonderfully legible hand, with his flourish of a signature at the end of each.

"The Thermidorian Reaction" followed: a period of months during which the National Convention sought to restore stability, partly by issuing a blizzard of decrees, many of which granted indemnities and pensions to compensate for the livelihoods that had disappeared when the revolutionary government abolished the institutions of the Old Regime. Lamarck succeeded in securing a pension replacing the one he had lost with the abolition of the Academy of Sciences, as well as an indemnity

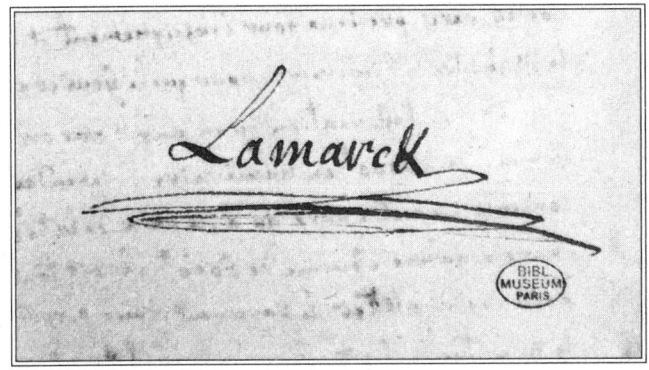

Lamarck's signature from a letter to Lucien Bonaparte, 1800

for his "contributions to the Republic." These funds, together with his salary as professor of insects and worms and Charlotte Reverdy's dowry, meant that he and his family enjoyed a modest and tenuous solvency. It was a good thing, too, since that fall, in October 1794, Lamarck's seventh child, and first with Charlotte Reverdy, was born. Like his two-year-old half sister, Cornélie, Aristide was named in revolutionary fashion: after the Athenian statesman Aristides, known as the Just, who defended the newly democratic Athens as a general during the Persian Wars.

TEN

Nemesis

B efore the end of his first year as a professor in the garden, Lamarck began making powerful and soon-to-be-powerful enemies. His first opportunity for enemy making came when he
took up the task of hiring an aide-naturalist for the Museum of Natural
History. He passed over the ambitious twenty-five-year-old Georges Cuvier in favor of a painter and taxidermist, Jean-Baptiste-Simon-Ferdinand
Desmoulins, on the ground that the job required an artist and that Cuvier was "more savant than artist."[1] Lamarck could scarcely have guessed
that Cuvier would become permanent secretary of the National Institute's First Class and in that capacity would have the task of writing Lamarck's official eulogy, which would in turn serve as the primary source
for all of his subsequent biographers. The moral of the story: never alienate someone who just might be in charge of writing your epitaph.

Cuvier, who appears in portraits as a rather pinch-lipped, hatchet-
faced person, was a Lutheran from Montbéliard, a town at the very edge
of the Jura Mountains that had become part of France only the previous
year when the revolutionary army had annexed it. Before then, Montbéliard had been a principality in the Holy Roman Empire. Over the course
of his long, extraordinarily productive and prosperous career, Cuvier

would accomplish foundational work in zoological classification, focusing the field on the arrangement of the internal organs, and in comparative anatomy and vertebrate paleontology. His studies of fossil bones would be instrumental in bringing about a general acceptance of the reality of extinctions.

When Lamarck first met him and denied him a job, Cuvier was just beginning a meteoric rise in Parisian naturalist circles. Two years after Lamarck passed him over, Cuvier was hired by Lamarck's colleague Jean-Claude Mertrud, holder of the garden's new professorship in comparative anatomy, as his own assistant. When Lamarck became a member of the National Institute's First Class in its botany section, Cuvier, too, became a member in its zoology and anatomy section. In 1800, Cuvier

Portrait of Georges Cuvier (1769–1832) by François-André Vincent, 1800

gained a lifetime professorship at the ultra-prestigious Imperial College (today the Collège de France), and two years later he succeeded Mertrud as professor of animal anatomy at the museum. A rumor had it that Cuvier had bribed Mertrud to name him as successor by promising to pay Mertrud's heirs a portion of his salary. In general, he had a remarkable aptitude for cultivating powerful people and maneuvering himself into positions of institutional authority. He would soon reign as the dominant figure in natural history across several political regimes and through the first third of the nineteenth century.

Even if he didn't much resent being passed over for the aide-naturalist job, Cuvier would have had ample reason to dislike Lamarck. In keeping with his Protestant faith, Cuvier was particularly unsympathetic to Lamarck's central, revolutionary idea that animals transformed themselves by their own agency and that minute changes, inherited and accumulated over generations, could explain their fitness to their environments.

All of natural history and zoology would fall apart, Cuvier believed, if living forms were fleeting. He was sure that species remained unchanging from creation to extinction. He understood the fossils of extinct creatures to reflect not transformations but rather a series of catastrophes, each eliminating the animals of a given time and place; they were then replaced by brand-new, equally fixed and unchanging beings through new acts of creation.

Cuvier had a powerful ability to translate religious dogmas into scientific principles. For instance, his central anatomical law, which historians generally refer to as his principle of correspondence of parts, held that one could infer all the parts of a creature from any single part since they all followed from one another by rational necessity. Implicitly, this must be the case because a divine designer had created them according to an optimal plan. "Every organized being forms an ensemble," Cuvier wrote, "a unique and closed system, in which all the parts correspond mutually . . . and consequently, each of them, taken separately, indicates and reveals the others." Just as one could take apart any artificial device, study it, and understand the function of each part as the engineer had intended it—and even predict missing parts based on the design of present ones—Cuvier boasted that he could do the same with the partial fossils of extinct creatures.

Studying a single "bud of a molar" from an extinct kind of hippopotamus, he had immediately known—so he said—what the rest of the bones would be like, and that they would bear the same relation to their counterparts in living hippos. He'd been gratified, once the bones were excavated, to see that they bore out "the infallibility of these general laws of zoology"—infallible because they had been established by an omnipotent lawgiver. Likewise, from half a lower jaw and metacarpal of an elephant-like fossil found in Italy, Cuvier inferred that the animal in question would have been "at least 15 feet tall" and asserted that this was an extinct species of elephant, the same species as the "*mammouth*" whose fossil remains were found best preserved in Siberia. Cuvier distinguished this European mammoth from a North American elephant-like creature that he said was also extinct and that he named—for the remarkable protrusions on its molars—"*mastodonte*" from the Greek roots *mastos* (breast) and *odon* (tooth), meaning "nippled teeth."

In each case, the parts of the animal made manifest a rational plan—as in any constructed device—and the job of the paleontologist was to uncover this plan by discerning the purpose of each part. The solid construction of an extinct rhinoceros's nose, for instance, was "doubtless destined to support the horn." Each structure, Cuvier assumed, had an intended purpose. Omnipresent, hovering between and around the lines of his text, is the divine origin of these purposes.

Regarding the cause that had rendered all these former creatures extinct, Cuvier concluded that it must have been "general," "sudden," and "aqueous"—very much, in other words, "as we imagine the flood" in the book of Genesis. All in all, Cuvier's core scientific ideas—the fixity of species, rational design, the catastrophic causes of extinctions—supported the doctrine of special creation: that God created each species for its niche in nature, and that each species then remained just as it had been created until God decided to wipe it from the earth and replace it with new forms of life. The beings themselves had no role to play in their own creations, transformations, or ultimate demise; they existed entirely at the mercy of a remote and supernatural power. Lamarck, for his part, scornfully dismissed the idea that some "so-called general catastrophes of the globe" could conveniently explain everything without requiring the naturalist to study ongoing natural processes.

. . . .

Fifteen students attended Lamarck's course on insects and worms during its second year, and the first for which a roster survives. They included Jacques-Antoine Creuzé-Latouche, a political economist and deputy to the National Convention who had refused to vote for the death of Louis XVI; after the fall of the Jacobins in July 1794, Creuzé-Latouche had taken a leading role in revising the constitution. How did he get on, we may wonder, with his classmate Jean Angélique Lemoine-Villeneuve, a "regicide" Convention deputy who voted for the death of the king? Lamarck's students in this year also included Nicolas Maréchal, a painter and illustrator at the Museum of Natural History; and Jean-Mathieu Artaud, a botanist from Arles. But they weren't all learned or politically powerful people. One student, a Mathieu Ferouillax from Lyon, must have been unaccustomed to writing numbers, because he initially wrote down his age as "91," then crossed it out and amended it to "61."

Just as the course reached the close of its second session, in the summer of 1795, two years after the political *bouleversement* that had turned Lamarck into a zoologist, another upheaval made him a botanist once more. The National Convention, having abolished all the learned academies two years earlier, re-created them as divisions of a single National Institute—which remains to this day France's state entity housing its various scholarly societies—and Lamarck was appointed to the botany section of the Institute's "First Class," its mathematical and physical sciences division. Nevertheless, in his main job at the Museum of Natural History and Garden of Plants, he remained a zoologist of insects and worms.

Using Charlotte Reverdy's dowry, Lamarck began proceedings to buy a small house in her native countryside, the pays de Bray, a region of pastures and woodlands in northwest France about a hundred kilometers from Paris. With the Thérain River running through it, the region is known for its clay soil, its cobhouses, and its butter and cheese. It was "very pleasant," according to a contemporary chronicler of the area, the Breton writer Jacques Cambry, "a mixture of knolls, hills, vast carpets of cultivation; but nothing could be more cheerful and fresh than the shores of the Thérain." The house itself was called Beauregard, and it was in the commune of Héricourt, a hub of the trade in spectacles and other optical goods.

The following autumn of 1795, the National Convention accomplished a new constitution and disbanded, giving way to a new republican government with a bicameral legislature that vested executive authority in a Directory of five directors. Lamarck was still struggling to acquire

Beauregard. As with other properties belonging to "émigrés," nobles who fled the country during the Revolution, the revolutionary government had declared the house to be state property and put it up for sale. But the family of the émigré owners had a different opinion and contested the republic's sale of what they regarded as their property. When that matter was at last resolved in Lamarck's favor, he confronted endless bureaucratic snafus in attempting to pay off his debt on the house by selling his magnificent shell collection—which had grown into one of the biggest and most beautiful in Paris—to the state.

At last, he prevailed, as Jacques Cambry's chronicle testifies: "There is a little country house in a delightful position; it is located on the slope of a hill whence the eye embraces the entire Thérain valley. . . . Citizen Lamarck, so famous for his knowledge of natural history and his meteorological observations, is the current owner."[2]

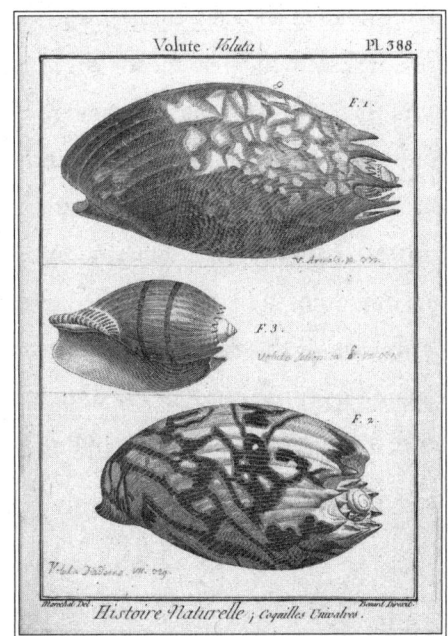

Shells by Nicolas Maréchal, 1827

. . . .

S ince the purview of Lamarck's new chair included mollusks, his collection of shells did in fact make him more of an expert in the area
than any of his colleagues at the Museum of Natural History. Also, according to Geoffroy, Lamarck had worked to learn a little bit about this
area of zoology in order to be able to talk to a good friend, Jean-Guillaume
Bruguière, whose discourse was altogether "limited to conversations
about shells." (This was in fact the very friend he'd originally had in
mind for the professorship of insects and worms.)

Nevertheless, Lamarck assumed the post with resignation rather
than enthusiasm. Later, he would confess to his students that he was initially dejected at being sentenced to study bugs and slugs for the rest of
his life. "What interest, I said to myself, can be inspired by the sight of a
mite that lives in cheese?" Or by the weevils in the grain bin, the larvae
eating holes in the furniture, the slugs destroying the garden? His assigned
creatures seemed to be mostly disgusting or at least tiresome household
pests. If he must become a zoologist, he wondered in frustration, why
couldn't he at least investigate grand and exciting creatures, like lions?
But then he said he'd been heartened by two inspiring thoughts. First, he
realized that his constituency comprised nine-tenths of the known animal kingdom, so he'd hardly been relegated to an unimportant corner of
nature: quite the contrary. And second, he'd come to hope that these
simplest of animals might reveal to him the basis of animal life—the essence of what it meant to be a living creature.

ELEVEN

The Invention of Invertebrates

Contemplating his insects and worms by the light of this gleam of hope, Lamarck went to work. As before with the royal herbaria, his penchant for making things neat and tidy asserted itself. The traditional taxonomic schemes for living beings ever since Aristotle had been linear, hierarchical, and static, each being in its assigned place along a spectrum of increasing complexity and perfection, with humans at the top, and every place in the natural order filled by a specially designed living thing. But when Lamarck looked at his newly assigned creatures, shoehorned into just two categories, "insects" and "worms," he saw not a rational order but instead a "chaos": a great profusion of forms that just didn't fit into the old categories. He set about straightening things out, and as he proceeded, a new picture began to emerge.

One big change he made was to rename the whole ensemble of his assigned creatures "invertebrates," creating a new taxonomic category. The invertebrates, Lamarck explained, were those animals lacking a vertebral column and articulated skeleton; which instead of red blood had a whitish humor; and whose bodies were soft and "eminently contractile."

The insects and worms, now "invertebrates," then took their place along-side plants as crucial creators of the world.

According to Lamarck's emerging theory, as we've seen, all the myriad forms of compound matter in the world originated in the incessant, creative activity of living things as they made and remade themselves and their surroundings; this productive dynamism, Lamarck said, was the essence of life itself. As a botanist, he had studied the primary role of plants in this process: synthesizing elements—sunlight, air, and water—into complex matter. Now, as he turned to his "invertebrates," he came to see them as embodying the generative essence of life in its most fundamental and powerful animal form. He didn't pretend to know how the very first animate beings had come to exist or what sorts of inanimate matter might have existed at the time. But looking around the world as he found it, he saw great mountains made by the activity of little invertebrates and boldly announced that the matter covering "all the parts of our globe . . . is entirely and solely due to the organic beings which are found there." Tiny coral polyps prepared limestone "in the bosom of the sea, by a continual excretion of their body." Even the very oldest mountains, made of the most primitive rock, must have been made by even older living beings.

Lamarck therefore swept aside Linnaeus's tripartite division of the world into animals, vegetables, and minerals, familiar to anyone who has played Twenty Questions, which descends from ancient traditions in natural history, notably Aristotle. He replaced it with a new and stark division turning upon life itself and its role as creator of the world: simply, the *living* and the *nonliving*. Next, following a procedure like the one he'd used in his dichotomous key for plants—the one that allowed him to turn a random passerby in the Garden of Plants into an amateur botanist—Lamarck divided the living into two categories, plants and animals; and animals, in turn, into two categories, vertebrates and invertebrates.

The vertebrates generally got all the respect, Lamarck acknowledged. They were the best known and the most complex, and they had the greatest number of faculties. They included mammals, birds, reptiles, fish—certainly all the creatures that now get called charismatic fauna. But Lamarck, warming up to the invertebrates, protested that they didn't merit the "scorn and indifference" they commonly received. Their class included more species than all the others combined. It included the insects, "so numerous, so singular, so elegant, even." Under Lamarck's pen, invertebrates metamorphosed from pests into fascinating creatures: "the most fertile in marvels of any kind, in the most singular and curious facts of organization, in piquant peculiarities." They had extraordinary "customs" and "habits," they were "amusing and very curious," they inspired "beautiful considerations." The invertebrates were also infinitely useful—Lamarck cited examples including the bumblebee (honey), the silkworm (silk), and the Mexican cochineal (crimson dye). On the other hand, invertebrates could also be extremely dangerous, which he counted as a further urgent reason to study them, since the best source of protection was surely understanding.

Above all, the invertebrates were important to Lamarck because he believed they revealed the essence and foundation of animal life. He wrote that by exhibiting the "astonishing gradation" of forms of life, they "conduct us insensibly toward the inconceivable limit of animalization," toward those beings "barely endowed with animality" with whom "nature perhaps began." This "limit" or "minimum" or essence of animality was none other than the power of generation. Simpler creatures have fewer animal faculties, Lamarck explained, but those faculties they do enjoy are much broader in extent than in more complex animals, starting with the power of generation. The very simplest animals, such as polyps, have the most widespread generative faculty, being able to reproduce an entire new organism from any part of their body.

. . . .

The first report of these astounding creatures, polyps, had come almost a century earlier, in 1702, from the Dutch microscopist Antoni van Leeuwenhoek, who was so fascinated by the worlds within worlds appearing under his microscope that he used it to inspect everything he could find. Having collected some water from a sluice near where he lived in Delft, he looked at it under his microscope and beheld, with delighted surprise, teeming throngs of what he called *"Animalcula."* In a letter to the Royal Society of London, of which he was a foreign member, Leeuwenhoek reported that he'd seen "a great many and different kinds" of these beings. Some had "long Tayls" and bodies "in shape like a Bell"—these were vorticellids, a type of ciliate—and they moved about vigorously, causing a commotion in the water; they "contracted their Bodies and Tayls in an instant, then softly extended them again," so that it was "very diverting to observe them."

Another particularly enchanting *"Animalculum"* that Leeuwenhoek observed was "a little Creature, the fore-part of whose Body was roundish . . . and presently from the same Rotundity proceeded 2 little Wheels that had a swift Gyration. . . . These small Wheels were as thick beset with Teeth, like the Wheel of a Watch." (This one is now classified as a rotifer.) Leeuwenhoek's "limner" (his illustrator) had gazed through the microscope in astonishment at this sight, hardly able to tear his eyes away.

Among the *"Animalcula"* Leeuwenhoek observed, one had a "Body sometimes long, sometimes contracted; and about the middle of its Body, which I conceived to be the lower part of its Belly, there was another of the same kind, but smaller, the Tayl of which seemed to be fastened to the other." These curious beings also had many "Horns [that] appeared of so wonderful a make to our Eyes, that it almost puzzled the Limner to draw them." The little animals could extend these horns so far that "looking on them through the microscope, you would think they were several fathoms long." This was the first description of what would later be called a hydra and then a polyp (for reasons that will presently be revealed). As he watched, Leeuwenhoek witnessed what he recognized as the birth of a "young *Animalculum*" of this variety, which started off as "a round little knob of Seed" on the body of the biggest adult. This knob grew ever larger, developed "Horns"—first two, then four—and finally separated itself off from its "Mother." The spellbound microscopist documented several more instances of this interesting mode of reproduction.[1]

Leeuwenhoek's tiny beings made a splashy reappearance several decades later, when the Genevan naturalist Abraham Trembley was working as a tutor to the two young sons of a Dutch nobleman, Count Willem Bentinck. Trembley was just the sort of tutor a child would love. His lessons involved taking the boys out to collect pond water in jars and trawl with nets; then they brought home what they collected to study it and draw pictures. They found plenty of Leeuwenhoek's "animalcula," including the one with the wonderful "horns," and in 1740, Trembley and his pupils made a new discovery about these. Not content to gaze at the animalcules through a microscope, Trembley decided—once again revealing his affinity with small boys—to see what would happen if he sliced the organisms up in bits. He found, to his astonishment, that they could grow a whole new creature from each section.

Even more fantastic was that, when he tried cutting them lengthwise from the head downward, but not entirely severing the two halves, they responded by developing two heads. Trembley fed each of the two mouths to fortify the creature, then sliced each of its heads in half again and got a four-headed animal, then a seven-, then an eight-headed one. He named his monstrous creation a "hydra" after the Hydra of Lerna, the many-headed monster of Greek mythology whom Hercules was supposed to slay as the second of his twelve labors, and who, whenever he cut off one of its heads, grew two new ones. "You can imagine," Trembley observed, "that having succeeded in making Hydras, I didn't stop there. I cut the heads of the one that had seven; and after several days I saw it to be a prodigy yielding nothing to the fabulous prodigy of the Hydra of Lerna. It had come to have seven new heads, and there's no doubt that if I kept cutting them as they grew, I could have caused others to grow."

Being able to regrow parts that had been cut off gave these organisms some resemblance to plants, Trembley acknowledged, and for a time he hesitated greatly over whether they were animals or plants. But he pointed out that the distinction wasn't always as clear as you might think. It might seem a simple matter to tell "a Horse from an Oak tree," but that was because each had many distinguishing features—hooves, acorns—that were specific to its particular kind of animal or plant. When you eliminated these, it became a much more difficult question. With regard to his hydras, Trembley felt it was impossible not to see them as animals "when one observes their progressive movement, & above all, when one sees them seize Insects, bring them to the mouth with their arms, swallow them, & digest them."

Trembley recounted his experiment to the French naturalist René-Antoine Ferchault de Réaumur, the reigning authority in entomology, who declared the marvelous animals to be "aquatic insects." Upon look-

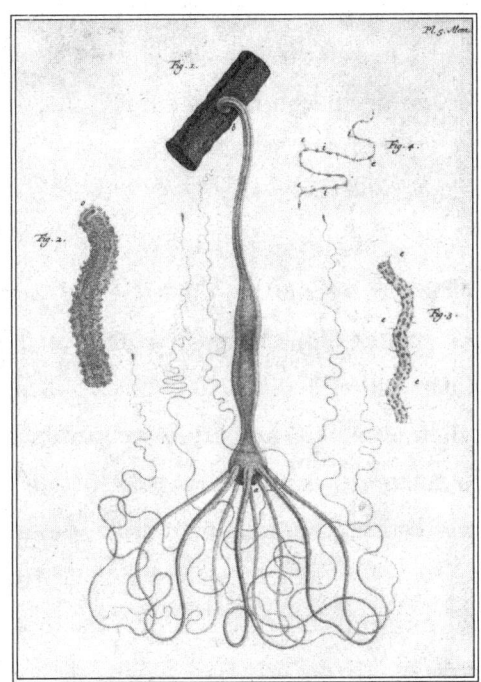

Trembley's Polyp by
Pieter Lyonet, 1744

ing into them himself, he proposed calling them "polyps" instead of "hydras" because their horns seemed to him more like the arms of an octopus or, in French, a *poulpe.* Trembley accepted the name suggestion in view of what he saw the voracious "polyps" do with their horns or arms. Catching other creatures barely smaller than themselves, they scooped these into their mouths octopus-style and swallowed them whole.

To Lamarck, the polyps represented an unbridled capacity to generate and regenerate themselves. Since they were among the simplest animals, they suggested that the power of generation was the essence of animal life. The life of each part of a polyp, Lamarck explained, was independent of the other parts, whereas in animals of more complex organization, the parts were interdependent, and the organism couldn't survive without certain parts. Moreover, lacking sex organs, or organs specific to generation, the polyps instead possessed a generative power distributed all throughout themselves, such that they could reproduce by separating off a portion of their bodies.

. . . .

Polyps were extraordinary beings, and so were insects, but still, Lamarck didn't follow Réaumur in classifying polyps as insects, nor did he adhere to Linnaeus by placing them among the worms. Taxonomy might seem like a humdrum occupation, but its organizing question is as foundational as you could get: What are the world's parts, and how are they constituted? In Lamarck's hands, taxonomy became a means of intellectual revolution. As he set about reclassifying the invertebrates in the museum collections, he was also teaching his students his new approach to living beings, remaking the way they understood the natural world.

To Linnaeus, things had been quite simple. Everything among the inferior orders of creatures that wasn't an insect had been a worm, and vice versa, hence the name of Lamarck's professorship: insects and worms. Up to his elbows in a diverse jumble of specimens all designated worms, an exasperated Lamarck lamented that naturalists had been so deeply in thrall to the late, great Swedish taxonomer that for so long "no one dared change this monstrous class of worms." Lamarck rolled up his sleeves and began tidying up once again, fetching things out into new categories of creatures. He reordered, rethought, and renamed, moving the puzzle pieces about and putting them back together to create a new kind of pattern.

TWELVE

A Can of Worms

F irst, Lamarck separated several classes from the worms. Cuvier had recently departed from Linnaeus by removing the mollusks and the echinoderms (such as starfish and sea urchins). Lamarck adopted these removals and added polyps (including Leeuwenhoek's hydras, sea anemones, and the little organisms that formed coral). That was better! Lamarck soon changed the name of the "echinoderms" to "radiolarians" in order to include creatures with radial symmetry but not spiny skin, such as jellies. Meanwhile, from insects, he plucked out the crustaceans and then created the "arachnids," from the Greek root *arakhne* for "spider." From two categories, Lamarck had now made seven . . . and counting.[1]

Heartened, Lamarck dove back into the diminishing chaos of worms and pulled out "cirripedes"—barnacles—from the Latin roots *cirrus* (curl) and *pede* (foot), meaning "curly-footed"; and "annelids"—such as earthworms—from the Latin root *anellus* (ring) for their segmented, ringlike structure.[2]

The annelids he extracted soon after attending Cuvier's lectures in the winter of 1801–2, where he watched as Cuvier showed that what he called "external worms"—those living outside in the world rather than

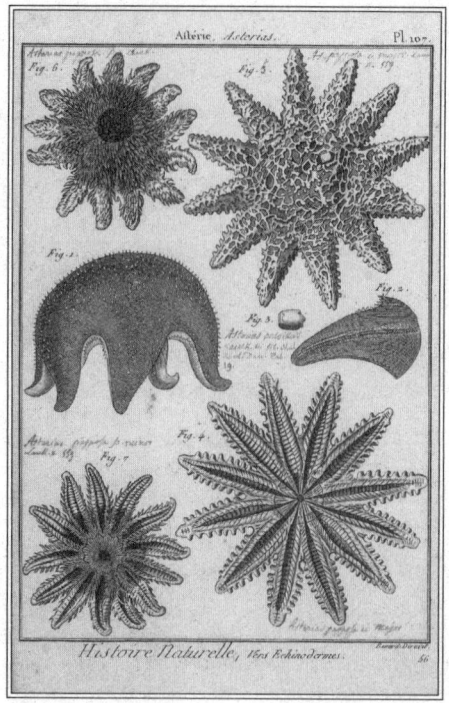

Echinoderms by Jacques Renaud
Benard, 1827

inside other animals—had circulatory systems of arteries and veins. This demonstration made a big impression on Lamarck. He had an "interleaved" copy of his own first zoological work on invertebrates, the *System of Animals Without Vertebrae*. This is a copy with blank pages inserted between the pages of text to allow the author to annotate the book for a potential revised edition. On one of the blank pages in Lamarck's interleaved copy of his *System* appears evidence that the earthworm's tiny circulatory system had a huge impact on his thinking, as Stephen Jay Gould discovered when this copy was sold at auction in 2000 and he got to have a look at it. Lamarck had made a careful sketch of the earthworm circulatory system revealed by Cuvier and added a note detailing the important differences between an organism with red blood circulating in arteries and veins and one with simply a white fluid flowing through its whole body.

Reflecting on this contrast, Lamarck decided that externally living "worms" in fact had little in common with parasitic worms other than soft, elongated bodies. Therefore, in his course the following spring, he introduced the new category of "annelids" for Cuvier's external worms, deciding to reserve the term "worms" for parasitic worms. This taxonomic revision would ultimately have dramatic implications for Lamarck's whole picture of animal life, as we'll see in chapter 20. Meanwhile, the annelids as he now defined them were those creatures such as earthworms that had long, annulated bodies, breathed by means of gills, and had circulatory systems and small brains with rudimentary nervous systems. Finally,

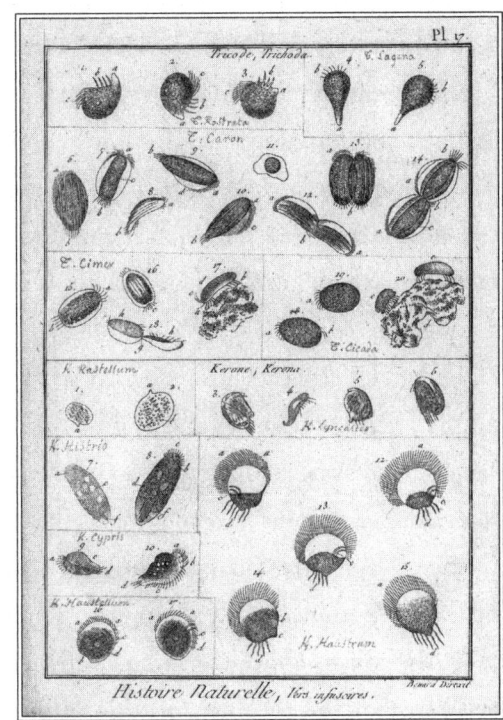

Infusoria by
Jacques Renaud Benard, 1827

he also separated from the polyps the freshwater microorganisms that German microscopists had been calling "infusoria," since they appeared in infusions of decaying matter (freshwater microorganisms such as Leeuwenhoek had spied in his sluice water).[3]

In this way, in place of "insects and worms," Lamarck established ten categories of "invertebrates": mollusks, annelids, cirripedes, crustaceans, arachnids, insects, worms, radiolarians, polyps, and infusoria. Added to the four categories of vertebrates—mammals, birds, reptiles, and fish— they made fourteen kinds of living beings. Overlaid onto these kinds, Lamarck also added a tripartite division of all animals: From infusoria through worms, he classed the invertebrates as "apathique" (insensible), having "irritability" but no sensation; from insects through mollusks, they were "sensible," meaning that they had sensation but no intelligence; and finally, all the vertebrates, Lamarck said, were sensitive, able to "acquire ideas," and were therefore "intelligent."

Current invertebrate taxonomy places Lamarck's categories on various levels, recombining some (for instance, cirripedes are currently folded back into crustaceans) but also adding a great many more. Invertebrate zoologists recognize 1.2 million living species and counting, making up well over the nine-tenths of animal kinds of which Lamarck boasted to his students.[4]

Lamarck's new categories opened up a true can of worms by destroying the ancient notion that living beings were arranged on a scale or chain of increasing complexity. Parasitic worms were more complex than jellies and sea urchins in some respects (for example, they had bilateral symmetry and directional motion) but simpler in others (for instance they lacked both nervous and circulatory systems). Without the annelids to lift the worms above the radiolarians, Lamarck found himself compelled to abandon the ancient chain of living beings. Already, as we've seen, he'd taken the ancient, static great chain of being and set it in thrumming, striving, living motion. But as we'll see in chapter 20, over the course of his next works in zoology, he would take the further and even more radical step of abandoning Aristotle's *scala naturae* altogether, replacing it with branching trees in continual growth and transformation. The simplest and most rudimentary of beings, the intestinal worms, had pointed the way from a single chain or ladder of complexity toward branching trees in which the branches didn't represent the progressive plan of a divine Providence but living beings' own creative responses to a variegated world that was also of their own making.

· · · ·

All the invertebrates possessed extraordinary powers of generation and regeneration, Lamarck pointed out: Insects metamorphose; crustaceans regrow their legs, starfish their rays, actiniae (sea anemones) their tentacles, and snails their *heads*. Growing a new head—how's that for a party trick? Certain worms, hydras, and polyps could multiply by partition of a single individual, as Trembley had shown; coral polyps also multiplied by a perpetual budding, forming plantlike stems. Lamarck explained that because these simplest animals had the greatest faculties of generation and regeneration, the invertebrates had an "immense" place in nature, far outnumbering the vertebrates, and he reckoned that we would likely never know their full extent because of the "infinite smallness" of the smallest among them. These tiniest beings displayed a "frightening fecundity" such that "it seems, so to speak, as if matter animalizes itself from every direction, so rapid is this astonishing fecundity."

"But what must be our astonishment," Lamarck continues, upon learning that some of the tiniest and simplest of creatures, the madrepore and millepora coral polyps, have been the chief producers of the great calcareous mountain chains and enormous banks of chalk found all over the earth? These minuscule animals create islands, fill up bays and gulfs, clog ports, and "completely transform the coastlines," all with just a little help from shellfish.

Shelled creatures, to Lamarck, also offered essential evidence in support of the theory he was developing. They revealed the animal creation of the world in action. Invertebrates with shells represented the living source of the infinitesimally slow processes that transformed the earth's surface. Shells were exquisite works of invertebrate production presenting "an astonishing diversity of forms" in an "almost infinite variety of colors," rivaling flowers, birds, and butterflies in their brilliance and magnificence.

Moreover, shells displayed the transition from organic to mineral matter, just like the partially mineralized vegetable matter that had captured Lamarck's interest during his trips with the Andrés and then with Buffonet. Any shell, Lamarck observed, was originally formed "of an intimate mixture of two materials of different natures, one of which is entirely animal, and the other purely calcareous." Both sorts of particles came from the same "viscous matter exuded by the animal" through special glands dotting the surface of its body. While the shell remained mixed in this way, partly animal and partly mineral, it could be considered alive; once its animal part had been destroyed and only the "stony part" remained, it was an inanimate fossil.

Lamarck scrutinized the muscles attaching shelled animals to their shells and observed that the muscle gradually gave way as the shell grew, releasing it bit by bit, and that the "tendinous" part of the muscle where it attached to the shell was "hard, horny, and is in this place completely inorganic." Once again, here was direct evidence of living organisms creating the inanimate world as organic matter became mineralized. He remarked too that shells grew in the same way as minerals, "by juxtaposition," through a continual deposit of new particles onto the interior surface or, more rarely, onto the exterior surface.

The discretion that animals employed in building their shells especially interested Lamarck. He noted, for instance, that as a shelled mollusk grows, it becomes ever more "constricted and uncomfortable in its shell" and therefore "seeks to make itself more comfortable." In order to achieve this end, "the animal moves a little in its shell, extends its body out of the opening, places itself in a stationary state, and remains motionless for some time." When it does this, it is patiently waiting for the uncovered part of its body to cover itself in the viscous shell-forming secretion, in this way expanding its shell. In spiral-shelled marine organisms, such as gastropod mollusks, "the fully grown animal, finding itself too uncomfortable in its shell," doesn't simply stick a part of its body outside the opening but gauges just how far to extend this part in order to continue the spiral in perfect proportion. "Depending on the species, the portion of its body that it exposes at each station is sometimes half a turn of a spiral, sometimes a third of a turn, sometimes a sixth of a turn, or even less."

Cowries from Senegal and Madagascar, Lamarck observed, build their shell by a two-part process. At first, they build from within, but later in their development they acquire "two soft and membranous wings which constitute the mantle of the animal," and which they extend outside their shell whenever they want to move around. Using these "wings," the animal adds layers of contrasting colors in various patterns—"spots, flames or bands, depending on the species"—to the external surface of its shell. It also uses its "wings" to transform the outer shape of its shell "in a remarkable way," adding "crenellations," forming "wrinkles, furrows, and . . . knobs," and "encrusting the turns of the whorl" with a kind of "teeth."

Murexes, a spiky, spiral-shelled marine mollusk, show an even greater degree of judgment. Lamarck described their procedure minutely. At first, they seem to block their own path with rows of spikes and fringes that obstruct the opening of the shell; these might seem to prevent the animal from sufficiently exiting the shell to be able to rotate around its axis and continue the spiral. "I felt this difficulty: nevertheless, by considering various individuals of the *murex tribulus*, the *murex ramosus*, etc., I convinced myself that the animal had known how to overcome it." He noted that on the already-formed spiral "we can still see the old rows of fringes or thorns which it did not need to destroy; but the thorns or testaceous [shell] fringes that it found in its path it knew how to make disappear." The animal therefore selectively "destroys the testaceous protuberances which hinder its exit."

. . . .

Insisting in this way on the agency and discretion of tiny creatures might seem like a small matter, but it was in fact an expression of Lamarck's radical view of life and of nature. Although in his early writing he described the origin and essence of life as impenetrable mysteries, he soon changed his tune. Life, he decided, consisted simply of the capacity for a kind of organic, self-directed movement.

Boldly, brazenly, Lamarck usurped God's monopoly on creation, ascribing it instead to mortal beings and down to the very humblest of them. Life originated, according to Lamarck, not when God created the world but when forces such as electricity and heat acted upon the elements of inanimate matter, spontaneously generating tiny, rudimentary beings—for plants, the simplest algae, and for animals, the globular in-fusoria visible under a microscope that he called "monads," using a phil-osophical term derived from the Greek *monas*, meaning "unit." A monad was the most elementary form of animal life, nothing but an "animated point." Lamarck wrote that he had nothing against religious belief and acknowledged that faith could be "a consolation for the good man who persuades himself of it." But faith was of no use to Lamarck. He all but declared his atheism. Souls were not physically knowable, whereas the dynamic, creative action of living organisms was apparent everywhere.

Life, as Lamarck now came to see it, was the manifest capacity to cre-ate in the face of all nature's forces of destruction. "There reigns through-out the universe an astonishing activity," he observed. He meant living things doing the work of living—growing, reproducing, eating, digest-ing, moving, propagating life among themselves—but not only that. To live, Lamarck thought, was to engage in a constant work of construction and combination, merging elements into compound matter and complex structure, in valiant defiance of the rest of nature's tendency in the oppo-site direction: to "destroy" all combinations, to decompose and dissolve. These "two powerful forces"—life versus the rest of nature—were locked in ceaseless combat, each perpetually struggling to undo the work of the other. "Without organic beings," therefore, "none of the compounds which are observed on our globe would ever have existed."

The science of life was the study of this capacity for heroic creation, this defining faculty of organic beings—to compose and complexify themselves and the world—to which Lamarck gave the name *pouvoir de la vie*, or "the power of life." The creative agency of living things set them apart from everything else and made them worthy of a science of their own. This agency was definitively present even in the most rudimen-tary of beings; Lamarck had seen it first in plants, and now he recognized it in the simplest animals. At its essence, to live was to create, and to create was to live.

It's a powerful image of living nature, and yet a distinct note of tragedy penetrates Lamarck's story, since of course all living things are doomed to ultimate failure. The inevitable end comes about when an animate being loses the "astonishing principle" that has hitherto allowed it to sustain itself against the onslaught of nature, continually dragging it toward its ruin. The vanquished being is now "completely at the mercy" of nature's destructive power. When living things die, Lamarck explained that they surrender all they have created in a lifetime: their bodies, secretions, and other productions. Natural forces such as heat and pressure then put these products through various destructive transformations, producing a succession of minerals that make up the clay of the composed world. "Thus, on all sides one perpetually sees nothing but an alternating succession of life and death, formation and destruction, effective movement and rest." It was a worldview of stark beauty that Lamarck presented to his colleagues, and also, to many, a scandalous one.

Lamarck's first use of the term "invertebrate" in print occurred in 1797 in his *Memoirs on Physics and Natural History*, although he had already been using it verbally with the students in his course. This work, despite its anodyne title, was in fact so polemical that when Lamarck tried to read it aloud in the traditional fashion to his colleagues at the National Institute, they interrupted and impeded him mercilessly during the first four sessions until he gave up right in the middle of reading the fourth of the seven memoirs, dismayed at "the contempt and odious rejection" he had suffered because of a few partisan members' determination to "deflect and prevent scientific discussions which do not suit them." Some years later, Lamarck vividly described the art of humiliation in sophisticated Parisian circles. With "a few monosyllables pertinently employed . . . a smile and a wink, an air of disdain," and a deft change of topic, practitioners of this art drowned anyone whose ideas they disliked in "ridicule and contempt."[5]

When Lamarck gave up on reading his *Memoirs*, he hadn't even gotten anywhere near the momentous "invertebrates," which come up only in the seventh memoir. He'd just been presenting his theory of the composition of matter. But the invertebrates were far from innocent in this regard, because as we've seen, Lamarck's theory of the composition of matter was as closely connected as could be with his theory of living things, and the invertebrates played a crucial role in both. Indeed, it would be no exaggeration to say that Lamarck remade the world itself around these creatures.

THIRTEEN

The Composition of Matter

L amarck's theory that mortal beings were the creators of inanimate matter departed from the traditional, religious view of creation; it also starkly opposed the National Institute members'
favorite theory of the composition of matter, which was the theory of
Antoine Lavoisier, a chemist, tax collector, and philosophical revisionist
who, along with the other members of the company of tax collectors, had
lost his head to the guillotine in the spring of 1794.

Lavoisier and his comrades had set out to do no less than remake the
general understanding of the nature of matter. The world, they said, was
not made of the ancient, traditional elements: earth, air, fire, and water.
Instead, the fundamental components of matter were a long list of unfamiliar, new elements whose strangeness these chemical revisionists emphasized by giving them Greek names (more on this naming practice
presently). These elements could exist as solids, liquids, or airs depending
on how much "caloric," or heat, they contained. This claim accompanied
startling experimental results. For instance, Lavoisier reported having
taken water apart into two different kinds of air: one highly flammable,
which he named "hydrogen," and the other excellent for breathing,
which he called "oxygen." As a general matter, Lavoisier and his fellows

claimed that the elements came together and apart simply as a result of temperature and atmospheric pressure. In their "new chemistry," which they also referred to as the "pneumatic chemistry" because of the central role it assigned to airs and their combinations, living organisms had no particular role to play.

Lamarck's countering theory that living beings were the agents of composition was especially offensive to two of Lavoisier's collaborators. One was the mathematician, astronomer, physicist, and statesman Pierre-Simon de Laplace, who was in his mid-forties and on his way up in the world. As a young man, Laplace had collaborated with Lavoisier to conduct a series of experiments on heat; the two had devised the first ice calorimeter, a tool that measures amounts of heat by quantities of ice melted. Using their device, Lavoisier and Laplace had established the equivalence of a burning piece of phosphorus or coal with a breathing

Portrait of Pierre-Simon
de Laplace (1745–1827)
by Jean-Baptiste Guérin, 1838

guinea pig. Both produced "fixed air" (carbon dioxide) in the same proportion to the amount of ice they melted. "Respiration is a form of combustion," they concluded, "very slow to be sure, but otherwise perfectly similar to that of coal." Living processes were no different from inanimate ones, they implied: a view sharply at odds with Lamarck's idea that living processes were the only creative forces in the world.

The other person Lamarck alienated with his theory that living organisms composed the world was a doctor and chemist named Antoine-François Fourcroy. Now in his early forties, Fourcroy had been a member of the Jacobin Committee of Public Safety and the Committee on Public Instruction; he had then handily survived the fall of the Jacobins to serve in the Council of Elders, the upper house of the legislature under the Directory. Fourcroy had been one of Lavoisier's coauthors on the treatise in which they had presented a new chemical language to suit their revisionary understanding of matter and its composition. As a basis for their new nomenclature, they assigned names to as many of their new elements as they could isolate in experiments. As mentioned above, they used Greek prefixes to emphasize the elements' great remoteness from daily experience and common language (Latin, they thought, would have sounded much too familiar and comprehensible). In place of the four trusty old Aristotelian elements, earth, air, fire, and water, the reformers presented a bewildering babel of Greek-derived names—"oxygen," "hydrogen," and thirty-one others—along with a promise that the list would grow even longer as more substances succumbed to chemical analysis.

To conduct their chemical analyses, Lavoisier and his fellow experimenters used an extremely precise balance to weigh the ingredients and products of each reaction—for instance, the hydrogen and oxygen into which they decomposed water and from which they then recomposed it. They assumed that matter was conserved, so the combined weights of the ingredients must equal those of the products. They then identified the lightest products they could make as the smallest, simplest units of composed substances. These were the true elements of matter, they said, and not the ancient, Aristotelian tetrad.

Aristotle wasn't their only victim. They also rejected a more current, popular approach to chemistry that, like theirs, featured Greek terminology, though in this case only a single Greek-derived name: "phlogiston," from a Greek root meaning "flame." "Phlogiston" was the name for what many philosophers assumed to be the matter of light, heat, and fire. Georg Stahl, a German chemist, doctor, and philosopher, had presented phlogiston chemistry in its most influential form around 1700. He described phlogiston as an active entity serving as the principle of all composition, life, and growth. Phlogiston gave plants and minerals their wholeness, gave animals their warmth, and was the spark of life in living things. The new chemistry featured oxygen and hydrogen but manifestly no phlogiston.

Rather than ascribing light, heat, and fire to a fiery fluid called phlogiston, the new chemists referred to the cause of heat as "caloric," from the Latin word for heat, *calor*, and remained resolutely agnostic about its nature. Did an object become hot because of the presence of a fiery fluid? Or simply because of the agitation of its own particles? It made no difference, Laplace and Lavoisier insisted in their jointly authored *Memoir on Heat* (1780). They could study the phenomena associated with heat without embracing either cause. Their interest in heat was not its cause but rather its effect on the composition of matter. According to the new chemists, composition and decomposition depended only on temperature and pressure.

Reduction was the new chemists' crucial principle. First, reduction was their chief experimental procedure. In experiments, they reduced various kinds of matter to their lightest constituents. Then reduction was the defining feature of their worldview. They saw the material world as reducible to invisibly tiny elements. Finally, reduction was their model of science. To them, explaining something scientifically meant breaking it down into its smallest (meaning lightest) parts.

Lamarck vehemently rejected this new approach to chemistry and the composition of matter that Laplace and Fourcroy had both been involved in establishing. He vigorously attacked the pneumatic chemists and their reductive science in a series of writings in the mid-1790s culminating in the *Memoirs* whose presentation at the National Institute provoked such enmity.[1] Fourcroy had defended the pneumatic chemistry in a manifesto titled *The Chemical Philosophy*, and it was this text that Lamarck chose to rebut, presenting a point-by-point refutation. A botanist turned zoologist, Lamarck nevertheless took on chemistry, announcing confidently that the pneumatic chemists had gotten everything utterly wrong. By assuming this apparently gratuitous position, he did his career such profound and lasting damage that for two centuries even his most sympathetic biographers have shaken their heads in rueful mystification.

But it wasn't actually gratuitous: Lamarck had a reason for his antipathy for the pneumatic chemistry, a reason connected with botany and zoology and, more generally, with the new science of life that he would soon name "biology." He was out to defend the creative agency and irreducibility of living things. According to Fourcroy, living things had no real agency. Nature (or Nature's God) merely lent them certain materials for a brief time, ultimately taking them back to continue her perpetual cycle of composition and decomposition. Lamarck retorted that nature could only aggregate, never compose anything. Only living beings could compose.

Against the new chemists' mechanical theory of composition by heat and pressure, Lamarck presented an entirely different theory of the composition of matter based on the agency of living beings and the principle that they were the unique force of composition. "The entire surface of the globe," he wrote with dramatic pathos, "the bosom of the waters and the whole atmosphere, are the vast field where nature ceaselessly destroys every composed substance which the principle of life does not defend or ceases to maintain."

At the same time, against the new chemists' reductive model of scientific explanation, analyzing all things including plants and animals into their lightest parts, Lamarck developed a markedly different kind of explanation; maybe we could call it compositional. He argued that living organisms and the things they composed were not "susceptible to analysis"; they were irreducible. No analysis could break them down into their component parts because, Lamarck said, they had no preexisting component parts. Organisms synthesized themselves and the world around them just from sunlight, air, and water. What Fourcroy took for the *constituents* of organic matter were in fact the *products* of "organic action." Therefore, "instead of reducing everything to immediate principles," Lamarck concluded, let's simply say that vegetable matter and animal matter are the products of organic action characterized by distinctive "remarkable properties."

By universal historical consensus, Lamarck was just plain wrong in rejecting Lavoisier's chemistry. But it's worth trying to see the situation from Lamarck's point of view because it might reveal something that the dominant tradition in modern chemistry has perhaps obscured. If we define the "elements" of matter not as the lightest components producible in a laboratory but as the primitive substances out of which living beings compose themselves and the world around them, it might make sense to say that these elements are sunlight, air, and water. It might even make sense, from this point of view, to say that oxygen is not an element but a product of organic action, since, as mentioned in chapter 6, its presence in Earth's atmosphere results from the activities of living beings. Nineteenth- and twentieth-century chemistry included extraordinary feats of scientific comprehension and world-changing engineering, but these were not by and large focused on discovering the role of living organisms in bringing about the geochemical conditions and surface mineralogy of the earth.

Historically speaking, of course, Laplace's and Fourcroy's way of defining elements triumphed, along with their reductive definition of science. In the wake of their success, Lamarck's vision of nature, of science, and of life became not just wrong but unthinkable, beyond the pale. Lamarck keenly felt the tenuousness of his position, both political and scientific. Soon after his catastrophic attempt to present his *Memoirs*, in the wake of a political scuffle in the Directory, he confided to his friend and publisher Henri Agasse that in the current climate he'd begun to fear being exiled—despite, as he protested, his fervent republicanism.[2]

Lamarck's science was dangerously radical, even in the midst of political revolution. And it wasn't just his opinions about pneumatic chemistry and the composition of matter that were impolitic. Lamarck seems to have lost no chance to alienate the influential.

FOURTEEN

Animal Collaborations.

Poor Charlotte Reverdy apparently never got to enjoy Beauregard, the house in her native countryside of Héricourt, because she died in November 1797, at the age of twenty-four, several months before Lamarck finally managed to take possession of it. She left behind three-year-old Aristide and a baby daughter, Eugénie, just eight months old, as well as her six stepchildren ranging in age from nineteen-year-old Rosalie to five-year-old Cornélie. Eugénie would have a formal education, going away to school like two of her brothers, but would survive barely longer than her mother had. She would die in 1822 at the age of twenty-five.[1]

A few months after Charlotte Reverdy's untimely death, Lamarck brought his children to Beauregard and had an experience there that deeply impressed upon him the creative, collaborative agency of living beings. One day in the spring of 1798, during a stay at the house with his family, he noticed a swallows' nest in one of the windows. Shortly afterward, a local child destroyed the nest, leaving the pair of swallows homeless just as the female was about to lay her eggs. Immediately, ten or twelve other swallows gathered around the wreckage. The birds were so agitated and noisy that at first Lamarck assumed they were quarreling, but he soon realized it was just the opposite. They had organized themselves to rebuild the nest collaboratively. He watched in amazement as some remained by the construction site while others came and went in rapid succession. Swallows generally take eight to twelve days to build their nest, Lamarck observed, but these worked so diligently that they finished it the following morning.

Lamarck later offered this story to his friend Geoffroy Saint-Hilaire for his collection of instances of mutual affection and helpful services rendered among animals, in so doing exhibiting precisely the generous quality that Geoffroy was keen to document. Geoffroy's collection, published in the *Annals* of the museum, featured the famous lion/dog duo we encountered on our prefatory stroll in the garden; cat mothers that obligingly nursed orphaned baby squirrels and rats; an eagle that declined to eat the roosters its handlers tried to feed it but instead befriended them; the "trochilus," a little bird that, according to Herodotus and Aristotle, blithely entered the ajar mouths of crocodiles to clean their teeth; the caracal, an African lynx that helped guide lions to their prey in return for a modest share; Senegalese lions and wolves that hunted in partnership; and pilot fish that acted as guides to sharks. This last phenomenon, Geoffroy himself observed while on board the frigate *L'Alceste* in the Mediterranean Sea between Cape Bon and Malta.

. . . .

Geoffroy was aboard the *Alceste* having temporarily left the Garden of Plants, where he now held the zoology professorship in quadrupeds and birds, in order to sail as a scientific member on Napoleon's expedition to Egypt. Napoleon was at that point still a general in the Directory's army; his mission was to seize Egypt from the Ottomans and

Barn swallows by Jacques de Sève,
1779, from Buffon, *Histoire naturelle*

the Mamelukes, who were the governing feudal order there, and use it as a stepping-stone to India, where he would retake possession of France's former Indian empire from the British. In the event, Napoleon won enough victories to establish tenuous control in Egypt, but he didn't go to India. Instead, in August 1799, he left most of his army in Egypt, sailed back to France, and staged a coup; two years later, British and Ottoman forces drove France out of Egypt.

At the French defeat, when the British claimed the French scientific collections for their own, Geoffroy swore to the British envoy, William Richard Hamilton, that he'd burn everything before surrendering it. Hamilton was the personal secretary of Lord Elgin, a diplomat and collector now known for having dubiously acquired the set of sculptures from the Parthenon ever since known as the Elgin Marbles, which remain, controversially, at the British Museum in London. Geoffroy promised Hamilton that posterity would never forget him if he insisted on seizing the French collections, because he too, like Julius Caesar, would have caused the burning of a library collection at Alexandria. Hamilton did manage to snag the Rosetta stone—also still at the British Museum—but Geoffroy's tactic prevailed, and he got to keep his natural history collections. He therefore sailed home bearing scientific plunder upon which he built much of his career. He was lucky even to have survived, let alone to return home with career-making booty. Long before the French encountered the various enemy forces, on the outward voyage, Geoffroy nearly drowned while crossing to another frigate to see some colleagues. The small boat carrying him capsized, and, unable to swim, he struggled in the sea but managed to catch hold of a rope ladder and pull himself to safety.

It was also on the outward voyage, on May 26, 1798, that Geoffroy witnessed the friendly collaboration mentioned above of some pilot fish and a shark. This took place during a lull, Geoffroy recounted, just as the passengers grew weary of languishing becalmed in the sea. Suddenly they caught sight of something to distract them. A shark was rapidly approaching the ship preceded by two pilot fish. The fish swam right up and inspected the ship from end to end while the shark followed their every movement at a distance. Having established that the ship held no interest for them, the trio turned and began to swim away. But meanwhile, a sailor had prepared a big hook baited with lard, which he threw into the water. At the sound, the three swimmers halted in astonishment, according to Geoffroy, and the pilot fish returned to investigate. The shark, while awaiting the results of their survey, cavorted in the ocean, rolling over on its back, diving down and resurfacing, but remaining always at a distance from the ship.

What's in it for the pilot fish? wondered Geoffroy. He had heard that they ate the sharks' droppings, but he found that these animals exhibited an obvious "intention to help one another." As the two pilot fish returned once again toward the shark, it apparently concluded they had nothing interesting to report and began swimming rapidly away, but they caught up with it and guided it back toward the ship so that it perceived, "thanks to the wisdom of his little friends . . . the fatal pray intended for him." The shark was caught and hauled aboard, where Geoffroy spent some happy hours dissecting it; he also conducted galvanic experiments on it, causing it to twitch and move, which were "for the whole crew a subject of astonishment and lively interest." Only belatedly did he think of examining the pilot fish too; the sailors immediately brought him one. Thus, the affectionate cooperation between shark and pilot fish ended in doom for both, due to the noncooperative spirit of that third species represented by Geoffroy himself.[2]

Once Geoffroy made it to Egypt, Napoleon appointed him to take charge of the natural sciences at the new Institut d'Égypte in Cairo, which the French were establishing in the service of what they saw as their *mission civilisatrice*[3] to impose French culture, science, and society upon the peoples of the earth. Ah, the comforts of empire. As Geoffroy's son, Isidore, later put it, in the midst of the "African nature, so rich and varied," Geoffroy now had all the pleasures of French academic life, even an institute to call his own. Thus ensconced, he set about conducting surveys of the fish of the Mediterranean, Nile, and Red Sea, including the torpedo fish and other electric fish. In Alexandria, he became ill and was struck with blindness for twenty-nine days, but despite all such travails he carried on, even through the British siege of Alexandria, during which he distracted himself from "the hubbub . . . the throwing of bombs, and the fires nearby, and the surprise attacks, and the plaintive cries of the victims" by focusing single-mindedly on natural history.

By "natural history," Geoffroy meant not only observing the Egyptian animals themselves but also studying ancient Egyptian treatments and representations of them. This was especially important to him because he believed the ancient Egyptians had been sensitive observers of something that particularly interested him, namely the behaviors, sentiments, and moral qualities of animals. Traveling to Thebes, Geoffroy examined the paintings of animals in the necropolis there, in the tombs of the pharaohs from the second millennium BCE. He was enchanted by depictions of a species of Nile catfish known as upside-down catfish due to their remarkable habit of swimming on their backs. The particular species—*Synodontis membranaceus*—is also called the moustache catfish for its long, drooping whiskers. And there it was, plainly represented in two different places in the tombs, doing its thing. "Amid several other fish, all placed in their natural attitude, this fish was represented swimming on its back." By talking with fishermen and observing the catfish for himself, Geoffroy confirmed that it did indeed almost always swim not only on its back but also often *sideways* and, moreover, that this curious mode of locomotion seemed to be purely for its own amusement, since "when he fears some danger, he immediately turns around, resumes the position usual for other fish, and flees quickly."

Animal customs and habits greatly interested Geoffroy, and he was frustrated by his contemporaries' commitment to what he called "ready-made ideas of psychology" that ruled out the possibility of animals having intelligence and moral sentiments. Modern authors persistently refused to recognize in animals "considered acts or judgements with the slightest appearance of morality"; their "proud pretensions" led them to place a "barrier" between themselves and other living beings. The ancients, in contrast, both Greek and Egyptian, had been keenly attentive to the nuances of animal behavior. They had placed all animals including humans on a continuum, assuming that intelligence varied with "organic structure," and had therefore carefully scrutinized the actions of animals in relation to human behaviors, a practice Geoffroy hoped to emulate and bring back into currency.

Herodotus's and Aristotle's reports of the intimacy between certain birds and crocodiles particularly intrigued Geoffroy, and he arrived in Egypt determined to confirm them. One day, he went out crocodile watching with some Egyptian guides who warned him to remain absolutely silent, because the creatures have extremely sensitive hearing. Imagine Geoffroy's delight when, crouching noiselessly on a bank of the Nile in northern Egypt near the ancient city of Thebes, he saw a plover hop right into the mouth of a crocodile! He watched with rapt attention and observed that the bird wasn't in fact cleaning the crocodile's teeth, as Herodotus had suggested—and which the crocodile could accomplish for itself by using its hind feet—but rather it was eating the crane flies that coated the inside of the crocodile's mouth to suck its blood. Lacking a tongue, the poor beast would have been at the mercy of the flies if it hadn't been for the services of the little plover. Geoffroy noted that observations in the French colony of Saint-Domingue of a similar crocodile, also lacking a tongue, showed that it received help keeping its palate clear of insects from a similar bird, known today as the Cuban tody. In each case, Geoffroy saw a situation of obvious "reciprocal obligations and mutual affection."

The crocodile, of all the creatures Geoffroy observed, was clearly his favorite. The primeval beast inspired a fascination in him. Never mind that, in making his observations, he became fodder for English caricatures representing him as foolishly trying to tame the monsters; these, according to his son, Isidore, were shamelessly inaccurate. He wasn't trying to tame them; he was trying to *understand* them. What riveted

Geoffroy's attention was—as he saw it—the intelligence, moral character, and advanced society of the crocodile. He admired, for instance, the crocodiles' arrangements for sleeping in safety on the banks of the Nile, which involved skill, foresight, prudence, and even "superior intelligence."

First, he said, they identified suitable locations along the bank, with a ramped area to get up and down the beach, and a cape projecting out into the deep part of the river to be able to immediately start swimming and not be slowed down by mud. Then they organized themselves according to a social division of labor. One crocodile served as guard, keeping watch over the whole troop; this guard crocodile always kept one ear partly pressed against the sand to listen "for the faintest sound." Otherwise, the social positions varied by age and size. The bigger ones who could jump farther into the deep water slept farther back; the smaller ones who didn't need such depth to be fully immersed remained near the shallower sides, with the elders "forming a sort of rampart" around them. Each crocodile had a designated place to which it would always return, constituting a kind of "property."

Yes, the crocodiles, according to Geoffroy, had a society sufficiently advanced to include a notion of real estate. He reckoned that no one had realized this before because the seasonal transformation of the river as it swelled and diminished over the course of the year required the crocodiles to vary their private property arrangements in concert with it, masking their "sociability." The same social arrangements were more easily visible in seals, whose "sleeping accommodations, made of flat stones or shaped parts of rock," were more stable. Seals, Geoffroy said, would banish any individual who trespassed on another's bit of rock, revealing "in these marine animals a real notion of property, a notion regarded as the product of a very advanced state of civilization."

. . . .

Among the main ideas with which Geoffroy returned to France was the notion that animals had emotional, social, and moral intelligence, which they expressed by forming cooperative relations both within and among species. He published his collection of examples soon afterward, including Lamarck's story of the collaborative swallows of Beauregard. Meanwhile, among the specimens Geoffroy brought back was a crocodile—albeit a mummified one. He had plundered the tombs at Thebes and blithely taken possession of many mummies—humans, a cow, many ibises, a mongoose—which he unwrapped and revealed to be perfectly preserved down to the smallest bones and hairs.

Lamarck, along with Cuvier and Lacépède, professor of reptiles and fish, composed the three-person committee who received and reported on Geoffroy's loot. Their report, written by Lacépède, reveals the growing tensions among the colleagues. It asserts that since Geoffroy's collection includes "animals from every century," it can shed light on the long-standing question "whether species change form over the course of time." But Lacépède then avoids answering this explosive question, although he does observe that the mummified ancient animals are "perfectly similar to those of today." Isidore Geoffroy Saint-Hilaire, Geoffroy's son, later wrote that Lacépède had been caught in an impossible position between his two coauthors. Under intense pressure from Cuvier, who insisted upon the fixity of species, Lacépède had nevertheless wanted to appease Lamarck, who as we know was developing his momentous theory of the transformation of living forms. Lacépède had therefore implied but not actually stated the conclusion that species were immutable. Further research had settled the matter in Lamarck's favor: Isidore wrote unequivocally that even if the animals of Egypt hadn't "sensibly varied" since ancient times, species did indeed vary.

Moreover, the variation of species was closely related to something Geoffroy had greatly emphasized in his reports from Egypt: that animals had a creative social and moral intelligence, which he saw them exercise through their collaborative behaviors and "habits." Geoffroy acknowledged that the habits of an animal had to correspond to its physical structure—to its size, shape, parts, and appendages—yet he observed that the physical structure nevertheless couldn't determine habits altogether. He pointed out that crocodiles resembled lions in many of their habits— for instance, when hunting their prey, they flattened themselves to be inconspicuous and waited motionlessly for the opportune moment to attack—even though the two animals had extremely different physical structures and environments. Here was where the brain and nervous system came in, allowing the animal to decide how to behave, to suit its instincts and habits to the situation and even to change them when confronting new circumstances. Hooded crows, when in France, timidly fled from people, having learned that humans were dangerous, whereas in Egypt the crows came right up to the farmers, who held them in high esteem, and even landed boldly on their plows. When Geoffroy made a move to shoot one such crow, the farmer begged him not to, explaining that it ate the insects that threatened his harvest.

Animals were not fixed and passive; they could change in active response to their circumstances. They could do new things. Lamarck's swallows were able to meet the moment. Encountering an unforeseen calamity, they pulled together and figured out how to build in a new way.

FIFTEEN

Fathers and Sons

Around the time his friend Geoffroy was setting sail on the *Alceste*, in the spring of 1798, Lamarck married for a third and last time at the age of fifty-four. His wife, Julie Mallet, aged thirty, had already been a sort of sister-in-law. She was the sister of Denise Mallet, who was married to Lamarck's brother Bazentin. Julie and Denise Mallet's father was a Freemason, like Bazentin, which perhaps provided the connection between the two families. Lamarck and Julie Mallet remained together for more than twenty years, until her death in 1819, but had no children together.

This was probably a good thing, given Lamarck's chronic shortage of funds and the trouble some of his existing children gave him. The year following his marriage to Julie Mallet, there at last came news of the missing André. Sometime in 1799, according to André's brother Auguste, Lamarck heard that André had arrived in Brest as a deckhand aboard another French ship. Lamarck rushed to find him, managed to secure his release from the engagement, brought him back to Paris, and— perhaps in an attempt to alter the boy's course by lifting him by the head, as had been successful in his own case—hired a math teacher for him. André, surprisingly, took up the study of mathematics with the same passion he had previously devoted to imperialist sea voyages (and that his father, post-head-lifting, had devoted to botany), quickly outstripping his teacher. He proceeded on his own with books and some kindly encouragement from the mathematician Joseph-Louis Lagrange, who was then living in Paris, working on his calculus of analytical functions and, to his own and his students' chagrin, teaching in his heavy Italian accent at the new École Polytechnique, founded as a military engineering academy in 1794. André soon gained admission there. Alas, the military air of the place restored his old stirrings, and he abandoned his mathematical studies for a position as ensign and then lieutenant in the marines.

Meanwhile, Antoine, André's next younger brother, was another worry to Lamarck. Following a childhood illness, he had become so hard of hearing that it was difficult to educate him or to imagine a future for him. Lamarck reasoned that perhaps a career in the visual arts would be possible, and in 1798 he wrote to Charles-Axel Guillaumot, director of the old and renowned tapestry factory the Manufacture des Gobelins, which had supplied a succession of monarchs with magnificent wall hangings, to ask if he would receive Antoine as an apprentice and pupil in the factory's drawing school. Guillaumot consented, but since Antoine had had no formal instruction, Lamarck sent him first to one of the new state art schools, where Antoine began to learn the art of drawing. A couple years later, in October 1800, Lamarck wrote once again to Guillaumot, explaining that because Antoine's classes took place in the afternoons, he might now come each morning to the drawing school of the Manufacture des Gobelins.[1]

Along with his biological sons, Lamarck came to have a sort of adoptive son in Pierre-André Latreille, a young man from the Limousin region in southwest France. Latreille anyway described himself as Lamarck's "adoptive son" and Lamarck as "the tenderest and best of fathers."[2] Latreille's actual parentage was ambiguous. He was officially an "orphan," but with a surprisingly well-connected and obliging network of benefactors. He was almost certainly the illegitimate son of the baron d'Espagnac (an aristocratic general and military strategist who distinguished himself in the conquest of Prague during the War of Austrian Succession), who financed and fostered Latreille's education. The Church was the best option for people in Latreille's equivocal situation; he became a clergyman but devoted all his time to his true passion, which was entomology. Already as a young child, he ran about the Limousin countryside trapping insects and pinning them to his clothing for want of proper entomological equipment.

This obsession, as Latreille recounted, later saved his life. In 1793, the Revolutionary Tribunal had him arrested as a cleric suspected of counter-revolutionary activities, imprisoned him in Bordeaux, and condemned him to be deported to Guyana. Luckily, his cell was rich in insect life. He also had a human cellmate, "a sick old bishop," who received daily visits from a doctor. One day, Latreille discovered a rare beetle, which so interested the visiting doctor that he brought it to Bory de Saint-Vincent, a fifteen-year-old aspiring naturalist and son of a local jurist. (Bory later became a naturalist, explorer, military officer, and agent of the French empire in Algeria.) Young Bory begged his father to intervene on behalf of the "clever Naturalist" who had discovered the beetle and also alerted the Bordeaux naturalist Raymond Dargelas. The rescue party arrived in the nick of time: Dargelas in fact had to take a boat down the Gironde estuary to intercept the pontoon that was already carrying the prisoners to the ship. Three days later, they learned that the ship had sunk almost immediately, and all the prisoners had drowned.

Latreille named the rescuing insect, a carnivorous beetle that feeds on the flesh of dead animals, *Necrobia ruficollis*, or "red-shouldered death-eater."

When Latreille died in 1833, the Entomological Society erected an obelisk over his tomb in Père Lachaise Cemetery in Paris bearing a likeness of the beetle with the inscription *"Necrobia ruficollis*, Latreille's savior." The monument is admirably poker-faced, bearing no hint of the irony that Latreille had presumably become his savior's lunch. Perhaps Latreille would have found it fitting to give back to the insect world after a life throughout which "the study of insects has delighted me, consoled me, saved me."

Latreille was around sixteen when Lamarck first encountered him in the late 1770s. He was a pupil of Haüy, the mineralogist and humanities professor who had been Lamarck's helper in writing the "Preliminary Discourse" to *French Flora*. Haüy brought Latreille over to the King's Garden to meet the flora and fauna residing there, including the naturalists, and Lamarck became the young entomologist's hero. A decade later, back in Paris as a deacon-entomologist in his mid-twenties, Latreille brought Lamarck an offering of some rare flowering plants. A few years after that, Latreille sent Lamarck various insects that were missing from the museum's collection—first a new species of wasp, then a whole box of carefully labeled specimens.

This courtship finally succeeded in July 1798, when Lamarck appointed Latreille as his assistant naturalist, in which position he would remain for three decades. He became one of Lamarck's closest companions and collaborators, and when Lamarck grew too blind and feeble to carry on lecturing, it was Latreille who took over his famous course in the garden's amphitheater.

SIXTEEN

The Creation of Biology

O n 18 brumaire year 8 (November 9, 1799), Napoleon brought the revolutionary era to an end and inaugurated the Napoleonic era by declaring himself first consul of France in a coup d'état. That year, forty-four students attended Lamarck's course on insects and worms in the garden's amphitheater. Among them was a seventeen-year-old boy named Louis-René Villermé who would later launch the public health movement in France. A doctor and political economist, Villermé worked tirelessly for prison and labor reform, for humane conditions in factories, and for limitations on child labor. His guiding principles throughout his career in science and political reform would be distinctly Lamarckian: the crucial importance of the environment in explaining differences among people in physical, intellectual, and moral characteristics including height, health, mortality, intelligence, and sobriety; and the endless potential of humans, among living things, for adaptation and advancement.[1]

These were not notably Napoleonic ideas. The emperor was interested in remaking the world himself and imposing an order of his choosing on its inhabitants, not in studying or fostering their own self-transformative and world-changing abilities. Lamarck's antagonists were all members of

Napoleon's scientific inner circle, and when Napoleon came to power, he appointed them to various important governmental positions and awarded them fancy honors. Laplace was briefly minister of the interior and then chancellor of the Senate, Fourcroy director of public instruction and a member of the State Council, and Bonaparte ennobled both in his new aristocratic order as counts of the empire. Cuvier for his part became inspector general of education and permanent secretary of the National Institute's First Class.

What Napoleon wanted from his scientific advisers was for them to help him establish a modern kind of imperial authority: one based on cutting-edge manipulative control of the natural world and its resources. It was important to him to distinguish this new mode of power both from traditional Catholic monarchy (though to be sure, Catholic kings too had done their share of plundering the natural world for valuable resources) and from the Jacobin republicanism of the revolutionary era. An important part of Napoleon's strategy for consolidating his power was to clamp down on the divisive radicalism of the Jacobin era, including atheism. At the same time, Napoleon certainly didn't want to cede authority to the pope or the clergy. So, he walked a careful line of appeasement but not concession.

An imperial approach to seizing hold of the natural world that was neither Catholic nor materialist—this was what Napoleon needed from his savants. Lamarck's vision of world-making organisms was no good at all. It was materialist, almost explicitly atheistic, and it invested agency in the living beings themselves. But Cuvier's Protestant mode of science, implicitly ascribing the patterns and structures of the natural world to the workings of a divine intelligence, perfectly fit the bill. Cuvier translated the authority of religion into science, producing a powerful modern concoction. He used paleontology and comparative anatomy to show that certain living beings had become extinct and new forms had replaced them. But he represented the beings themselves as purely passive throughout these revolutionary changes. Each transformation had been imposed upon them by catastrophes, such as floods, and ultimately by the divine author of these events. Each of the new forms of living beings that repopulated the earth owed its existence to the same implicit designer. In other words, here was a scientific doctrine of revolutionary change that was at the same time profoundly conservative, and the whole thing was guaranteed not by the Church but by Napoleon's own scientific inner circle. What, from the emperor's perspective, could be better?

In contrast, Lamarck's natural world of inherent active powers, its substance literally generated and shaped by its own living denizens, did the opposite of consolidating power. It distributed creative power throughout the living world. By association with the likes of Fourcroy and Cuvier—and with them, against Lamarck and the sort of natural philosophy he represented—Napoleon claimed for himself and his political order the absolute authority of religion translated into modern science. In relation to that supreme, modern, and scientific authority, Lamarck's view of nature and natural science was plainly subversive.

. . . .

On 21 floréal year 8 of the republic (May 11, 1800), an international group of sixty-four students—one from as far away as Brazil and another from the United States—assembled for what promised to be a routine event: the opening lecture of Lamarck's course on the new class of organisms he called "invertebrates."[2] Instead, they became participants in an extraordinary moment, as Lamarck presented for the very first time his radical idea that living beings could create and transform not only the inanimate world around them but even *their own selves*. Polyps, the simplest of the invertebrates, he told them, embodied the lower limit of animal life and the rudimentary essence of animality: the unbridled power of generation. He described the extraordinary generative abilities of invertebrates. Lamarck then explained that nature created living forms using two principal means: "time and circumstances." The first of these, time, had "no limit": Departing from the traditional time frame of the Bible, Lamarck described time as unimaginably deep, a bottomless well. As we'll see in chapter 17, this idea of deep time played a crucial role in his theory of life. Just minutes into what they might have expected to be a dry lecture on the humblest and most mundane of creatures, the students were plunged into an endless abyss of time.

"Circumstances," meanwhile, included climate, temperature, and the particularities of place, but the crucial circumstantial factor was the myriad, creative responses of living beings to the varying conditions in which they found themselves—their habits, their ways of living and surviving. By responding to their environments, animals dynamically shaped and reshaped themselves, expanding and strengthening their bodies across generations. For instance, the waterbird created webbed feet for itself by its habit of "spread[ing] its toes when it wants to hit the water to move along the surface"; birds that perched in trees developed hooked talons by curving their toes "to embrace the branches"; and shorebirds, unable to swim, elongated their legs as they waded deeper into the sea and stretched to keep their bodies above the water.

The momentous moral of the story Lamarck told his students that day was that the evident fitness of animals' bodies to their circumstances arose "entirely" from their own actions. "I will be able to prove," he promised, "that it is not the form, either of the body or of its parts, that gives rise to the habits," but just the reverse: It was a creature's habits that gave rise to the form of the body. Through their behaviors and ways of life, animals created—and were continually re-creating—themselves.

Lamarck said all this aloud to his sixty-four students in the class of year 8, and the following year his dramatic opening lecture became the introduction to the book in which he first published this sea change of an idea in print, his *System of Animals Without Vertebrae* (1801). He had set Aristotle's ancient, static Scale of Nature not just in motion but in a great commotion as, rather than being arrayed along it, living beings were themselves, by their own incessant activity, ever making and remaking it, and in many directions at once.

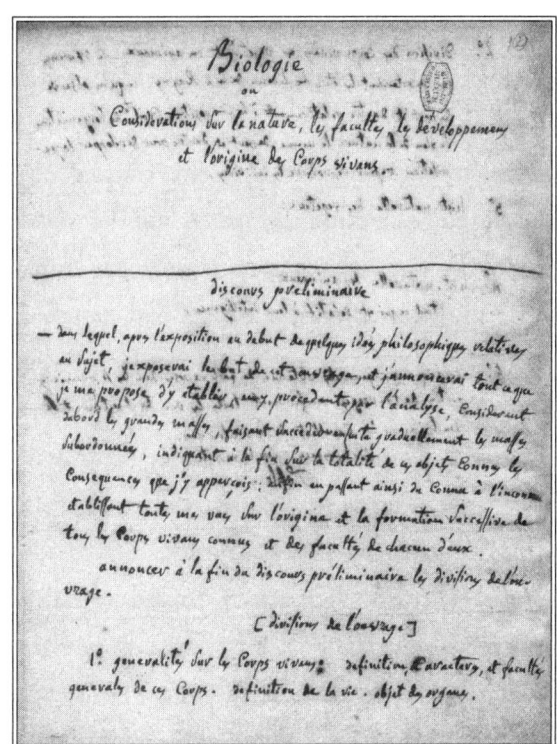

Biologie manuscript,
Lamarck, no date

. . . .

At work in his garden study, Lamarck wrote down for the first time—in bold, declarative letters across the top of a page—another new word of his own coinage: "Biologie." Under it, he added in a smaller and more pensive script, "Considerations on the nature, faculties, development, and origins of living bodies." This small, private act marked the beginning of a new science: a science of *life*.[3]

Such a thing had never before existed. There had been *natural history*, which studied everything in nature, whether animal, vegetable, or mineral. There had also been *anatomy*, a science of animal bodies; and *botany*, a science of plants. But there was not a science of life: of all and only living things. Lamarck now forged one, carrying over ideas he had begun developing as a botanist into his study of invertebrate zoology. The organizing principle for his new science was the radical idea that living beings were not the passive objects of a divine creator, the recipients of creative power. Instead, they *were* themselves the cosmos's creative power, eternally making and remaking themselves and the world around them. With the help of the unassuming yet extraordinary creatures for whom he coined the term "invertebrates," Lamarck had launched his own revolution.

His *Researches on the Organization of Living Bodies* (1802) was, he said, a kind of sneak preview: a first, brief presentation of the new science of biology. As if that weren't enough—reassigning creation from God to insects and worms—Lamarck also here mentioned in passing that he thought humans probably descended from some "quadrumane" (four-handed) animal, using a term coined by Buffon to denote nonhuman primates by the fact that they have an opposable thumb-like big toe on each foot, making them seem to have four hands and no feet.[4] (More to follow on this daring suggestion in chapter 22.)

People now came thronging to Lamarck's course to hear him unfold his extraordinary new science of a world eternally in the making by its own mortal inhabitants, who were constantly remaking also their own selves. On May 19, 1802, when Lamarck arrived at the garden's amphitheater to teach the opening lecture of his annual course on invertebrates— the first iteration of the course since he had published his new vision of the world in his *System of Animals Without Vertebrae*—he found 131 students waiting eagerly to enroll. One of these was Jacques-François Bissy, aged forty-five, a former member of the Jacobin Club, then of the Council of Five Hundred—the lower legislative house under the Directory—and now, under Napoleon, a member of one of the Courts of Appeals. Bissy took notes throughout the course—or tried to, though he often had trouble keeping up—which perhaps explains why he returned to take it again the following year.

Bissy's script—firm and legible but at a hasty upward slant—carries a whiff of barely controlled excitement where, at the top of his notebook on the first day of class, he wrote, "Wednesday 29 floréal year X, La Marke. animals without vertebra. 1st lesson." After that, things went downhill fast. There followed a torrent of information in which Bissy was apparently tossed about, grabbing at any firm object he could find to steady himself. Lamarck began by listing the seven classes of invertebrates (as he then saw them), but apparently much too quickly, because Bissy somehow managed to miss the very first class, the mollusks, and also everything after the third one, merely entering a desperate "1. . . . . 2. the annelids 3. the crustaceans &&&." But he seems to have gotten better at keeping up. On Friday, June 11, in the midst of taking down notes about the "very pretty and infinitely common" shells of a mollusk called *Columbella rustica*, he reached the end of his first notebook and hastily got out a second one mid-lecture, writing at the top "2d notebook of invertebrates. Continuation of the lesson of 22 prairial year 10."

It happened again on Monday, July 12, while Bissy was hurriedly taking down information about a genus of crabs called *Grapsus*, which includes the delicate red rock crab colloquially known in English as the Sally Lightfoot. This time Bissy broke off mid-sentence after noting that "it has 4 legs situated on its back with which it seizes what it wants to take," and quickly switched notebooks, continuing in the new one, "3rd notebook of invertebrates/ continuation of the lesson of 23 messidor/ and devour.—it covers its back with the first shell it finds convenient and thereby covers itself as under a shield—its legs resting on its back are— powerful. these crustaceans are small and flat."

On the last day of class, Friday, 25 thermidor (August 13), when Lamarck arrived finally at the simplest creatures, the microscopic infusoria—he proceeded always from the most complex to the simplest, tracing life to its origin—Bissy was right there with him. These beings "offer us the limit of animality . . . Are as old—as the World." At the end of the final class, Lamarck arrived at the most basic of infusoria: "the monads," Bissy wrote. "This is the limit of animality. you can no longer see either the intestinal canal or the mouth. feed by absorption, through their lateral pores.—it is plausible that there are others still below: but which we do not see because of the imperfection of our senses and our instruments." After a tumultuous spring and summer, he'd arrived with Lamarck at the origins of the world, at the edge of life, at the limit of human sensation and knowledge. At the bottom of the page, Bissy added "end—end. end—" and we can hear the repeated word sounding in his mind, as though he needed to persuade himself: Is this the end? The final dash already pointed his way back to the class the following spring.

. . . .

In December 1804, Napoleon summoned Pope Pius VII to Notre Dame Cathedral in Paris to watch him crown himself emperor. In fact, Napoleon bestowed upon himself not one but two crowns for good measure: a Roman-style golden laurel wreath, which he arrived at the cathedral already wearing, and a replica of Charlemagne's crown, which he briefly brandished aloft over his wreathed head. The Revolution was over; the Church and the Crown were back, albeit in a new, modern configuration: the Church clearly subordinate, the Crown claiming the authority of science, nature, and reason.

Napoleon's savants were his apostles, and the doctrine they preached was scientific support for a nominally Christian, authoritarian state. Cuvier, for his part, began to argue in his public lectures that geological evidence confirmed the story of creation in the Bible, and to represent his colleague Lamarck as a fool who proliferated species in his taxonomic work and yet—by describing them as being in a state of constant transformation—denied their real existence.[5] Lamarck shot back obliquely, remarking in his course that although "certain authors" (he meant Cuvier) continued to confuse crustaceans and arachnids with insects, and annelids with worms, these categories were importantly distinct.

It was a bleak time for Lamarck. Soon after the coronation, René, his fourth child with Rosalie de la Porte, died at the age of eighteen. Lamarck's new science of biology, with its world made by mortal beings, seemed dangerously subversive with the return of church and crown, and his enrollments fell off a cliff. Only seven students arrived to take his course in the spring of 1805. He hastily arranged to have his youngest children, Aristide (almost eleven) and Eugénie (aged eight), baptized at the Saint-Médard Church, a few minutes' walk from the garden, correcting their revolutionary names: Aristide became Jean-Louis and Eugénie became Julie Joséphine Eugénie, gaining as protection both her stepmother's and the empress's names.

But there were some happier developments. All the members of the National Institute including Lamarck became knights of Napoleon's new Legion of Honor, which remains the French governmental entity in charge of awarding the state's highest honors, and which Napoleon founded on the principle that men could only be led by enticing them with "baubles." This membership brought Lamarck a pension and allowed him to educate his children at the state's expense. Auguste, now thirteen, would go to board at a lycée in Paris; Aristide would attend the lycée of Rouen, the Norman capital city on the Seine; Eugénie would be one of the first pupils at the new Legion of Honor boarding school for girls in Écouen, just north of Paris; and Auguste, at thirteen or fourteen, would go to board at a lycée in Paris. With André again at sea, this left at home Rosalie, in her mid twenties; the hard-of-hearing Antoine, in his late teens, studying art at the Manufacture des Gobelins; and Cornélie, age twelve.[6]

Students began returning to Lamarck's lectures as it became clear that he was in fact surviving the transition from revolution to empire. Many were foreign students, and some of these were illustrious and even royal, such as three young German princes: the future Georg II, Prince of Waldeck, then nineteen, and two of his brothers, Friedrich, eighteen, and Franz Ludwig, who punctiliously noted his age as "13½." In 1808, the same year as the German princes, the very first woman students attended Lamarck's course: Henriette Dumas, née Petit, from Hamburg, aged sixty-seven, and her sister, who signed herself merely as "S. Cole née Petit."

Before this diverse crowd of students—women and men, young and old, foreign and domestic, humble and royal—Lamarck continued to develop his revolutionary worldview. The idea that living beings created and transformed themselves was an outrageous assertion in a world where people generally believed in divine creation; even today, Lamarck's claim might seem ludicrous to anyone who has taken high school biology, since it's at odds with the dominant tradition in evolutionary biology, according to which living beings are the passive objects of natural selection. But although he scandalized people and drew plenty of opprobrium, he carried on. For a quarter century, walking across the garden from his residence to the graceful, neoclassical amphitheater where he gave his lectures, Lamarck unfolded his radical theory of mortal rather than divine creation.

. . . .

Meanwhile, not long after hiring Latreille, in fructidor of year 7 (September 1799), Lamarck had published the first in a decade-long series of meteorological yearbooks: a yearbook for year 8, otherwise known as 1800, the first year of the new century. In a fraught situation, the weather is usually a safe topic. Not so in this case!

....

The Weather

In which our Hero takes up the study of Meteorology when, as a poor Student living in a Parisian Garret, he can see nothing but Clouds. He becomes the First European to Categorize and Name the Clouds; and he establishes the First Government Weather Service. His meteorological Efforts incur the Wrath of the Emperor and his inner Circle by running counter to their vision of Nature, Science, and Authority.

From the garret window of the room where Lamarck lived when he first moved to Paris in his mid-twenties, he could see nothing but clouds and sky. The clouds therefore became his companions and source of entertainment. Watching them, he began to notice how they formed, gathered, and dispersed. They didn't behave randomly, he observed, but exhibited types and patterns. He began to watch more carefully, and in this way he became the first person to classify the clouds, producing a veritable cloud atlas for his first presentation to the Academy of Sciences in 1777. He divided clouds into five types: (1) veiled (*en voile*), (2) gathered (*attroupés*), (3) dappled (*pommelés*), (4) sweeping (*en balayures*), and (5) grouped (*groupé*). But Lamarck became busy with plants and then invertebrates and waited twenty-five years before actually publishing his classification of clouds, finally including it in his meteorological yearbook for year 10 (1802).

The following year, a young English chemist named Luke Howard introduced the basic categories of the current taxonomy: "cirrus" (made of parallel fibers, from the Latin word for a tendril of hair), "cumulus" (conical heaps, from the Latin word for a heap or pile), and "stratus" (horizontal sheets, from the Latin word for a spread). Lamarck and Howard apparently never heard of each other, which is a shame since they might have enjoyed comparing notes on the clouds. On the other hand, they would probably have disagreed about some things too. For instance, Howard said he used Latin terms to name the types of clouds because he intended his system to be adopted by the "learned of different nations." This strategy was successful: Howard's terminology established itself internationally and is still in use today.

Lamarck, on the other hand, used French terms, not because he wanted to speak only to French people—after all, he drew people from all over the world to his classes—but because he hoped to create a participatory community of meteorologists that would include the non- "learned." Only an educated minority would have understood Latin terms, and Latin names would have indicated that meteorology was an elite, scholarly pursuit, which was exactly the opposite of what Lamarck intended. To be sure, when he named plants and invertebrates, he often used Latin roots, but in those cases he was working within traditions that had existed for centuries before him. The clouds, in contrast, had no traditional nomenclature in Latin or any European language. Embarking on an entirely new science, Lamarck was free to define its terms as he liked. And, just as he boasted that he could turn any passerby in the Garden of Plants into a botanist, he also meant to turn everyone in France, and beyond, into a meteorologist. He addressed his readers fondly and familiarly as "friends of nature."

In September 1799, two months before Napoleon declared himself first consul of France in a coup d'état, Lamarck published the first in his decade-long series of meteorological yearbooks, offering forecasts for the year 1800. In this inaugural yearbook, he issued a special "invitation to amateurs of meteorology," encouraging readers to annotate their copies of the book by recording their own observations in them and to send him their marked up yearbooks at the end of the year. He offered careful instructions for how to make these annotations.

In order to understand how you'd have entered notes in your year-book if you'd been one of Lamarck's participatory readers at the turn of the nineteenth century, you first have to know that each book includes a calendar arranged in columns. From left to right, there's a column for the day; then for the month; then for "natural periods useful to observe" such as the migratory arrivals and departures of certain birds, the blooming of particular plants, and the fall of leaves from various species of tree. Next comes the most important column, which is somewhat mysteriously labeled "epochs of changes of constitution." This column in fact contains the positions of the moon: its apogees, perigees, and the northern and southern extremes of its orbit. In the course of a lunar month, the moon's orbit carries it not only around the earth but also north and south, above and below the equator. Lamarck explained that when the moon was traveling through the six northern signs of the zodiac, he called this period the moon's "boreal constitution"; when it was traveling through the six southern signs, he called it the moon's "austral constitution."

The column containing the moon's constitutions is the crucial one, because Lamarck's central meteorological principle was that the moon imposes cyclical effects upon the weather by exerting a gravitational pull on the earth's atmosphere: pulling to the north, causing south winds to prevail, or to the south, bringing north winds. (Lamarck's essential idea remains present in current meteorological science; today, meteorologists refer to the effects of the moon's gravitational pull on the earth's atmosphere as atmospheric tides.[1])

After the column of lunar constitutions comes one for the "meteorological season," and here Lamarck divides each of the four regular seasons into two parts: the winter and summer into "solstitial" and "median" phases, and the spring and fall into "equinoctial" and "median" phases. Next comes a column for the time of the moon's passage over the meridian line through Paris, where Lamarck was conducting his observations, and finally, one for its declination (its angular distance above or below the equator) at noon. A second calendar, arranged in paragraphs rather than columns, describes what the weather will probably be like for each successive "constitution" of the moon. According to Lamarck's theory, boreal constitutions, pulling northward, draw winds from the south, and therefore warmer temperatures, lower pressures, humidity, rain, storms, and tempests; in contrast, austral constitutions, pulling southward, draw winds from the north, and therefore colder temperatures, higher pressures, and clear, dry weather.

In his invitation to his readers, Lamarck explains how to record their observations of the weather day by day in the column containing lunar constitutions. He also offers detailed instructions for "how to judge the state of the atmosphere." First, he says, you have to examine whether the weather is "simple" or "mixed." Simple weather either is calm or has wind blowing in only one direction; mixed weather has upper and lower winds blowing in different directions. The lower winds are the familiar ones that turn weathervanes and windmills and fill the sails of ships; the upper winds are those we can see through gaps in the lower clouds as they blow around the upper clouds. Sometimes, even when the sky is heavily overcast and you can't see its upper regions, you can still discern that the weather is mixed. For instance, if a south wind is blowing the clouds northward, and yet the barometer is steady or rising, there must be a cold north wind up above bringing the higher pressures; if it's rainy even though a north wind is driving the lower clouds southward, you can be sure there's a warm south wind up above.

Having explained how to observe the atmosphere and how to annotate the yearbooks, Lamarck encouraged his readers to send him their marked-up books at year's end and said he'd be especially grateful to hear from any whose observations differed significantly from the probabilities he described. He also included a thorough explanation of the new metric system of weights and measures that the National Convention had introduced in April 1795, even though the system had no immediate use for annotating the yearbooks. The metric system was provoking widespread exasperation and resentment, but Lamarck believed in it; here was a further expression of his purpose to spread scientific literacy and a participatory feeling among his readers.

. . . .

True to Lamarck's romantic nature, he really loved storms. The participatory feeling wasn't all about measurement systems, charts, and tables. He wanted his readers to pay attention to the weather not only by recording it but also by relishing it. Storms, he said, were "the most imposing and the most beautiful" of meteorological phenomena. One summer while he was up at Beauregard, his country house in Héricourt, he observed a hurricane, which he described rapturously in the following yearbook in a special section titled "Observations on the Hurricane of July 31, 1808, and on a Large Rotating Cloud." The storm's great diversity of cloud shapes and colors had fascinated him, their rapid dance across the sky so "magnificent" that "I couldn't tire of admiring the beautiful spectacle before my eyes."

Science and the appreciation of beauty were inseparable, since it was while he stood rapt in the midst of the storm, watching the drama unfold, that Lamarck made a surprising observation: "Suddenly I perceived in a large cloud placed to the southwest a singular movement that I had never noticed." The cloud in question took up much of the horizon and was oddly shaped, resembling "an enormous pyramid" with its summit pointing downward toward the earth. "Soon the sides of this pyramid were torn into shreds" while its base took on "a gyratory movement," turning on its axis "with great slowness." This stately, circular motion was portentous: "A most violent wind" and squall soon followed, so powerful that Lamarck was obliged to flee back to the shelter of Beauregard and "abandon my observations."

He had witnessed a similar phenomenon in Paris just a couple years earlier, in 1806, which he had also reported in the following yearbook. "On the 16th of May at about one hour thirty-five minutes in the afternoon, the sky being loaded with various clouds on the point of becoming stormy, the barometer being at 28 inches three quarters and the thermometer . . . at 17 degrees, I heard a rather loud clap of thunder, which had been preceded by a flash of lightning." Turning to look toward the southwest, where he reckoned the storm would be coming from, he saw "a large blackish cloud" and beneath it "a whitish and hanging portion of the same cloud having the shape of an inverted cone or imitating that of a funnel." It seemed a bit like a waterspout, except that the funnel-shaped part was quite small compared with the rest of the giant cloud above it. As he watched, he became most interested "to see the misty parts of the waterspout in a continual movement. Those of the funnel or the inverted cone turned very distinctly as if around an axis but rising and forming a spiral." As with the other sorts of cloud forms, so too for funnel clouds, Lamarck seems to have been the first to assign them a name in any European language.

Lamarck also loved the "spectacle of the sky" in all its myriad moods—majestic, sublime, terrible, magnificent—at sunrises and sunsets with "the clouds adorned successively in the most beautiful tints, like the tender pink, the fiery red, the most brilliant purple, finally the beautiful violet," or else at other times when the clouds were "grouped into mountains, their faces shining, illuminated with the brilliance of silver or purest gold." The "meteorologist's sky," he explained, was the portion of the atmosphere that generated the weather, an "immense laboratory" overhead.

The emotional, the lyrical, and the scientific were all one sky, one weather system, one world. And the observer—Lamarck himself and each of his readers—was right there in the thick of it, not outside looking in. He emphasized the direct action that the atmosphere exerted—and the moon through its action upon the atmosphere—upon all living beings: "the intimate relations that each particular state of the sky exercises on us, as well as on all the bodies on the surface of the globe." Anyone who has witnessed a total solar eclipse has experienced dramatically just what Lamarck meant. During totality, as the moon moves across the face of the sun, the awesome thing is that you feel the astronomical event viscerally, directly in your body: the chill, the darkness, the spreading hush. Even if we only notice it at certain moments, especially during extreme experiences like eclipses or storms, we're always being acted upon by the atmosphere, which is always being acted upon by the sun and the moon.

Rather than lifting the observer out of the world into a position of external, objective witness, Lamarck's science resided in the thick of the storm. It was all about the intimacy of witness with world.

. . . .

Lamarck's yearbooks generated lots of interest and commentary. In the dailies, they were announced, attacked, defended, and their probabilities compared with the actual weather as it came to pass, by Lamarck himself and by others.[2] With all the attention they attracted, they often sold well enough to require a second printing.

One particularly dashing defender of Lamarck's was the Parisian balloonist and parachutist André-Jacques Garnerin. Captured by the British during the Revolutionary wars and then handed over to the Austrians, Garnerin had spent three years as a prisoner of war in Buda (the now-Hungarian city directly across the Danube from Pest), passing the time by dreaming up designs for parachutes with the idea of escape. After his release, on the clear autumn evening of October 22, 1797, in a park in Paris (the Jardin de Mousseau, which is now the Parc Monceau), Garnerin made the world's first descent by parachute. He first rose to a height of nine hundred meters and then audaciously cut the cords attaching his basket to his balloon. The crowd craned its collective neck and held its collective breath as Garnerin initially plummeted like a rock; then, when his parachute opened, he began to pitch about violently in the wind, finally dropping out of view. Aghast, the spectators rushed together in the direction of Garnerin's fall, then cheered riotously when they saw him come galloping back into the garden on horseback, perfectly intact except for a mildly sprained ankle.[3]

A young woman in the crowd named Jeanne Geneviève Labrosse was especially inspired by the performance. She made her way to Garnerin to introduce herself and soon became his pupil, then his wife and the first woman parachutist. Before starting her lessons, Garnerin had to persuade the authorities that women should be allowed to go up in balloons alongside men. This had been disallowed for a couple reasons. First, the members of the Academy of Sciences warned that women's fragile organs might not tolerate air travel; then the government authorities had judged it unseemly for men and women to occupy the same balloon basket.

Garnerin's arguments prevailed, however, and he soon received the following remarkably considerate response from the city government: "Citizen, according to the complaint that you have addressed against the order of the central office, which forbids you to travel in an aerostat with a young *citoyenne*, we have consulted the minister of the interior and that of the general police. . . . [We all] think that there is no more scandal in seeing two persons of different sexes rise together into the air than in seeing them get into the same carriage, and that moreover one cannot prevent an adult woman from doing in this respect what one allows men to do, and from giving, by rising into the air, a proof both of her confidence in the practice and of her intrepidity."[4]

In the autumn of 1805, on the occasion of Garnerin's thirty-ninth aerostatic ascent and fifth descent by parachute, he cited Lamarck's yearbooks as the most useful thing to him in his aerial adventures. Since his balloons were ungovernable, leaving him entirely at the mercy of the weather, he said the best he could do was to know as much as possible about the forecast. "Never oars nor rudder, wings nor sails, bird's flight nor carp's leap, will direct a balloon," waxed Garnerin poetically. "Mr. Lamarck's meteorological directory is worth more . . . than all the calculations of modern Icaruses."

But Lamarck's colleagues were less enthusiastic. Once again, when he tried to read his meteorological theory to the members of the First Class of the National Institute in 1802, as had happened five years earlier with his work on the composition of matter, he encountered vehement opposition, particularly from Laplace.[5] To Lamarck's chagrin, his adversaries cast his meteorological work as outdated, superstitious nonsense. They constantly referred to the yearbooks as "almanacs" and to the probabilistic reasoning they contained as "predictions," insinuating that Lamarck's meteorology was essentially astrology. As with Lamarck's other struggles with his critics, this dispute with Laplace was fundamental, having to do with the nature of science, and even of the world itself. What counts as a scientific explanation? Is it legitimate science if you acknowledge, as Lamarck did, that the causes are not absolutely determinative of certain outcomes? Is the world deterministic?

Lamarck didn't see the world as deterministic or science as the description of deterministic mechanisms. In the case of the moon's influence on the weather, he emphasized that many different factors affected it, rendering its effects uncertain. Lamarck identified two sorts of complicating factors: those that acted in a regular and predictable manner, and those that acted irregularly. The regular complications included the fact that the moon's elliptical orbit carries it away from and back toward the earth during each lunar month. From its farthest point, its apogee, its gravitational pull is weakest, while from its closest point, its perigee, the influence is strongest. There are also the lunar "syzygies"—its "oppositions" with the sun, when it is on the opposite side of the earth, and its "conjunctions" with the sun, when it is on the same side.

Solstices and equinoxes—and in general the positions of the earth relative to the sun—were also regularly occurring situations that affected the influence of the moon on the atmosphere. Finally, in terms of complicating factors that act irregularly, Lamarck mentioned clouds and storms; in other words, the weather has a feedback effect upon itself. Added to all this complexity was the fact that the earth itself was in a state of continual transformation: According to Lamarck, nothing in nature remained the same even "for two seconds in a row." The oceans were always modifying the surface of the globe; the moon was continually moving the sea basin and the earth's center of gravity; and the planet was always absorbing solar light, which transformed its matter, mass, and volume. All in all, the earth's landscape, atmosphere, and climate were a great commotion of interacting elements.

Others who had sought connections between the moon and the weather had tended to simplify the situation by focusing on certain points in the moon's orbit such as full moons, new moons, apogees, perigees, oppositions, and conjunctions. For instance, in 1770—while Lamarck had been garrisoned with his regiment in Provence admiring the wildflowers—an Italian priest and professor of geography and astronomy at the University of Padua named Giuseppe Toaldo had published a "Meteorological Essay on the True Influence of the Stars upon the Seasons and Changes of Weather." Toaldo wrote that the moon acted upon the earth and its inhabitants by means of its light, heat, and gravitational attraction and that its influence became important at certain points such as perigees and syzygies. Lamarck drew upon and appreciatively cited Toaldo's observations in developing his own theory, but, he said, it was a mistake to focus upon certain points of the moon's orbit. The moon acts upon the earth's atmosphere all the time, in a continuous way, and always all mixed up with many other influences, not "in indivisible and determinable instants."

This complex mixture of factors made the moon's role hard to identify, and Lamarck confessed he had almost given up hope. "For more than twenty years," he recalled, "I alternately resumed and abandoned this interesting research. I often spoke about it to my friends." But try as he might, he couldn't find patterns to confirm his conviction of the moon's influence on the weather until he finally realized that this influence wasn't absolute but rather was subject to many complicating factors, and yet it remained perceptible. Overall, he estimated that the weather corresponded to the probabilities derived from his theory about five-eighths of the time. This was far from perfectly predictive, but he thought it was enough to be useful. The probabilities he offered were grounded in theory and observation, he insisted, and therefore, whatever Laplace might say, "it is not an opinion that I am presenting here, it is a fact that I am announcing; it is an order of things."

To Laplace—whose work in mathematics included pivotal contributions to the field of probability theory—Lamarck's "probabilities" were no better than astrological prognostications. Lamarck offered descriptive forecasts such as "we have reason to expect strong winds, maybe violent or tempestuous. . . . It would be imprudent to set out to sea or to engage in any enterprise that requires calm weather." These were not probabilities in Laplace's estimation. His probabilities were mathematical expressions, not conversational descriptions. This divergence was connected to a deeper difference between the two. Laplace's world, even including the most minor and insignificant events, was as deterministic "as the revolutions of the sun." Humans' impression of having free will and making unconstrained choices was just an "illusion of the mind." In Lamarck's view, in contrast, the world was made up of many small agencies acting in concert, its future open to whatever they might bring about.

To express his idea of an utterly deterministic world, Laplace imagined an infinite intelligence that could comprehend all the interacting forces in the universe. For such a being, he said, nothing would be uncertain, "and the future, as the past, would be present to its eyes." The human mind could gain but a "feeble idea" of such an intelligence. The best sort of approach we could make, Laplace thought, was through astronomy and mathematics, which would lead the mind "back continually to the vast intelligence . . . from which it will always remain infinitely removed."[6] I don't suppose Laplace's hypothetical omniscient being might remind you of anyone?

Isaac Newton, whose science provided Laplace with his starting point, came right out and said it in 1713 in the second edition of his *Principia*, the magnum opus in which he laid out his physical system of the universe: "It follows, that the true God is a Living, Intelligent and Powerful Being; and . . . that he is Supreme or most Perfect. He is Eternal and Infinite, Omnipotent and Omniscient." The "French Newton," as the popular press reverently nicknamed Laplace, showed Newtonian tendencies not only in his system of celestial mechanics but also in some of the theological ideas he attached to it.

God controlled everything in Newton's universe. The world operated according to God's "original perfect idea by the continual uninterrupted exercise of his power and government." All the parts of the world were God's "Creatures subordinate to him, and subservient to his Will," and there were "no powers of nature at all that can do anything of themselves." In fact, according to Newton, the world machinery functioned only through God's immediate presence all throughout and would otherwise grind to a halt.[7] In Laplace's cosmology, likewise, nothing enjoyed its own power to act. The deterministic mechanism of the divinely created universe left no room for other agencies.

This might seem like a surprising thing to say about Laplace, who has often been invoked as an early champion of atheism. But in fact, Laplace wasn't an atheist. While he left behind the Catholic orthodoxy in which he grew up, and especially rejected the possibility of miracles, he consistently expressed belief in a supreme power behind natural processes. He referred to a divine intelligence in his published writings and to God in private letters. According to an often-told story, Napoleon asked Laplace why he made no mention of God in his work on celestial mechanics, and Laplace replied that he had "no need of that hypothesis." The anecdote, a favorite of popular science writers, seems to have originated with Napoleon's doctor, François Carlo Antommarchi. In Antommarchi's version, Bonaparte was displeased by the response, but then Antommarchi's purpose was hagiographic propaganda, and he offered the story as evidence that the emperor was respectably devout.[8]

According to the astronomer William Herschel, who witnessed the exchange between Laplace and Napoleon, Laplace had simply wanted to show that "a chain of natural causes" could account for the "wonderful system" of the heavens. The question of God's existence didn't come into it. This is perfectly consistent with what Laplace himself wrote: that the supreme intelligence might well have worked entirely through material causes, with no need for direct supernatural intervention of the sort Newton insisted upon. If indeed Laplace said he had no need of a particular theological hypothesis, it was likely Newton's idea of constant, direct divine intervention in nature that he rejected, not the idea of God's presence behind the world machine. In fact, Laplace boasted that his own findings regarding the stability of the solar system would have confirmed Newton's conviction that the universe could only be the work of an intelligent and all-powerful being.[9]

In short, the deterministic mechanism of the world, in Laplace's science, was the manifestation of omnipotence: a total consolidation of power, with no room for even minor assertions of choice or will, and no such thing as contingency or uncertainty. These were but figments of human ignorance, illusion, and folly. How different a world—how even *opposite* a world—from Lamarck's churning commotion of beings and forces in continual self-transformation.

. . . . .

S oon after publishing his first yearbook, at the end of 1800, Lamarck went to see the chemist Jean-Antoine Chaptal, whom Napoleon had just appointed minister of the interior. Lamarck wanted to persuade Chaptal to create a government service that would receive meteorological observations and information from all over France: the world's first national weather bureau. Chaptal was enthusiastic; the very next day, he sent off letters to departmental prefects in every region of the country asking them each to name someone to the new position of government meteorological correspondent. Chaptal charged Lamarck with the welcome task of directing this first government weather agency, which entailed sending out exact instructions to the correspondents regarding how to conduct and record their observations, receiving their responses, and overseeing the process of organizing all the information into general tables. This job brought Lamarck to the Interior Ministry several times a week as weather information began pouring in from all quarters.[10]

Alas, the agency was short-lived. In August 1804, three months after Napoleon had crowned himself emperor, Chaptal resigned as interior minister. In his letter of resignation, he said he wanted to spend more time with his chemistry. He told a more complicated story in his journal, having to do with the ins and outs of foreign and domestic diplomacy in the wake of Napoleon's self-promotion to emperor. Meanwhile, a rumor circulated that Napoleon had provoked a falling-out with Chaptal over Chaptal's lover, Marie-Thérèse Bourgoin, a stage actress at the Comédie

Française, with whom Napoleon was also having an affair. According to the rumor, Napoleon had deliberately summoned Bourgoin to his residence one evening while he was working there with Chaptal just to rub his own relations with her in Chaptal's face. When she was announced, Chaptal brusquely gathered up his papers and stormed out. This coup de théâtre, if it took place, must surely have been designed to provoke Chaptal's resignation, which at any rate came at a moment when Napoleon was eliminating all the more independent-minded administrators and surrounding himself with absolutely loyal followers.

With Chaptal's departure, the meteorological correspondence service was abolished. All the records were confiscated and sent to the Bureau of Longitudes (a government agency founded in 1795 by the revolutionary National Convention to improve astronomy and navigation), which was just coming under the directorship of—you guessed it—Laplace. That was the end of official French meteorology for a time. Only after the passage of several decades and political regimes did a Parisian doctor, botanist, and geologist named Charles Martins, together with two colleagues, restart the practice of a meteorological correspondence and annual yearbook. That same year, 1848, saw a new revolution: The Second Republic was establishing itself just as Martins and his colleagues were launching their meteorological project. But history repeated itself. By the time they published their third volume, in 1851, the Second Republic was already giving way to the Second Empire under Napoleon's nephew Napoleon III (*Napoléon le petit,* as Victor Hugo devastatingly nicknamed him, or to Karl Marx, the farcical and grotesquely mediocre echo of his formidable uncle[11]).

Martins was an admirer of Lamarck. He also published a new edition of Lamarck's *Zoological Philosophy* to which he added a long biographical introduction, proclaiming grandiosely, "The hour of justice has sounded, and the posthumous glory of Lamarck casts an unexpected glow on France." In the first of their yearbooks, for 1849, Martins and his colleagues echoed Lamarck by modestly referring to themselves as "three friends of Meteorology" and promised to compile the information gathered by "zealous and disinterested" observers from all over France. In 1852, they cofounded the Meteorological Society of France, giving rise to the governmental Central Meteorological Bureau in 1878, the forebear of today's Météo-France.[12]

But that was all decades after Lamarck's death. In his own time, after the self-emperor-ification of Napoleon, the resignation of Chaptal, and the cancellation of the meteorological correspondence, the opposition to Lamarck's own yearbooks grew ever more concerted. The nastiness of the campaign against him was so painful that decades later, following Lamarck's death, Auguste would express deep bitterness on his father's behalf. No science could be more useful and important, Auguste reflected, and yet although people had done plenty of observing and recording, no one had embarked on an actual science of meteorology: "my father was the first to try it." Instead of praising and supporting his pioneering effort, those with power and influence had subjected him to ridicule and "persecution plotted in the shadows."

Lamarck ignored it for as long as he could and carried on compiling and publishing annually, but at last, in 1809, as the tenth yearbook appeared, Napoleon himself instructed Lamarck "to immediately cease all publication of his observations on the atmosphere," as Lamarck later recalled. "What a strange thing," he lamented, to ban his work even though he wasn't "writing at all about politics," and devoted himself exclusively to "studies of nature." It does seem extraordinary that Napoleon the great modernizer, the mastermind of rational, scientific approaches to governmental administration and authority, should have taken such a hatred to meteorology, of all things. But perhaps, upon reflection, it's not so surprising after all. Meteorology resists a reductive, deterministic model of science. And the weather defies control. Perhaps that's at least partly why Lamarck found it so fascinating.

At any rate, Lamarck was quite mistaken to suggest that "studies of nature" are separate from "politics." Lamarck's natural science was overtly materialist, granted creative agency to the humblest of living beings, combined poetry with observation and measurement, insisted upon the embroilment of the observer—sensory, physical, emotional—in the world observed, and promulgated a participatory model of scientific and governmental authority. To Napoleon and his scientific advisers, that was all plainly subversive.

PART THREE

Birds and Mammals

In which Lamarck presents his Zoological Philosophy *to Napoleon, then to the World, showing that Living Beings are Self-Making and that Birds and Mammals can even make themselves by Acts of Will. We witness the Emperor's Hatred of Lamarck's Science; and the authoritative Verdict that it is not Science but Poetry. Despite suffering Imperial Scorn, Lamarck continues to develop his Theory and to teach his Internationally Renowned Course on Invertebrate Animals. His Tree of Life leaves behind the Old Linear, Hierarchical Schemes Entirely, becoming a Representation of Creative Living Agency in Action. Elderly, Blind, Infirm, having survived Three Wives and Three of his Children, our Hero is aided by his Surviving Daughters; Rosalie writes the Last Volume of his* Natural History of Animals Without Vertebrae. *Lamarck dies and is buried in a Pauper's Grave.*

SEVENTEEN

The Zoological Philosophy

A cold November day in Paris, 1809 (across the English Channel in Shropshire, Charles Darwin is nine months old); a salon of the Palais des Tuileries, where some of the members of the National Institute—those who are newly elected or who have new publications to present—are restively waiting to be received by Napoleon, who is attending Mass. The savants and men of letters are dressed in their formal *habit vert*, black coats embroidered with a golden-green leaf pattern; the military aides-de-camp wear big gold epaulets. François Arago, aged twenty-three, newly elected, finds himself standing near Lamarck, now sixty-five, frail, and in ill health (though he'll live another two decades). Lamarck holds a copy of his *Zoological Philosophy*, his magnum opus in which he presents his full theory of living beings as self-making and world-making, the culmination of decades of work. Arago wryly observes the anxiety of each of the men around them that he might fail to receive the emperor's notice.

Napoleon at last finishes Mass and enters the room. He approaches Arago as he would a new conscript, fixing him with a stare and barking, "You're very young! What's your name?"

Portrait of François Arago
(1786–1853) by Charles de Steuben,
1832

Before Arago can answer, his neighbor on the right, eager to be noticed, replies, "His name is Arago!"

Napoleon demands, "What science do you cultivate?"

Arago's neighbor on the left, not to be outdone, pipes up with "He cultivates astronomy!"

"And what have you done?" demands the emperor.

"He has just measured the meridian in Spain!" announces the right-hand neighbor, seizing back the spotlight.

While Arago ruefully reflects that Napoleon must take him for a mute or an imbecile, the emperor turns to Lamarck, who presents his *Zoological Philosophy*. Napoleon snatches it, growling, "What is this? . . . It is your absurd Meteorology . . . this yearbook which dishonors your old age; . . . I take [this] only out of consideration for your white hair.—Give it here!" And without so much as glancing at the book, Napoleon thrusts it into the hands of an aide, while Lamarck, tears of humiliation rolling down his cheeks, tries to explain what's in it.

But had Napoleon known what was in the book, with its revolutionary understanding of animal life, he probably would have liked it even less. Given what he and his scientific inner circle—Laplace, Fourcroy, and Cuvier—already thought of Lamarck's earlier claims that living organisms were the creators of the inanimate world, the emperor would likely have taken a dim view of his new theory that animals were constantly making and remaking themselves as well as the world around them. The materialism and all-but-explicit atheism wouldn't have suited Napoleon, who was trying hard to distance himself from the materialist, atheist revolutionaries as part of his ongoing bid for the support of the

Lamarck presenting the *Zoological Philosophy* to Napoleon
by Mikhail Dmitrievich Ezuchevsky, ca. 1921–25

conservatives who wanted to restore the monarchy and the Church. Lamarck gave an occasional perfunctory nod to "the supreme author of all that exists," but it was invariably to insist on his total irrelevance to the study of living nature.

"As a naturalist and physicist," Lamarck admonished, "I am only concerned" with observable things. God was not an observable thing, and regarding the question of an immortal soul, "all we can ever say on this subject is baseless and purely imaginary. If we are studying nature, it alone should occupy our attention." All in all, Lamarck's science was uncomfortably ungodly and his natural world of inherent active powers, its substance generated and shaped by the agency of its own living denizens, decidedly uncongenial to authoritarians.

. . . .

What could be more interesting than the study of animals?" demands Lamarck disarmingly on the first page of the *Zoological Philosophy*. The book is the proverbial wolf in sheep's clothing: an immensely daring work that presents itself unassumingly as just a set of modest conclusions based upon years of minute research and conscientious teaching in the Garden of Plants. True, the book did expand upon what Lamarck had already been telling his students in his courses over the previous decade. Initially, in fact, he said he'd just been planning to write a textbook to use when teaching his course on invertebrates, offering his students some basic principles for the study of animal life. Invertebrates provided an excellent foundation for thinking about animal life in general, since they were the simplest living beings, and also since they comprised the vast majority of animals, both in variety and in sheer numbers.

But as Lamarck set about the task of writing his textbook, he found himself drawn into a more profound reflection. Why, he wondered, did the forms of animals exhibit such a "singular" and striking gradation from one to the next across the whole array of animal kinds? As he contemplated this gradation, it began to move, to transpose itself from space into time. What he was seeing, it now seemed obvious, was in fact the encapsulation in a single moment of a process taking place over millennia: the forms of animals arising gradually, "in an extremely remarkable manner," developing "little by little and successively" from simple to ever more complex.

The reason people hadn't realized this, Lamarck explained, was that it happened too slowly. We can't witness such infinitely slow changes, which take place over inconceivably long expanses of time, so we're "naturally prone" to think that things have always been as they appear to us now. This was among the factors that rendered life the very hardest scientific subject, in Lamarck's estimation, even harder than astronomy with its inconceivably remote objects. "It seems to me," he mused, "that it was much easier to determine the course of the stars in space, and to ascertain the distance, magnitudes, masses and movements of the planets of our solar system, than to solve the problem of the origin of life." At least in the case of the stars and planets, their motions took place on an observable timescale.

The idea of a deep abyss of time departed dramatically from the traditionally established temporal framework, the time frame of the Bible, which is measured in human generations and graspable in proportion to a human lifespan. The idea of eternity did have ancient precedents, though; for instance, the Roman philosopher-poet Lucretius invoked an "infinite time past," and in this fathomless depth of time, he said, "the generations of living things are changed and like runners pass on the torch of life."

When Lamarck was writing his *Zoological Philosophy*, the age of the world had been a matter of controversy for more than a century. During the mid-seventeenth century, certain writers, such as the polemical French author and diplomat Isaac de La Peyrère, had rejected the time-scale of the Bible and endorsed the much greater antiquity that ancient pre-Judeo-Christian peoples—such as the Egyptians, the Chinese, and the indigenous Americans—had claimed for their civilizations. In response to this radical divergence from Judeo-Christian historical time, other writers heatedly refuted the claim, drawing on evidence from the Bible. Isaac Newton was a leading example. He devoted enormous energy to biblical chronology and to arguing that the world was at least as young as tradition suggested. In fact, Newton calculated that ancient Greek history was actually *shorter* than generally described by half a century, and ancient Egyptian history by a full millennium.

This adamantly defended short time frame suffered new challenges during the first half of the eighteenth century. A diplomat and natural historian named Benoît de Maillet produced a scandalous interview with a wise "Indian philosopher" named "Telliamed"—discerning readers quickly spotted the ploy, as this was in fact the author's own name spelled backward—claiming that the earth was billions of years old. Neither Diderot nor Buffon went quite so far, but they both argued for a much longer timescale than the traditional scriptural one. Diderot imagined the earth and its living inhabitants transforming over "several million centuries." Meanwhile Buffon, on the assumption that the earth had started out molten and based on his estimate of its rate of cooling from the outside in, reckoned it must be seventy-five thousand years old. And why, he demanded, should the human mind boggle at an expanse of time more than of space, or in fact at any other measure? Why should it be harder to conceive of a hundred thousand years than of a hundred thousand pounds of silver money?

Around the time Lamarck started working with Buffon in the Garden of Plants, during the mid to late 1780s, a Scottish farmer and geologist named James Hutton presented to the world—beginning with a lecture before the Royal Society of Edinburgh—his theory of the earth's fathomlessly deep history. Hutton described the earth as a perfect "machine" governed by an "almighty power, and supreme wisdom." In such a machine, any waste must be rectified, any decay repaired, any excess trimmed. In keeping with this principle of perfection, Hutton described the earth's history as an endlessly repeating cycle in perfect balance and harmony with itself. The solid land was formed in the depths of the oceans, from the remains of "animal and vegetable bodies"; then the oceans wore the continents away and swallowed them, and then they re-emerged by the same process as before, and so on, forever, eternally. Poetically capturing the shiveringly awesome aspect of earth's fathomless history, as he saw it, Hutton wrote that there was "no vestige of a beginning,—no prospect of an end."

In some ways, Hutton and Lamarck were kindred spirits. Hutton regarded the earth as one "living world" made up of "an almost endless diversity of plants and animals." He described this world as existing in a constant state of lively commotion: "Where so many living creatures are to ply their respective powers . . . we are not to look for nature in a quiescent state; matter itself must be in motion, and the scenes of life a continued or repeated series of agitations and events." On the other hand, for Hutton the agitations and events of the living world were all imposed by external design, in accordance with "Divine wisdom," for the purpose of propagating and maintaining perfect forms of life. These forms didn't shape or transform themselves. They were produced in a state of perfection by the divine machinery and remained just as they were created until the same machinery eliminated them from the earth and replaced them with other beings.[1]

In Hutton's bottomless well of time, although everything was in constant motion, in another sense nothing changed at all. The divine machinery eternally executed its cycles. Lamarck's version of deep time was entirely different in this regard. His living world was ever developing itself in new directions, though at such a slow pace that it was invisible to the tiny and short-lived organisms within it.

Reaching for a way to make this profoundly unfamiliar idea intuitive to his readers, Lamarck thought of an analogy. "To explain what I mean," he wrote, "let me make the following supposition." Say that each human life lasted only one single second. Imagine, though, that clocks such as our actual ones still existed in this alternate world. If one of our clocks were somehow wound up and set going, a person watching it would detect no movement in the hour hand over the course of their whole life. In fact, the "observations of thirty generations would furnish no clear evidence" that the hour hand was moving. Perhaps certain very radical thinkers, drawing on even older observations going back more than thirty generations, might begin to argue that the hour hand was changing its position. Those hearing them would surely refuse to believe them at first, Lamarck said, "since they had always seen the hand at the same point of the dial." Nevertheless, the revisionists would be quite right: The hour hand is moving.

EIGHTEEN

The Power of Life and the
Inner Feeling of Existence

L amarck invoked two different sorts of internal, organic agencies propelling the development of living beings from within. The first was a natural force like the forces of contemporary physics, such as gravity, electricity, and magnetism. For some time already, in his lectures and writings, Lamarck had taken to calling this force the "power of life": an upward-striving tendency that drove living matter to compose, elaborate, and complexify its organization over time.

Tendencies of one kind or another were very common in physics (as indeed they still are). A little over a century earlier, Isaac Newton had described a universal tendency of common matter by which the planets gravitated toward one another and toward the sun, and by which objects on earth moved downward toward its center.[1] Gravity was a central force in physics during Lamarck's time, and, of course, it remains so today. Also, during Lamarck's lifetime, people had been discovering the dynamic and fiery tendencies of electricity, another force that retains its place in current science. Electricity would move through some substances but not others, following metal or water, avoiding glass or wax; it could cause objects to attract and repel one another; and it glowed, sparked, and even created explosions as it moved among different substances.

While Lamarck was a child in the 1740s and 1750s, a Parisian experimental physicist named Jean-Antoine Nollet had performed spectacular feats with electricity in his public courses in Paris. He had channeled the electrical fire into and through all kinds of things including living people: an apparently willing boy suspended from the ceiling by silk cords, 240 members of the royal guard at Versailles standing and holding hands before King Louis XV, hundreds of Carthusian monks obediently doing the same in their monastery, and at the Collège de Navarre, at the top of the Montagne St. Geneviève in Paris, more than 600 people in a linked, rapt crowd. According to an observer, it was "singular" to see "the multitude of different gestures and to hear the instantaneous exclamations of surprise" from hundreds of people at once as they received the electric shock.[2]

Around the same time in Philadelphia, Benjamin Franklin had noted that electricity tended to move toward and through pointed objects, leading him to devise one of the first lightning rods. And in 1785, while Lamarck had been engrossed in his botanical research, a military officer, engineer, and physicist named Charles Coulomb had studied magnetic attraction and repulsion and demonstrated that magnetism followed the same pattern as electricity and gravity. All the forces decreased as the square of the distance between the objects attracting or repelling one another.[3] Magnetism, like electricity and gravity, remains an established force in physics.

If Lamarck's "power of life" had endured in biology in the same way as gravity, electricity, and magnetism in physics, it's hard to know how science might have looked over the past two and a half centuries, but surely it would have differed interestingly from the reductive, mechanistic theories of the later nineteenth and twentieth centuries, in which living beings were fundamentally passive, shaped only by random variation and natural selection.

Lamarck of course read Newton, and also Franklin, whose works he kept in one of the bookcases in his garden study, and all the recent research on electricity and magnetism.[4] If inanimate matter exhibited these various behaviors and proclivities, why shouldn't living matter have its own tendency? The power of life in organic matter began, he said, with the very simplest plants and animals: tiny, fragile, translucent algae and monads. Lamarck wrote that these infinitesimal beings were continually being spontaneously generated: Nature "begins again every day." He judged that conditions favorable for spontaneous generation existed "exclusively in water or very moist places," where these most rudimentary organisms were always found, so that "water is the true cradle of the entire animal kingdom."

Starting with the tiny, spontaneously generated organisms appearing in water, over an "incalculable series of centuries," driven by the power of life to tend toward ever greater complexity, the simplest forms of life gave rise to every kind of living being. Just as living things were always dying, inanimate things were always becoming animate and embarking on the long process of transformation; it was an immense circle of changes, Lamarck said, these "passages from life to death, and from death to life."

From the simplest living beings through the most elaborate forms of life, Lamarck described a continuous spectrum of increasing complexity, but at the very beginning, between inanimate matter and the simplest organisms, he identified a threshold. Only living bodies had "individuality," a kind of integral oneness spread throughout themselves, by which all the parts derived their purpose and character from one another. It was therefore a real mistake, he thought, to try to reduce animate bodies to inorganic ones.

I n addition to the power of life, in animals another kind of agency is
always at work according to Lamarck: responsiveness, an ability to re-
act to the environment. In the simplest animals, infusoria and polyps,
this responsiveness takes the form of a primitive "irritability," a contractile motion in response to stimulus. This basic capacity distinguishes even
the most primitive animals from plants, which—lacking irritability—
are incapable of performing "repeated sudden movements" of the sort
even elementary animals can make. Moving slightly upward in complexity, Lamarck thinks there might possibly be rudimentary nervous systems in certain animals—such as "radiates" (like starfish) and "fistulides"
(such as sea cucumbers)—with nerve fibers allowing for a more elaborate
kind of movement, muscular motion. But these animals still have no
brain, and therefore no capacity to feel.

The distinction between irritability and sensation is important, Lamarck says: Animals can move without sensation and vice versa. For this
reason, he rejects an utterly grisly set of experiments performed by the
doctor and physiological researcher—and, apparently, wanton sadist—
César Legallois, in which Legallois decapitated baby rabbits, or sometimes severed their spinal cords at the neck to fully paralyze them, then
forced air directly into their lungs. From the fact that their limbs moved,
he concluded that the artificial respiration had fully restored "sensation
and voluntary motion." Legallois describes in detail how the rabbits'
mouths gasped for air while he repeatedly extinguished and reawakened
movement in their bodies. The experiment is useless, Lamarck writes.
The movements tell you nothing about the animals' sensations, only
about the irritability of their bodily fibers. No experiment that triggers
involuntary motion in an incapacitated animal, he says, can reveal any-

thing about sensation or voluntary motion. If the animal's nervous system is intact, we can assume it is suffering, while the experimenter is learning nothing.[5]

Whether or not there are animals having muscular motion but lacking sensation, Lamarck writes that all animals starting from the level of insects have both. This is because they have nervous systems in which the nerve threads come together in a single "medullary mass," or brain, which he calls the "special organ of feeling": It endows animals who have it with sensations and an "inner feeling of existence." Insects can experience "simple and fugitive perceptions when any object affects them" and perhaps also some rudimentary "inner emotions." The more complex the animal, the more developed the inner feeling. Birds and mammals have not only a brain—a medullary nexus of nerves—but what Lamarck calls a "hypocephalon," which he identifies as the "organ of intelligence" (which today goes by the name "cerebral cortex"). It consists of "two wrinkled, pulpy hemispheres" enveloping the brain. This organ brings the "faculty of will": the ability to form complex ideas, produce judgments, and act upon them.

In its most developed form, the hypocephalon allows a being to "think, reason, invent, and perform various intelligent acts." When the environment creates a need, animals with a hypocephalon respond to it by willfully undertaking new actions. They form "habits" and "penchants." Whenever they decide to perform a given action, the fluids in their body move into the necessary organs and limbs to bring the action about. By many repetitions of this process, animals can strengthen, extend, and even create the necessary parts. Therefore, by acts of will—exercising or failing to exercise their various parts and organs—the more complex animals transform their bodies. These transformations are infinitesimal, Lamarck acknowledges, but he believes they can be inherited by the animals' offspring, accumulate over thousands and thousands of generations, and so account for how animals have achieved their current forms. By adapting their behavior to their environments, they are always gradually making and remaking themselves.

"When my dog is out for a walk," Lamarck reports, "and sees in the distance another animal of his own species," he clearly experiences a visual sensation that emotionally affects his inner feeling. It then guides his nervous fluid into his muscles, propelling him immediately and impulsively toward the fellow dog he sees. At other times, watching his dog dreaming, "barking in his sleep, and giving unequivocal signs of the thoughts which agitate him," Lamarck writes that he becomes "convinced that he too has ideas."

The desires and fears of animals are the driving forces in Lamarck's theory of transformation, and he thinks these should be central topics in zoology. It's obvious that "many animals possess feelings and that some also have ideas and perform intelligent acts." Lamarck finds it to be the most "curious and interesting" subject, and also the most difficult, to try to understand how material parts can produce ideas and feelings. After all, intelligence is surely a natural phenomenon—sensations and thoughts are no more miraculous than other natural things—and so they must result from physical causes. The inner feeling of existence is at their very crux.

. . . .

One day, Lamarck's left eye was sore and especially painful in the sunlight. Going out, he covered it with his pocket handkerchief. All of a sudden, as he walked along the street, a horse and rider fell down quite close to him on his left side. He didn't consciously see them, because of the handkerchief, but apparently sensed their presence since "I instantly found myself transported two steps toward the right by a movement or bound, in which my will had not the smallest share." This was his "inner feeling of existence" manifesting itself. Surely, Lamarck writes, everyone has had an experience like this. The same inner feeling is in play when we're moved "by the sight of a precipice, [or] of a tragic scene, either real or on stage or even in a painting," or while listening to a fine piece of music, or when we feel joy or sorrow upon hearing good news or bad.

The inner feeling of existence, in other words, is the basic current of feeling underlying all other thoughts and emotions. Nothing else is as important as this "intimate and continuous feeling," one of the chief causes of animal organization, and yet constantly overlooked. We generally don't notice it in ourselves precisely because it's so fundamental and constant. But we can get a "sort of hint of the existence of this interior power" if we think of moments of shock or agitation. Who hasn't had the experience of being startled by a loud and unexpected noise, which "makes us start or give a sort of jump"? This reflexive response is the inner feeling asserting itself directly, overcoming or bypassing the conscious will.

Lamarck characterizes the inner feeling of existence as an "all obscure" and yet "very powerful" feeling arising from the confused jumble of inner sensations that accompany the movements of life in our internal parts. It's the "me" possessed by all sensitive animals, although only the intelligent ones can notice it and become conscious of it, and it is suspended during sleep. The inner feeling is the origin of the emotions, both the moral emotions that result from thoughts and the physical emotions that result from sensations. These emotions in turn provoke the "active motor" that allows animals to deliberately perform movements by dispatching nervous fluids to the muscles, exciting them into action.

Watching a deaf and mute young woman performing on the piano, Lamarck admired her ability to keep perfect time (he pronounced her playing "far from brilliant, but passable"). As she played, he observed that "her entire personality was stirred by regular movements of her inner feeling," which he thought must compensate for her lack of hearing. Her teacher explained to Lamarck that he had had her practice keeping time using visual signs. These signs, Lamarck was convinced, "stirred within her the feeling in question." He concluded that the musical capacity we generally attribute "to the highly trained and delicate ears of good musicians belongs rather to their inner feeling." The young woman's performance also reflected the fact that education, habits, and temperament all contribute to modifying the capacity of the inner feeling to experience emotions.

The most important thing about the inner feeling of existence is that it is the culmination of a transfer of power. For the most rudimentary organisms, Lamarck explains, the power to cause their movements and actions is entirely external. They exist only at the mercy and under the influence of their environment. But by preserving in offspring the changes the parents underwent, nature has "transported this power into the animal itself" until finally, in the most complex animals, it is "at their disposal." With the inner feeling of existence, animals can cause their own actions, and with intelligence and will they can do so systematically, forming habits and ways of life.

Such a process must be the only way to account for the existence of sentient living beings, Lamarck is sure. "If nature had confined itself to its original method, that is, to a force entirely external and foreign to the animal," he writes, then animals would never have been anything more than "purely passive machines," and there would never have existed living beings with the wonderful capacities of "sensibility, the intimate feeling of existence, the power of acting," ideas, or "the most astonishing of all, thought, in a word, intelligence."

Lamarck's theory was the exact opposite of the established view, held by "just about everyone," as he lamented, that God had designed each animal for its particular environment and that animals behaved in accordance with their divinely given structures. This theory assumed that each animal had a fixed and unvarying form, and that the environments inhabited by animals remained equally static and changeless, an idea Lamarck rejected as obviously false. The world and all its inhabitants were clearly in constant flux. No, he maintained—as he'd been telling his students for the past decade—it's just the reverse of what everyone says. It's not the shape of the body that dictates the animals' habits and ways of life, but on the contrary it's the habits and ways of life that give rise to a bodily form to suit them. Animals are their own creators.

Along with this reversal went another, closely related one pertaining specifically to human beings: Lamarck writes that people have focused a lot of attention on the "influence of the physical on the moral," the ways in which our bodily organization shapes our character, but haven't paid enough attention to the "influence of the moral on the physical," the ways in which our feelings shape our body. (It's important to understand that Lamarck doesn't mean "moral" in the sense of good or virtuous; he uses "moral" to mean having to do with thoughts and the feelings they cause.)

The year before the *Zoological Philosophy* was published, Pierre-Jean-Georges Cabanis had died; Cabanis was a medical school professor and polemical materialist famous for having remarked that the brain digests sensory impressions to secrete thoughts, just as the stomach digests food. Thought, in other words, is a bodily function like any other. Cabanis observed that melancholy people often suffered from abdominal problems, and no doubt these were the cause of their gloom. Lamarck thought Cabanis was wrong to assume that the causality flowed in only one direction, from physical states to emotional results. Surely, Lamarck argued, it must go both ways. Some original grief might have damaged the melancholy person's abdominal organs, permanently inclining them to sadness even after the original cause was gone. Cabanis's mistake was to overlook "the strength of the inner feeling" to exert changes in the body.[6]

After all, we know that a moral feeling, when very powerful, can temporarily extinguish physical sensation and can disrupt the functioning of the essential organs. Lamarck offers examples. Distressing or joyful news can lead to minor effects, such as digestive troubles, or in weak or elderly people can have dangerous and even fatal consequences. In religious fanatics who are subjected to torture, the moral feeling can be so overdeveloped as to overcome even extreme pain. Normally, the moral emotions are more powerful than the physical, Lamarck thinks, but when physical feelings are very strong, they can sometimes overwhelm the intellectual faculties, causing delirium. Still, overall, he finds that moral emotions have a greater influence on bodily organization than physical sensations. Thoughts are the means by which animals of sufficient complexity, including humans—slowly, infinitesimally, over countless generations—shape and transform themselves from generation to generation.

NINETEEN

The Giraffe and Company

Evidence of animals' self-transformation is everywhere. Lamarck begins with cases of the diminished use of a part or organ resulting in its weakening or even disappearing altogether. For instance, right whales, birds, and anteaters can swallow their food without chewing it, so they no longer have a developed set of teeth, but they apparently once did: Geoffroy Saint-Hilaire found teeth concealed within the jaw of a right whale fetus and also saw grooves for teeth in the beaks of certain birds. Another category of examples concerns eyes. Moles, which live in the dark, don't use their eyes much, so these have become tiny and barely visible. The same is true for the *Spalax*, a kind of mole rat that lives underground, and the olm, an aquatic reptile that inhabits deep underwater caves. Then there's the case of legs, or rather their lack. Snakes, Lamarck says, must once have had four legs like their closest amphibian relatives, frogs and salamanders. But having "adopted the habit of crawling on the ground and hiding in the grass," through "repeated efforts" to pass through narrow spaces, snakes elongated their bodies, such that "legs would have been quite useless to these animals."[1]

Ears are the clincher. Unlike teeth, eyes, and legs, ears are common to all vertebrates. In this contrast, Lamarck finds proof of his claim that it was disuse that caused the teeth, eyes, and legs of certain vertebrates to shrink or even disappear. He reasons that while light doesn't penetrate everywhere in the world—there are inhabited places that are absolutely dark—there are no inhabited places without sound, which "penetrates everywhere and passes through any medium." Some vertebrates have no use for teeth, eyes, or legs, and so they've gradually lost these parts, but all vertebrates use their ears, at least to some degree, and so they've retained them.[2] Use it or lose it.

Then there's the opposite process in which the increased use of an organ or part develops and strengthens it. Beginning with his momentous lecture course for the year 1800, as we saw in chapter 16, Lamarck cited to his students the case of water birds versus birds that perch in trees, and he develops the example further in the *Zoological Philosophy*.

Jacana bird by Jacques de Sève, 1781, from Buffon, *Histoire naturelle*

By making "efforts to swim, that is to push against the water so as to move about," not only ducks and geese but also "frogs, sea turtles, otters, and beavers" have slowly stretched the membranes between their toes and acquired webbed feet. In contrast, from the custom of clasping branches, a bird's "claws in time become lengthened, sharpened and curved into hooks" such as the talons of eagles. Lamarck also includes another of his favorite examples from teaching his course, the shorebird's long legs and neck. This bird, he says, dislikes swimming yet finds itself obliged to wade into the water to catch its prey. Continually sinking into the mud, the bird stretches its legs in distaste, trying to keep its body above the water. This habit has resulted in the long, skinny, featherless legs that make these birds look as if they were on stilts. The same fastidiousness has shaped their necks. Wanting to fish without wetting its body, the bird stretches and cranes its neck, resulting in "a remarkable lengthening" in the necks "of all shorebirds."

The proof that these extravagant bodily forms have followed from the animals' tastes and habits, Lamarck says, is the comparison of shorebirds with swans and geese, which have long necks but short legs. These swimming birds, being perfectly happy in the water, have felt no need to stretch their legs, which have therefore remained short. But from "the habit of plunging their head as deeply as they can" in search of aquatic larvae, plants, and other edible creatures, swans and geese have greatly elongated their necks. In other animals, such as anteaters and woodpeckers, it's the tongue that has grown long from repeatedly sticking it way out to slurp up their insect prey. The tongue has become not only elongated but also forked in hummingbirds, which use it to collect nectar from flowers, and in lizards and snakes, which "use theirs to palpate and identify objects in front of them" and can feel "several objects at the same time" thanks to their divided tongues. For greater convenience in using their tongues, snakes have even developed "an aperture at the extremity of their snout to allow the tongue to pass without having to separate the jaws."

While some animals crane their legs, necks, and tongues, others make strenuous efforts to see in the needed direction that ultimately end up repositioning their eyes, according to Lamarck. He contrasts the perpendicular fish of the deep ocean, which have one eye on each side of their head, maximizing their lateral vision, with the fish who swim in shallow water near the beach. These are obliged to swim on their flattened surfaces, receiving more light from on top than beneath; they also have a particular need to focus on what's happening above them. The constant effort of looking upward has directed their inner fluids so as to slowly but inexorably force "one of their eyes to undergo a sort of displacement, and to assume the very remarkable position found in soles, turbots, [and] dabs." In these species, Lamarck says, the eyes have arrived on the upper surface, but they're still asymmetrically placed because the mutation is "incomplete." In skates, it's fully accomplished. These fish are altogether flat, with their eyes symmetrically placed and directed upward. Snakes, too, since they crawl on the ground, mostly need to look upward and to the sides; their eyes have accordingly gravitated to the "lateral and upper parts of their head." As a result, they can barely see what's right in front of them. This could cause injuries if they weren't able to compensate with their tongue, "which they are obliged to thrust out with all their might."

Darwin, several decades later, would comment similarly on the "remarkable peculiarity" of the placement of the eyes in flatfish, and would attribute it—like Lamarck—to the inherited effects of habit. Young flatfish, Darwin noted, had one eye on each side of their heads. But because of their shape, the size of their fins, and their lack of a swim bladder, these fish find it hard to remain upright in the water and soon grow tired, flopping over onto one side. They then "often twist . . . the lower eye upwards, to see above them; and they do this so vigorously that the eye is pressed hard against the upper part of the orbit. The forehead between the eyes consequently becomes, as could be plainly seen, temporarily contracted in breadth." As they grow, they become increasingly flat, and the lower eye "begins to glide slowly round the head to the upper side" as a result of this habit. Darwin surmised that the changes in eye position were inherited to some degree so that the eyes were shifting not only in individual young fish but from generation to generation.

Herbivorous mammals present "remarkable" examples of the effects of habit, according to Lamarck. For instance, certain grazing quadrupeds such as elephants, rhinoceroses, buffalo, and oxen do a lot of standing around. They don't move much, or climb trees, but mostly just eat. This habit has rendered them heavy and massive, with thick horns enveloping their toes. In fact, their toes get so little use that they've dwindled from five to four to three in pachyderms (elephants, hippos, and rhinos, respectively) to two in ruminants, to just one in horses and donkeys.

Exotic animals offer yet more wonderful examples. Lions and tigers, as a result of using their claws to climb trees and tear into their prey, have grown large, curved claws that would impede them when walking or running, but their efforts to pull the claws back out of the way have resulted in the claws being retractable. In contrast, the sloth's big, hooked claws are not retractable. Encumbered by such great claws, along with extra-long arms and short little bowlegs, the sloth's bodily structure might appear to have "forced it into the habits and miserable state in which it exists." But Lamarck says it's just the other way around. Some set of dangers forced sloths "to take refuge in the trees, to live there habitually and feed on their leaves" so that their main needs became clinging to branches, dragging themselves from branch to branch, and otherwise remaining "in a state of inactivity in order to avoid falling off." By acting on this restricted set of needs, they shaped themselves into the strange, slow, sleepy creatures they became.

. . . .

In the *Zoological Philosophy*, Lamarck summarized the effects of animals' responses to their environments by asserting two new "laws of nature":

1. The frequent use of an organ gradually strengthens, develops, and enlarges it, while disuse imperceptibly weakens and deteriorates it, until it finally disappears.

2. The results of the use or disuse of an organ are inherited in the next generation as long as they are common to both parents.

Six years later, in the *Natural History of Animals Without Vertebrae*, Lamarck expanded his laws to four and also modified one significantly:

1. Life, by its own power, tends to grow, increase, and extend its parts.

2. New organs arise from animals' actions in response to their needs.

3. The continued strength and development of organs is due to their use (and by implication, if they aren't used, they weaken and disappear).

4. Everything individual organisms acquire in their
lifetime they transmit to their offspring.

Note that he has removed the requirement that changes be "common
to both parents" in order to be transmitted to the offspring. Perhaps he realized that the requirement was incompatible with sexual dimorphism—
the differences in traits between the males and the females of species—a
phenomenon he studied extensively, especially in insects (more on this in
chapter 23). He now elaborated that in sexual reproduction, when the
two parents had undergone a modification "unequally," it was not always
transmitted to the next generation or might be transmitted only partially.

In deriving his laws of transformation of living forms, he had the advantage not only of the dead specimens in the collections of the Museum
of Natural History but also of the living animals of the menagerie, in
which he was keenly interested. For a time, he was in charge of overseeing the menagerie. He fretted over an elephant's bout of indigestion and
oversaw its treatment with elephantine quantities of honey. Soon he was
able to reassure his colleagues. The minutes of an assembly in the spring
of 1802 record that "M. le chevalier de Lamarck reports on the state of
the elephant and calms any anxieties the assembly may have conceived;
the usage of honey in sufficient quantities has produced a very good effect." Lamarck studied the breeding of the goats and noted the resemblance of a newborn bull calf to its Scottish Highland father. He urged
his colleagues to address the need of the growing lionesses for more
space, and under his tenure as director the menagerie developed a successful technique for soothing the teething of the lion cubs so that after
the loss of three males they were able to nurse a young female through
the dangerous period.

Seal by Jacques de Sève, 1782, from Buffon, *Histoire naturelle*

Lamarck also oversaw a concerted and ultimately successful attempt to acquire kangaroos. After several disappointments and failures, a pair of kangaroos finally arrived from a collector in England during the summer of 1802; alas, they remained only until the spring of 1804, when Joséphine Bonaparte wanted them for her estate outside Paris at Malmaison.

When a seal arrived in the menagerie of the Museum of Natural History at the end of June 1809, Lamarck watched it with great interest. The *Zoological Philosophy* was already in press, but he hastily wrote up his observations, and the reflections they inspired, as a last-minute and very important "addition" to the end of volume 2. He noticed that the seal could join its hind flippers together, separating the digits to spread out the membranes between them, forming a large paddle "which it uses for traveling about in the water in the same way as fish use their tail." When it was on land, the seal could move speedily "by means of an undulatory movement of the body," not using its hind legs at all—these it held stretched out behind—but supporting itself on its forelegs "on the arms up to the wrists," and not really using its "hands." The seal seized prey with either its hind feet or its mouth. When it remained underwater for an extended period, it completely closed its nostrils "just as we close our eyes."

One lesson Lamarck drew from his observation of the seal was that, from the habit of using its hind legs as a fin or paddle, the legs had shifted in position so that they extended "in the same direction as the axis of their body." However, the legs had not fully joined together and become a fin because the seal still used them for seizing prey. In other words, by their distinctive uses of their hind legs, seals had produced a structure that could work either like a paddle or like a hand.

The seal's obvious kinship with land mammals confirmed Lamarck's sense that all mammals, and indeed all animal life, had originated in the water. The transition between land and sea mammals also led him to reflect on another transition of mammals from land to air. Birds must have come about by a gradual passage from land mammals through "those animals that can make but a very prolonged leap," such as flying squirrels. By their efforts to extend their jumps, these animals had extended their skin to form flaps, which exemplified structures partway from arms to wings. In bats, you see the transition from land mammal to flying mammal completed. Watching the seal, Lamarck began to think in more detail about a branching structure of animal life, originating in the water and branching into more complex sea organisms, amphibious creatures, and animals living on land and in the air. He drew a "chart serving to show the origins of the various animals," beginning with two main

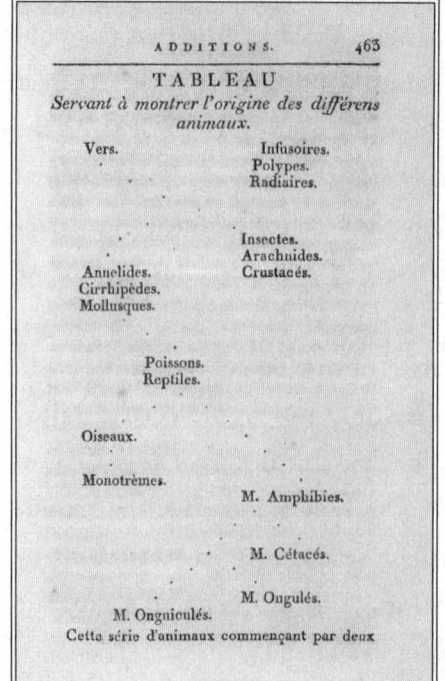

"Table showing the origins of different animals," Lamarck, 1809

trunks, worms and infusoria, and dividing into ever smaller branches as the tree of life spread through the ocean, onto the land, and into the air. (More on this chart in chapter 20.)

. . . .

And finally, the moment you've been waiting for (drumroll please): the giraffe! More than any other creature, the giraffe came to symbolize Lamarck's theory, although it occupies only a couple brief passages in all his voluminous writings. The giraffe does indeed present an arresting example. "Concerning habits," Lamarck writes, "it is curious to observe their results in the particular shape and size of the giraffe (*Camelo-*

pardalis)." Inhabiting the arid and largely barren African savanna, giraffes need to graze the leaves of the trees, "and to struggle constantly to reach these. The result of this habit, maintained over a long time in all individuals of the race, is that the front legs have become longer than the hind legs, and the neck has elongated itself such that the giraffe, without standing up on its hind legs, lifts its head and attains a height of six meters (almost 20 feet)."

Lamarck hadn't seen a living giraffe when he wrote these lines, and he never would get to see one, although he would have one for a neighbor during his last couple years of life. The giraffe we met during our prefatory stroll, which Geoffroy Saint-Hilaire escorted from Marseille to Paris, arrived at the garden in the spring of 1827, at which point Lamarck had been blind for almost a decade. But had he gotten to see her in action—a young creature transplanted halfway across the world who took Paris by storm and drew eager throngs to the garden and museum—surely, she would have confirmed his conviction about animal agency. Giraffes with their improbable, undeniable grace seem clearly to be beings of their own creation.

Giraffe from Buffon, *Histoire naturelle*, 1781

TWENTY

A Very Branchy Tree

T he overall picture of the living world that emerges from the process Lamarck describes is a "branching series": crooked and irregular, but—at least in principle—unbroken and continuous. He acknowledges that we don't yet have enough knowledge to fill in all the gaps among living structures, but the more we develop our zoological collections, the more continuous the picture becomes, until we're forced to base distinctions between species on the minutest and most trivial of differences. At that point, the distinctions start to look entirely arbitrary, as the categories appear to merge and blend together.

Take the platypus and the echidna, for example, two trophies from the natural plunder that English and French traveling naturalists had been amassing in the land then known in Europe as New Holland (the aspirational designation by a seventeenth-century Dutch cartographer would linger until 1817, when the British imperial government renamed it Australia from the Latin *Terra Australis*, "southern land"). These strange creatures seem to indicate an intermediate form between birds and mammals. They're quadrupeds covered in hair or bristles, but they're unlike mammals in having no visible teats and in laying eggs; on the other hand, they're also unlike birds in having no winglike limbs. They seem to Lamarck to represent a new form that fills the space between birds and mammals, rendering the passage between them closer to continuous.

There may well be plenty more extraordinary intermediate creatures like the platypus and the echidna, Lamarck suggests, that European naturalists don't yet know about, and that fill in the apparent gaps among living forms. In general, he rejects the idea of discontinuities caused by major, catastrophic extinctions of the kind Cuvier describes. Rather, fossils that don't appear to correspond to any known living organism must belong to species that have undergone big transformations.

In current science, the question whether scientists should consider a particular kind of animal extinct or just radically transformed often remains a matter of uncertainty. In 2004, I happened to be a regular at the American Museum of Natural History in New York City (I had two small children at the time) as the curators were changing all the exhibits to explain that while for almost two centuries paleontologists had considered dinosaurs extinct, they had now come to regard current birds as living dinosaurs. "You probably saw lots of living dinosaurs right outside the museum on your way here today," one of the guides told his young, rapt audience, "and some of you might have had a dinosaur sandwich for lunch." More recently, the question came even closer to home as research in evolutionary genetics suggested that probably at least half of the Neanderthal genome lives on, spread throughout the current human population (though only about 1–4 percent of the genome of any individual is Neanderthal). The Swedish geneticist and Nobel laureate Svante Pääbo, one of the leading researchers in this area, has said that "in effect Neanderthals are not really extinct at all, they are in us."[1]

What makes an animal extinct rather than just radically transformed? It seems sometimes to be a matter of definition. For Cuvier, since he didn't believe that living forms transformed at all, any fossil that didn't correspond to a known living animal must be of an extinct species. Lamarck writes that "naturalists" (whom he leaves conspicuously unnamed but everyone, including Cuvier, knew he meant Cuvier), having failed to perceive that animal forms change over time, have offered the "facile" and purely imaginary explanation of a global catastrophe that wiped out all the animals at a given moment. Lamarck rejects the idea of such catastrophic events resembling those described in the Bible, such as the Great Flood. The only cause of extinctions that he deems plausible are human beings. In those parts of the earth where humans exercise absolute power, they might quite possibly have destroyed whole species for which they had no use; Lamarck wouldn't put such an act of extermination past humans.

The irregularly branching structure of forms of life that Lamarck describes raises the question of whether there are better and worse animals, more or less perfect, higher or lower? In the *Zoological Philosophy*, he is inconsistent on this point. He regularly describes animals as more or less "perfect" or "imperfect," and as "higher" or "lower." But he also emphasizes the difficulties in making this kind of reckoning, because animals are constantly adapting to their immediate needs, and what's better or worse, more or less perfect, is entirely relative to environment. Whales and snakes both lack legs not because they're imperfect but because they have no need for legs and would indeed be impeded by them.

Is the process of transformation Lamarck describes a steady progress toward perfection or something altogether different? This question preoccupied him during the decade following the publication of the *Zoological Philosophy*, and by the end of his life he had arrived at a startling answer. Stephen Jay Gould, following his discovery of Lamarck's marginal sketch of the earthworm's circulatory system, traced how Lamarck's reclassification of annelids ultimately led him to revise his whole picture of animal life.

Remember that Lamarck had taken Cuvier's distinction between "internal" and "external" worms and hijacked it for his own purposes. After witnessing Cuvier's demonstration during the winter of 1801–2 that "external worms" had circulatory systems with arteries and veins, Lamarck had decided that the internal, parasitic worms were too different from the external worms, such as earthworms, to share the same category. So, for his course that spring, he had renamed the external worms "annelids." This seemingly modest taxonomic revision had had momentous implications. It had undermined the idea that living forms were arrayed along a ladder of increasing complexity. Without the annelids, there was no compelling way to rank what remained as "worms" with regard to radiates such as jellies and sea urchins. They weren't higher or lower, just different.

This messiness of worms threatened to undermine a principle that Lamarck had held dear from the start: that thanks to the power of life, living forms arrayed themselves along a continuum of increasing complexity. In his early works—the *System of Animals Without Vertebrae* (1801) and *Researches on the Organization of Living Bodies* (1802)—and on through the first volume of the *Zoological Philosophy* (1809), he clung to this principle. In each work, he presented a chart of animals forming a continuum from simplest to most complex.

But while the *Zoological Philosophy* was in press, Lamarck's worm-related misgivings got the better of him. He tried solving the problem by creating two separate series of animals: unarticulated ones, beginning with spontaneously generated infusoria, and articulated ones, beginning with spontaneously generated worms. The infusoria, he thought, had given rise to polyps and radiates. Meanwhile, the worms had produced all sorts of other beings. Lamarck now thought that at least one kind of worm, the *Gordius*, commonly known as horsehair worms, could live outside other animal bodies; no doubt there were other such worms. Some of these, leaving the water and living in air, must have become insects, arachnids, and crustaceans, while others, remaining in the water, had given rise to annelids, cirripedes, mollusks, fish, and reptiles. The reptiles had in turn given rise to two branches: birds and mammals.

Creating two separate series had the advantage that Lamarck could retain his principle of a continuum of increasing complexity within each series while still accommodating the fact that worms seemed neither higher nor lower than radiates, but just different. While he was at it, and contemplating the newly arrived seal, he also used branches to avoid ranking birds and mammals, and to avoid ranking the mammals that descended from amphibious mammals such as seals: the cetaceans, the ungulates (hoofed), and the unguiculates (having nails). Despite these concessions to messiness and the unrankable nature of certain organisms, in the first volume of the *Natural History of Animals Without Vertebrae*, Lamarck still retains his two separate series of animals, each essentially a continuum of increasing complexity. He still clings to his original idea that such a continuum is the primary pattern of the animal world, while acknowledging that animals' responses to their various environments constitute a secondary cause that messes up the continuum, introducing gaps and branches.

Five years later, blind and dictating his last work to Rosalie, the *Analytical System of Human Knowledge*, Lamarck at last relinquished the continuous hierarchy of animal beings. In his last general picture of animal life, it all begins with spontaneously generated monads; living beings then develop by their power of life and grow continually more complex, but they also constantly ramify themselves outward along branches as they transform in response to their circumstances. The result of this process isn't a single, continuous hierarchy, nor two or more continuous hierarchies, nor even several continuous hierarchies with gaps and branches thrown in. Lamarck now simply sees an "extreme diversity" in living forms resulting from their adaptive responses to their environments. The order of life, he says, is "very branchy," its branchiness as important as the increasing complexity of each little twig.[2]

The *Analytical System* presents Lamarck's growing and branching tree of life in its most radical form. In his earliest conceptions of the tree, he retained a remnant of the old, linear, hierarchical arrangements that had characterized natural history ever since Aristotle. In the final version of the tree, to be sure, humans remain the most complex form of life, superior in intelligence to all others (with a decidedly ugly side to their superiority, as we shall see in chapter 22). But there is no single scale of perfection. Living beings transform and vary by following their own creative impulses in response to varying circumstances, branching ever outward as well as upward. The tree is creative agency in action.

TWENTY-ONE

The Natural History of
Animals Without Vertebrae

F ifty-nine students gathered in the garden amphitheater in the
spring of 1810, the year after the publication of the *Zoological
Philosophy*, to take Lamarck's course on invertebrates. They
were an exceedingly international crowd, including Russians, Germans,
a Spanish geologist, and an American professor of natural philosophy
named Robert Maskell Patterson who would later be the sixth director
of the U.S. Mint and president of the American Philosophical Society.
There were a couple of Belgian aristocrats, one of whom was the twenty-
seven-year-old Jean-Baptiste-Julien d'Omalius d'Halloy, a frequent visitor to Paris; he went on to an illustrious career in geology, cartography,
and politics.

The French students that same year of 1810 included an assortment
of naturalists and professionals, including doctors, engineers, writers, politicians, lawyers, and bankers. Among them was Lamarck's young colleague Cyprien-Prosper Brard, a geologist at the museum, just embarking
on a career as a mining engineer and mineralogist. Brard was twenty-
three years old when he took his seat in Lamarck's course; he had been
appointed two years earlier as an aide-naturalist to Barthélemy Faujas de

Saint-Fond, holder of the chair in geology. Another geologist in the group was Constant Prévost, who took the course along with his close friend, the devout Catholic and misanthropic aristocrat Henri Marie Ducrotay de Blainville. Prévost and Blainville, who shared lodgings for several years and often traveled together, had an established practice of enrolling together in the garden's renowned natural history courses.

They constituted a rather hostile presence in the crowd. Blainville was a student and protégé of Cuvier, with whom he collaborated on an examination of Saartjie "Sara" Baartman, a Khoikhoi woman from the Dutch Cape Colony (now a part of South Africa), to compare her body with those of European women and orangutans. Cuvier's fascination with this subject remained unfulfilled until Baartman died, whereupon he dissected her body in the name of science and preserved the parts in the museum's natural history collection (more on that dreadful story presently). Blainville, who wrote the first report of the examination of Baartman, was by all accounts ill-tempered, touchy, bitter, difficult, and unsociable. "Ask Mr. de Blainville his opinion on anything, or even just say hello to him," Cuvier observed, "and he will answer: No." How did Prévost tolerate him? They seem to have bonded over their shared belief in the continuity and constancy of the natural world, both geological and organic: no transformation of living forms over time such as Lamarck described, but also, no catastrophes or extinctions à la Cuvier.

In the 1822 article in which he coined the term "paleontology," Blainville dismissively relegated the use of fossils for distinguishing geological strata to "a sort of school of geology" of which Cuvier was the chief representative. This kind of thing might have contributed to the souring of his relations with Cuvier. But despite such impolitic tendencies, Blainville's professional ascent seems to have been unstoppable. After Lamarck's death, he was appointed to one of the two chairs into which

Lamarck's position was divided, the chair in "Mollusks, Worms, and Zoophytes." Latreille, Lamarck's "adoptive son," held the other one, in "Crustaceans, Arachnids, and Insects." Blainville had actually connived to cheat Latreille out of the position altogether by petitioning Lamarck, both in person and in writing, to name him, Blainville, as his successor. Latreille's health was "too dilapidated" for the job, Blainville said, and furthermore, should he receive the position, he promised to pay any one of Lamarck's children a pension of 1,800 francs from his salary, a ploy he claimed Cuvier had used with Mertrud. Blainville's machinations succeeded at least to the extent of securing him half the position, and shortly afterward he succeeded Cuvier in the chair in comparative anatomy. Willful ignorance and malevolence rarely go unrewarded.

. . . .

In the winter of 1814, a coalition of countries formed against Napoleon, including Austria, Prussia, Russia, Great Britain, and Spain, and defeated the French army in Russia and Germany, then advanced through France. On March 30 and 31, these allied forces stormed Paris and forced a French surrender, sending Napoleon into exile on the island of Elba, midway between his native Corsica and the Tuscan shore. Elba is only about six miles from the Italian mainland, today a forty-five-minute ferry ride. But exile is exile, and it made way for the Restoration of the Bourbon king Louis XVIII, a brother of the guillotined Louis XVI (whose son Louis XVII never actually ruled as king and died in prison in 1795 at age ten). Apart from Napoleon's momentary return to power, there would be Bourbons on the throne for the rest of Lamarck's life: Louis XVIII would rule until his death in 1824, when he would be succeeded by his brother Charles X. The king restored the Royal Academy of Sciences, though under the auspices of the National Institute, and in the membership list the *citoyen* Lamarck became once again the *chevalier* de Lamarck.

Lamarck's youngest son, Aristide, was also in a kind of exile. He was interned at the Charenton mental asylum on the southeast outskirts of Paris. After attending lycée in Rouen, Aristide had served briefly in the army, but now he was committed at Charenton. There's no telling why. People were sent to such hospitals for cognitive and emotional symptoms but also for homosexuality, libertinism, prostitution, and venereal disease among other pretexts. Charenton had a reputation for compassionate practices, and many musicians, composers, artists, and writers were interned there. Among Aristide's fellow inmates was the marquis de Sade, who spent his last years confined at Charenton for "libertine dementia."

The asylum was then operating under the enlightened directorship of François Simonnet de Coulmiers, who supplied Sade with writing paper. Coulmiers, a priest and member of the legislature during the Revolution and under Napoleon, instituted various humane reforms at the asylum. He replaced physical restraint with gentler methods and encouraged artistic expression, in the service of which he created a theater in the asylum, and enabled Sade to produce plays casting the other inmates alongside Parisian actors. Perhaps Aristide even performed in one of these. It became the height of fashion to attend a play featuring the "lunatics" at Charenton. But Aristide's residence there was tragically brief; he died in 1814 at age twenty, the same year as the seventy-four-year-old Sade, both perhaps victims of the typhus epidemic that Napoleon's *Grande armée* carried lethally across Europe.[1]

Working steadily through the political turmoil and personal travails, Lamarck continued teaching his world-famous course and conducting his research on invertebrate zoology. The following spring of 1815, when

Lamarck's course began in mid-May, Napoleon had sailed and then marched his way back to Paris two months earlier and reseized the throne; France was in the midst of Bonaparte's "Hundred Days" back in power. But when the course was only about a third of the way through, on June 18, another coalition of forces led by Austria, Prussia, Russia, and Great Britain re-defeated Napoleon, this time definitively, at the Battle of Waterloo. By the time the course was about halfway done, the victors had re-dispatched him into exile. Having learned their lesson, they sent him to the island of St. Helena, then a British colony, right in the middle of the South Atlantic Ocean between the coasts of Brazil and Angola, almost twelve hundred miles from the African mainland.

While his nemesis was being routed at Waterloo, in March 1815, Lamarck was publishing the first volume of his ultimately epic seven-volume *Natural History of Animals Without Vertebrae.* This work actually rivals the *Zoological Philosophy* for the position of magnum opus. The *Zoological Philosophy* was the book in which Lamarck introduced his fully developed revolutionary theory to the world, but the *Natural History of Animals Without Vertebrae* was the first comprehensive compendium and classification of invertebrates, including hundreds of new species previously unnamed. In the Catalogue of Life, an up-to-date listing of all the world's known species, Lamarck has 6,301 entries, almost half as many as Linnaeus, the founder of modern taxonomy (who has 13,647, and as a further point of comparison, Cuvier has a mere 2,660).[2]

The first volume of the *Natural History of Animals Without Vertebrae* also includes a 382-page introduction that is essentially an expanded edition of the *Zoological Philosophy*, which had sold disappointingly few copies. This was a clever move: Lamarck's exhaustive taxonomy of invertebrates was essential to naturalists everywhere, and by inserting his theory into its first volume, he stealthily snuck it into libraries, cabinets, and laboratories all over the world. Here, he declared, are *seven volumes of evidence* for my theory, which deserves "great attention," because it's the first and only general theory of animal life. At least, adds Lamarck, "I know of no other."

To create his vast compendium of invertebrates, Lamarck drew upon the museum's collections, those in his own cabinet, and notes and descriptions sent from all over the country and all over the world. Félix Lamouroux, a professor of natural history from Caen, up north on the English Channel, sent notes on polyps, starfish, and jellies. From the same area of Normandy came a description of a single-celled bioluminescent marine organism that glows blue at night. This was the contribution of a man named Suriray, chief physician at the hospital in Le Havre, the port city on the Channel at the mouth of the Seine. The doctor had grown curious about something he noticed on summer evenings in Le Havre. The seawater in the port often glimmered with a sparkling, vivid blue light. Examining some of the water under a microscope, he discerned a tiny, glowing organism and named it "Noctiluca," meaning "nightlight."[3]

Suriray often explored the seashores of Normandy with two friends, François Péron and Charles Alexandre Lesueur, who contributed many of the descriptions and specimens in Lamarck's compendium. Péron and Lesueur had both been members of an expedition to Australia between 1800 and 1803, led by Nicolas Baudin, an especially greedy and villainous sea captain, cartographer, colonialist, and sometime slave trader. Péron, who had lost his right eye as a soldier in the Revolutionary wars, had been discharged from the army and begun medical school in Paris, but then he made an abrupt decision to travel and got a spot as an assistant zoologist on the expedition. He received a promotion when the expedition's main zoologist, René Maugé, died of dysentery in Tasmania. (The expedition had an unusually high mortality rate, probably due to Baudin's practice of selling everything at each port for his own personal profit, including the ship's provisions, medicine, and scientific supplies.) Together, Péron and Lesueur assumed Maugé's duties. Lesueur, for his part, had signed on as an assistant gunner, but had been promoted to artist when the expedition's appointed artists—one of whom was Lamarck's old friend André Michaux—abandoned the voyage, presumably in disgust at Baudin's nefarious practices, in Île de France (now Mauritius).

Péron returned to France as an established naturalist, despite the dubious nature of some of his reports. Cuvier cited his fantastical description of the vaginas of women in certain populations in Africa at the beginning of his report on his dissection of Saartjie Baartman's body. Péron had not based his account on observations of his own but had cited descriptions produced by the governor of the Dutch Cape Colony, Jan Willem Janssens, during a special "tour undertaken while he was governor." Lamarck, for his part, was interested only in Péron's descriptions of invertebrate marine animals.

Péron and Lesueur's specimens might well not have made it back to France at all, since Baudin had sold all the ship's supply of preserving alcohol, but they resourcefully used arrack, an Indonesian drink made from fermented coconut sap and sugarcane, and this did the trick. Their contributions to Lamarck's compendium included corals, sea sponges, hydrozoans—small, predatory animals related to jellyfish and corals—and bryozoans, often called moss animals for the mossy appearance of their colonies. Lamarck especially admired some "extremely pretty varieties" of these that the pair had brought back and that were now in his own cabinet: purple and tawny white, funnel shaped and tubular, with systems of delicate branches. He delighted in the rich colors of the marine organisms, noting another bright purple one, an organ pipe coral described by Péron, with "fringed tentacles of a beautiful green."

One of the most remarkable creatures from Lesueur and Péron, in Lamarck's estimation, was a jelly known as the Venus girdle, a long, gelatinous, transparent, ribbonlike being with iridescent edges, all milky white with glints of blue. Also from the duo, he had in his cabinet several specimens of a starfish-like echinoderm known as a brittle star that uses its long arms to crawl along the seafloor. And from the Southern Ocean, Péron and Lesueur had brought back a *Spirula* preserved in fluid (presumably arrack!) that Lamarck liked to show his students during lectures. The *Spirula*, colloquially known as the ram's horn squid for its horn-shaped internal shell, or sometimes the taillight squid because of a light-emitting appendage on its rear end, is a kind of cephalopod rarely seen because it lives so deep in the ocean. Lamarck was apparently the first to capture a likeness of the elusive *Spirula*. He made sketches of it in pen for his lec-

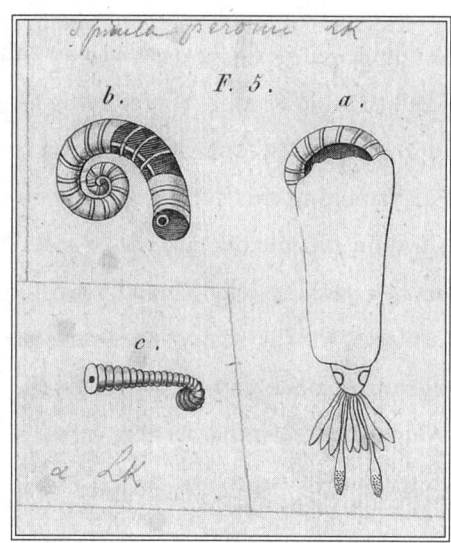

Lamarck's drawings of *Spirula*, 1827

tures, then had them engraved for the mollusks section of Panckoucke's *Encyclopedia*, which he was supervising, and which was by then being published by Panckoucke's daughter Antoinette-Pauline Agasse.[4]

Another contributor of creatures for the compendium was Jules-César Savigny, a former student from Lamarck's lecture course, who sent information about varieties of polyps and also about sea squirts: mostly sessile, potato-shaped marine creatures that hang out on the undersides of ships, the pilings of piers, big rocks or seashells, or even the backs of large crabs. Savigny was from Provins, a medieval city about a hundred kilometers southeast of Paris, and had first arrived in the capital at age sixteen in 1793 to study medicine; he enrolled at the School of Health that replaced the old Faculty of Medicine under the National Convention around the time Lamarck began teaching his course on insects and worms. Savigny had a lot of things in common with Lamarck. He too began as a botanist and later remade himself into a zoologist in order to take a job, in Savigny's case a place on Napoleon's expedition to Egypt. He too lived in a garret apartment as a newly arrived, poor student, and his garret happened to have been in the rue Copeau, where Lamarck lived with Rosalie and their six children. Also like Lamarck, Savigny would later in life be afflicted by debilitating eye trouble, perhaps the delayed result of an illness contracted in Egypt.

Being right next to the garden in the rue Copeau, Savigny began attending courses of lectures there, including Lamarck's. He made a habit of being the first to arrive at the amphitheater, then positioning himself as close as possible to Lamarck, following every word with rapt attention. In this way he made an impression on his teacher, and one day when Lamarck was feeling overwhelmed with work on the botanical volumes of Panckoucke's *Encyclopedia*, he asked Savigny to help him by drafting the article on sorrel. The next morning, Savigny arrived to deliver the text.

"Goodness," exclaimed Lamarck, surprised at the speed and quality of the effort, "is it really you who wrote this? In that case, no false modesty: sign this article and from now on I expect you to have the honor of your work. It's only fair." It was the beginning of a long and productive friendship reflected in the warmth and importance of Lamarck's frequent references to Savigny throughout the *Natural History of Animals Without Vertebrae*.

Savigny's observations of sea squirts led Lamarck to create an entirely new branch of animals, the "ascidians," branching off from the polyps in one direction, leading to mollusks, while the radiates branched off in another. In this way, the sea squirts became an occasion for Lamarck to return to the question that had been preoccupying him: the apparent messiness that animal structures presented, refusing to fall into a single line of progression from simple to complex. The new branch of ascidians, Lamarck writes, "made me feel the necessity" of distinguishing between the simple series that naturalists develop for their own convenience and the real "order of the production of these beings," which is "far from being simple," but rather a jumble of branches.

"In whatever way we go about it," he writes, "I am convinced we will never succeed" in fitting all the animals of each series into a simple order of progression with no branches. He recommends therefore creating two separate tables: one showing an abstract, simple order of animals, for use in writing and teaching, to help students and readers understand the relative complexity of different animal structures; and the other telling it like it is, showing the actual, unruly state of affairs in the animal world. This second chart also reflects a world in boisterous motion. The mistake naturalists have always made, writes Lamarck, has been to see in the distribution of animal forms only a static order based upon relationships among structures, "and yet it obviously represented *an order of formation*," the results of a ubiquitous and ongoing activity.

To Lamarck's great admiration, Savigny also revealed the full structure of the mouths of Lepidoptera (butterflies and moths). Previously, only their tongue had been documented: a spiral with two layers that they use for sucking fluids such as nectar, water, or the juices from rotting fruit, dung, or carrion. Savigny was an astoundingly astute and dedicated observer of tiny structures. Using powerful magnifying glasses, microscopes, fine tweezers, and scalpels, he dissected the mouths and sensory organs of hundreds of species of insects and drew each part separately. Through "extraordinary sagacity and patience in observation,"

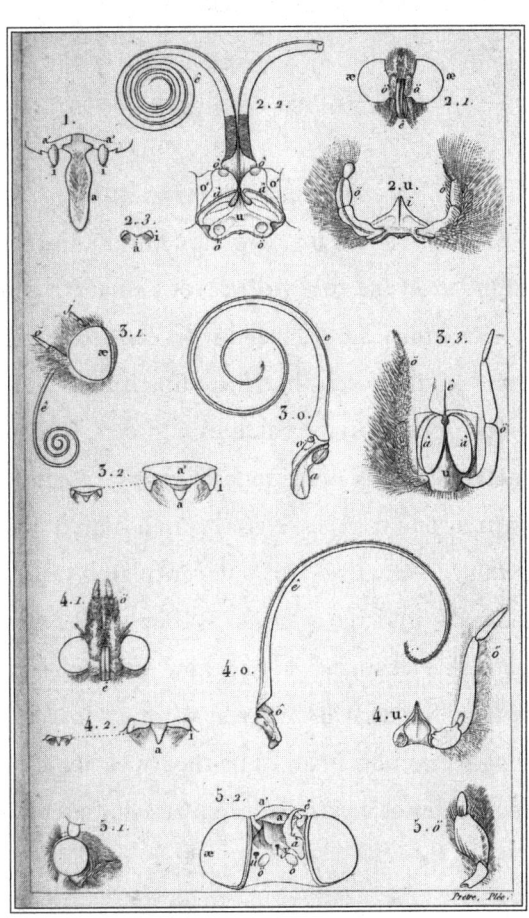

Savigny's drawings of the parts of Lepidoptera mouths, 1827

Lamarck marvels, Savigny managed to show that Lepidoptera also have two lips, two mandibles, two jaws, and four palps, sensory appendages used for detecting and manipulating food.

The important point for Lamarck is that since these animals only need to suck, they don't use most of these instruments—just their jaws and tongue. Unused, the other parts are underdeveloped and tiny, which is why they had previously escaped notice. In contrast, Hymenoptera (such as bees, wasps, and ants) use their mandibles for chewing and grinding, so these are larger and more developed. The development of the parts of the mouth, Lamarck concluded, depends upon their usage.

. . . .

The *Natural History of Animals Without Vertebrae* is of course very much about invertebrates, but not only invertebrates. They are the foundation and the evidence for Lamarck's general theory of animal life. The volumes are threaded with connections joining invertebrates with every kind of vertebrate including humans. A particularly charming example appears in the volume on insects, when Lamarck observes that the metamorphosis undergone by many insects, such as when caterpillars turn into butterflies, is really just a kind of puberty. All animals that reproduce sexually—and one can plainly tell he's speaking as one such animal—undergo a "crisis" as they reach sexual maturity producing "remarkable changes." Here's how we can make sense of the otherwise strange fact that in his great work on invertebrates Lamarck includes a long section on "man." His theory of life is universal, and he unflinchingly extends it even to encompass human beings.

TWENTY-TWO

The Self-Made Primate

T he radical idea that humans evolved from earlier primates, though it is now firmly associated with Charles Darwin in the popular imagination, did not in fact originate with him. Darwin himself emphasized that the "conclusion that man is the co-descendant with other species of some ancient, lower, and extinct form is not in any degree new. Lamarck long ago came to this conclusion." Actually, even before Lamarck, the idea had been abroad in the land. The brazen materialist naturalist and journal editor Jean-Claude Delamétherie published it in 1780. "We can suspect," he suggested, that some earlier version of "man" had become "bipedal by habit; which must have produced great changes in his body." In 1802, Lamarck advanced this idea, as we saw briefly in chapter 16, referring to nonhuman primates including human precursors as "quadrumane," or four-handed, borrowing Buffon's terminology. (Humans, in contrast, are bimanous and bipedal, having two hands and two feet, while quadrupeds, for their part, have zero hands and four feet.) Lamarck wrote that the differences in organization distinguishing humans from other primates—for instance, the relative position of the head and neck, the dexterity of the fingers, the muscles of the lower extremities, the lack of a thumb-like toe in the foot—might have

Gibbon by Jacques de Sève, 1766, from Buffon, *Histoire naturelle*

been "acquired little by little over a long period of time," and he mused, "What a subject of meditation for those who have the courage to delve into it!"

As it turned out, Lamarck himself had that courage. In the *Zoological Philosophy,* he returned to the dangerous topic and developed it further. He again surmised that humans had indeed developed gradually over time from some earlier kind of primate. According to Buffon, quadrumane animals filled the taxonomic interval between humans and quadrupeds. Lamarck now suggested that certain quadrumane animals had actually propelled themselves across that interval by transforming themselves into bimanous animals. "If some race of quadrumane animals," he observed, were to abandon "the habit of climbing trees and grasping branches with its feet ... there is no doubt ... that these quadrumane animals would at length be transformed into bimanous."

It wasn't just that humans had developed from older primates, but that these older primates had actively begun re-creating themselves, and their offspring had carried on the project of self-transformation from generation to generation. Acting on their own projects and desires, some quadrumane creatures had struck out in a new direction. "If the individuals of whom I speak, driven by the need to dominate, and to see both far and wide, strove to stand upright," Lamarck wrote, "and continually adopted that habit from generation to generation, there is again no doubt that their feet would gradually acquire a shape suitable for keeping them in an upright attitude." This was essentially the view that Darwin later adopted. "As soon as some ancient member in the great series of the Primates came to be less arboreal, owing to a change in its manner of procuring subsistence, or to some change in the surrounding conditions, its habitual manner of progression would have been modified," Darwin wrote, and so a bipedal primate would have emerged, its hands free to use tools and weapons.

Lamarck pointed out that although an upright posture is characteristic of humans, it still isn't restful for them. It generally takes more than a year to learn, continues throughout life to require energy to maintain, and grows tiring after a time. The strain of standing upright, he thought, was an indication of humans' common origin with other mammals and served as evidence that humans had not only made themselves human in the distant past but continued to do so through their striving activity in the present.

The self-making rise of humans, as Lamarck tells it, is not an entirely beautiful story. Humans, he says, are "in some ways incomprehensible," capable of extremes of both goodness and cruelty. Their highly developed intellectual faculties "confer the means for doing harm as easily as good." Lamarck imagines that the earliest quadrumane pioneers would have gained mastery over the other quadrumane animals and taken possession of all the best parts of the earth, exiling the others to less desirable areas, holding back their growth and arresting their progress. In rising to dominance, the members of this special group would have continually acquired "new wants" that would have driven them to develop new "abilities and faculties," including the power to form articulate sounds and ultimately to speak. This ability would have given rise to many languages as usage changed from person to person and country to country. And by the use of language, humans would have expanded their intelligence.

In these ways, the rising bimanous primates would have asserted a sharp distinction between themselves and their quadrumane former comrades. One result would have been that the exiled quadrumane animals had come to live "a wretched, anxious life, incessantly compelled to take refuge in flight and concealment." Even among the bimanous humans, the domination of a wretched and oppressed majority by an aggressive minority would be the normal state of affairs, because humans made themselves human by asserting their will, by pursuing their desires and increasing their powers, to the marked detriment of their fellow creatures.

. . . .

In the spring of 1815, while Lamarck was putting out the first volume of his *Natural History of Animals Without Vertebrae*, Saartjie Baartman arrived at the Garden of Plants for three days of scientific inspection arranged by Cuvier, Blainville, and Geoffroy Saint-Hilaire. Baartman had been born in the Dutch Cape Colony, which had become a British colony when she was five or six years old. Orphaned as a child, she became an indentured servant—essentially enslaved—following the 1809 "British Hottentot Proclamation" formalizing the indenturing of the Khoekhoe and San people (for whom "Hottentot" was the colonial, racist term). Soon afterward, the head of the household where she worked hatched a plan with an associate to make their fortune by exhibiting Baartman in London, and so she appeared first in England and then in France under the name "Hottentot Venus" between 1810 and 1815, when she was in her early to mid-twenties.[1]

For several days in March 1815, Baartman suffered the scrutiny of Cuvier, Blainville, Geoffroy Saint-Hilaire, and a crowd of eager viewers. Lamarck seems to have been exceptional in his absence, at least as far as we can know, from this great event at the Paris Museum of Natural History. Among the group were Pierre Flourens, a physiologist from the Academy of Sciences who studied the brain, and three artists: Pierre-François De Wailly, Nicolas Huet, and Jean-Baptiste Berré. Cuvier's brother Frédéric, who was then in charge of the garden's menagerie, might also have been present; he and Geoffroy later included Baartman at the beginning of the first volume of their coedited *Natural History of Mammals*. These men were all desperately eager to see Baartman naked—until now, she had always appeared in skintight flesh-colored clothing to give just a suggestion of nudity—and they especially wanted to see her vagina, having heard extraordinary things from Péron.[2]

When Baartman arrived, the illustrious group of men accosted her, begging for a look, until she reluctantly agreed to take off her clothes; then these prestigious gentlemen of science at the height of their profession implored the now-naked young woman to open her legs and show them her vagina, Blainville even offering her money; to their disappointment, she angrily refused. Blainville nevertheless did his best in the face of adversity, publishing a report in which he minutely describes what he could see of Baartman's labia as she stood, bent over, and walked. Meanwhile, the three artists painted Baartman, in Berré's case including beneath the portrait a close-up of her vagina done from his own lurid imagination.

Within the year, Baartman died from an unknown illness, whereupon Cuvier performed—no, not an autopsy to determine the cause of death—a *dissection* focused particularly on Baartman's genitals, which he describes in an obsessively detailed report. After her vagina and pelvis— which he says show her, like other "Negresses and female Bushmen," to be closer than European women to monkeys—he pays the most attention to Baartman's head and brain. "I never saw a human head closer to a monkey than hers," he declares. Certainly no "negroes" could ever have created the sophisticated ancient Egyptian society; Cuvier's measurements of the heads of mummies brought back from Napoleon's expedition show him—surprise!—that the ancient Egyptians were "caucasian." These findings appear, along with de Wailly's depictions of Baartman wearing only an expression of grave resolve, at the beginning of the first volume of Geoffroy Saint-Hilaire's and Frédéric Cuvier's *Natural History of Mammals*; Baartman is the first mammal, followed by a langur monkey.

Cuvier for his part frequently returned to his pet theme throughout his major works, repeating that "the caucasian [race], to which we belong," is the most beautiful, and the best, while the most degraded race is that of the "negroes," the closest to monkeys and other brutes. Comparative anatomy itself, as its founders practiced it, was a science largely built upon what Lamarck identified as a mistake: a reduction of the moral to the physical. Blainville, Cuvier, and the others were sure that all living beings including people were made of fixed, physical parts that absolutely determined their moral behaviors, and also, of course, that their own highly esteemed parts and behaviors represented perfection. If Lamarck was right that it was really more the other way around, one has to wonder whether Blainville's, Geoffroy's, Cuvier's, and their accomplices' physical parts bore any imprint of their perverse and repugnant behaviors.

. . . .

W hy did Lamarck remain remote from these proceedings and from the essentially universal tendency among his fellow naturalists to engage with joyful abandon in producing scientific theories of human races? It's not because he avoided the subject of humans and their place in nature; he boldly discussed this subject often, including in his two greatest works, the *Zoological Philosophy* and the *Natural History of Animals Without Vertebrae*. But in all his voluminous writings, and throughout a period in which the scientific characterization of human races was a general cultural obsession, Lamarck seems to have had just four very brief things to say about race in humans.

In his 1802 *Researches on the Organization of Living Bodies*, Lamarck remarks in passing, citing no evidence of any kind, that "the head of the Negro man, less flattened in front than that of the European man, necessarily has a less central occipital hole." What can we make of this statement? The "occipital hole," or foramen magnum, is the opening at the base of the skull through which the spinal cord passes. Decades earlier, Daubenton, whom we met as Buffon's old friend and coauthor, who consulted with Lamarck about the preface to *French Flora*, had introduced the idea of an "occipital angle" that distinguished humans from other animals. In humans, Daubenton said, the occipital opening is at the very base of the skull so that the skull sits on top of the spine, consistent with an upright posture. In quadrupeds, the occipital opening is situated more toward the back of the skull, consistent with walking on four legs. In monkeys, who alternate between upright and horizontal postures, the occipital hole occupies an intermediate position. Daubenton compared the positions of the occipital holes in various animals by means of the angle between lines drawn through it to the eyes and to the jaw. Nowhere did he mention any variation in this angle among human beings.

Next, though, things took an ugly turn. Petrus Camper, a Dutch doctor and comparative anatomist, was inspired by Daubenton's measurements to construct a theory of facial angles between forehead and jawline among human races and in comparison with monkeys. Camper had in mind only aesthetics, not mental capacity. "Negroes," he said, were aesthetically closer to monkeys and the furthest removed from the ideal of Grecian beauty. He helpfully included drawings of ancient-statue-looking heads alongside "European," "Negro," monkey, and orangutan heads, so you could see what he meant.

Then came our friends Cuvier and Geoffroy Saint-Hilaire, who co-authored an article in 1795 that was primarily about orangutans, but in which they cited Camper and incorporated his "facial angle," along with the size and shape of the skull, into a calculus of cognitive ability and cultural sophistication among humans. Human races showed "the same series of relations, as in the various species of animals, between the protrusion of the skull and the degree of intelligence," they opined, adding, "We do not see, at the least, that any of the peoples with depressed foreheads and prominent jaws have ever furnished individuals equal to the average European in terms of the faculties of the soul."

Lamarck's observation about the placement of the occipital hole in

"Negro men" now takes on a chilling resonance, especially since it immediately follows a discussion of the placement of this hole in orangutans. It plainly echoes Camper's and Cuvier and Geoffroy's application of Daubenton's original, innocent idea to the creation of pernicious rankings of human races. At the same time, the particular claim Lamarck makes about the occipital hole placement—that in people with "less flattened" faces than Europeans the occipital hole is necessarily situated less centrally in their skull—seems to have no bearing on posture, beauty, cognitive capacity, or anything else. In fact, Lamarck attaches no significance to it of any kind. Why does he include an unsupported assertion with no significance? It's impossible to know; perhaps it's to show that he's not ignorant of the conversation about occipital holes, even while refraining from joining it.

In the *Zoological Philosophy*, Lamarck makes two remarks about race in humans. One is a definition and taxonomy of "bimanous animals," or humans: "The Bimanous Animals: mammals with separate limbs, unguiculate [having nails or claws]; with three kinds of teeth, and with thumbs opposable to the hands only [that is, not the feet]. Varieties: Caucasian, Hyperborean [far northern], Mongolian, American, Malay, Ethiopian, or Negro." These varieties, he says, are superficial and fluid: "We may be sure that if men were not kept apart by the distances of their habitations, the crossing in reproduction would soon bring about the disappearance of the general characteristics distinguishing different nations."

Finally, a lapse into rote conventionality. In his last work, the *Analytical System of Human Knowledge*, Lamarck writes, "The human species offers different varieties to which the name of *races* has been given; and . . . each exists in a particular region of the globe. Probably, the oldest of them is the Caucasian race, which is at the same time the most perfected." Immediately after this, with a striking lack of logic, he describes the state of European civilization and its people as very far from perfected. In the "civilized" world, he says, there's "an immense disparity" in people's situations and means. A small minority have seized power and dominate the others, keeping "the multitude in a state of inferiority, by skillfully inspiring in them prejudices and fancies that keep them chained." The impoverished masses are forced to live "in unhealthy places, breathing only foul air, irregularly and poorly nourished," in a state of diminished intelligence, while the rich minority, leading indolent and overfed lives, are equally unhealthy, subject to an increasing number of endemic diseases.

Insofar as Lamarck drew significant distinctions among categories of people, they were founded in circumstance and environment, not anatomy or physiology. If we want to understand human beings, he wrote, we must keep in mind that "as they change in age, situation, state, fortune, or power, they also constantly change in their way of feeling, considering, judging." People aren't fixed by their physical parts but are continually remaking themselves. For this reason, Lamarck rejected "phrenology," the theory of two German doctors, Franz Josef Gall and Johann Spurzheim, that the parts of a person's brain reflect by their sizes the degrees of the person's mental powers, and that one can evaluate these by examining the shape of the skull (Spurzheim assigned the name to the theory and made it official, but as we know, Cuvier and Geoffroy, among others, had already published versions of the idea).

Gall and Spurzheim presented their phrenological theory to the world in the same period that Lamarck was developing and publishing his *Zoological Philosophy*, and phrenology became the basis for decades in which people used minute measurements of forehead bumps and head shapes to rank human beings by race, sex, and social class. Unsurprisingly, their findings always confirmed that men were innately smarter than women, rich people smarter than poor people, and white western Europeans smarter than everyone else. Lamarck, for his part, judges that Gall has been guilty of "an all-too-common abuse of the imagination." The moral feelings and behaviors of a person shape the physical structures and not vice versa. "It is . . . not true that any of our intellectual faculties are innate," Lamarck insists. They depend on the exercise we give them and the particular set of circumstances of our environment. If some people are smarter than others, it's because they have more of a chance to exercise their capacity to think.

To regard the intellect as innate is a "dangerous opinion and moreover a real error." Lamarck admits the possibility that a much-exercised intellectual organ could grow noticeably large over time, like the kangaroo's hind legs or the elephant's trunk, but that would of course take place across many generations and characterize the whole human species, not certain individuals. He doesn't absolutely rule out the possibility that extraordinary intelligence might have externally visible signs in an individual brain, but that would be a result of extreme exercise, not innate genius.

Lamarck also exceptionally abstained from invoking nature and science in the service of cultural bigotry. For instance, while people all around him were busily demonstrating the supremacy of French among languages—the most natural of tongues, the most suited to clear thinking—Lamarck was adamant that "no language is more natural to humans than any other; there is no *mother language*." In his unshakable view, all social institutions were equally natural, and all biological forms also social, as people and other animals remade themselves by their habits and behaviors.[3]

The stark contrast between Lamarck's approach to the human animal and that of seemingly everyone around him is dramatically apparent if you leaf through the *New Dictionary of Natural History*, published by Jean-François-Pierre Déterville. Like Panckoucke, Déterville was a publisher of massive, multivolume works, especially in natural history. He published works by Buffon and Cuvier; he published Lamarck's *Natural History of Animals Without Vertebrae* and *Analytical System of Human Knowledge*; and he roped Lamarck into contributing twenty articles to the second edition of his thirty-six-volume dictionary of natural history, appearing between 1816 and 1819. Often, rather than replacing the entry from the first edition of 1803, Déterville simply placed Lamarck's contribution after it in the new edition so that the two entries sit side by side in eternal confrontation; this helped sell copies as readers were drawn into the drama of conflicting views. Such is the case with the article on man.[4]

The first article on man, retained from the 1803 edition in an updated version, is a 270-page screed by the chief editor of the *Dictionary*, a pharmacist and naturalist named Julien-Joseph Virey. It includes a full-throated rendition of the occipital-hole perversion. In "negroes," Virey writes, the position of the occipital hole is closer to where it is in monkeys, which is why they don't stand up nice and straight like Europeans: "[Negroes] have their lower backs thrust backward, in order to establish a sort of counterweight to their muzzle. . . . In monkeys, this tendency is even more pronounced." Virey also divides humans into six races, and further divides these into two distinct species, by skin color and "facial angle." White- through copper-skinned people with facial angles of eighty-five to ninety degrees make up the first species. They have a moral character and an expansive intelligence, and they're skillful, industrious, and courageous. Brown- through black-skinned people with facial angles of seventy-five to eighty-two degrees form the second species. They tend to stand around with their knees jutting out looking disaffected, they have only a limited intelligence, they're lacking in skill, industry, and courage, they prefer sensual pleasures to moral affections, and they're generally similar to brutes.

Virey also describes the examination and later dissection of Saartjie Baartman. He explains that the proximity of "Hottentot bushmen" to monkeys, and of "the negro race" in general to quadrumanes, has already been well established; therefore, he'll just focus on "the cause of the configuration of these female bushmen, of whom we have had before our eyes a tame one, so to speak." He thinks she and other women of her kind have what he describes as a monkey-like form associated with sitting around in the hot sun "in the manner of baboons, mandrills, and Barbary apes."

Arriving queasily at the end of Virey's virulent 270 pages, we encounter Lamarck's economical treatment of the same subject in just 7. The thing to know about humans, he says, is that they're natural beings, like other animals. They have intelligence, instincts, and sentiments. They're driven by certain "inclinations": the tendency to seek well-being and avoid suffering, a feeling of self-love, a drive to dominate, and a repugnance at the idea of their own demise. These inclinations are universal: They're the same in human beings "of all countries and all times," whether in "civilization" or in nature. The entry is pretty banal in itself, far from the most interesting or momentous of Lamarck's writings. Or rather, it's banal until you consider its juxtaposition with the one that comes right before it by Virey, and with Cuvier's writings about humans, and Geoffroy's, and Péron's, and Camper's, and Gall's and Spurzheim's, and the musings of Lamarck's mentor Buffon, and of his precursor in botanical and zoological classification Linnaeus, both of whom had promulgated influential rankings of human kinds, and . . . essentially *everyone else* in Lamarck's insular, powerful world. His short, unassuming dictionary entry on man then becomes truly remarkable, both for what it *doesn't* say—it doesn't say that some humans are nearly gods while others are practically monkeys—and for what it *does* say: that humans in all times and places are essentially the same. The universal human essence according to Lamarck is irreducible to parts. It's a way of being in the world.

TWENTY-THREE

The Birds and the Bees

(AND THE KANGAROOS)

L amarck had even less to say about human sexes than about human races, which is to say essentially zero. Invertebrate sexes, on the other hand, he had plenty to say about. He carefully documented their distinctive features, emphasizing that in some cases they're dramatically different from their counterparts. In red-tailed bumblebees (*Bombus lapidarius*), for instance, observers had mistaken the male and female for two different species. In ordinary fleas, he mentions, the female is larger than the male. In dragonflies, the sexual organs are placed differently according to the sex: In the female, it's at the very bottom of the abdomen, while in the male it's under the first segment. Lamarck paid careful attention to the distinctive features and behaviors of female animals, describing the stingers of female wasps, the borers of female sawflies, the antennae of female beetles, the winglessness and glow of female glowworms, the egg-laying procedures and habits of various female flies, the nests built by different sorts of female bees. He details how the *Andrena* bee constructs her nest, lays her eggs in it, and stocks it with provisions necessary to the larvae. In general, he observes, as soon as female insects are fertile, they look for a suitable place to lay their eggs "where the little ones, upon being born, can find the food they need."

How does one sex an insect? In many cases, Lamarck says, you'll see at the end of the abdomen some "tail-shaped filaments," or appendages, or perhaps a stinger, which might also be retractable. These are "almost never common to both sexes." Sometimes they serve as a borer for the female to pierce into a bit of wood or the body of an animal to deposit her eggs, or as a weapon to attack others or to defend herself, or sometimes they serve the male as a pincer to hook and hold the female during mating. When beetles mate, he notes, the male is almost always on the back of the female, and their union usually lasts several hours, sometimes a day, sometimes two. Even if the distinguishing parts aren't readily apparent, it's still "easy to ascertain the sex of an insect." Lamarck offers careful instructions. The relevant parts "are usually placed at the end of the abdomen and hidden in the anus. . . . [Just] press its belly enough to make these parts come out; then one will easily recognize those of the male by the hooks which accompany them, and those of the female by a kind of auger at the end" (he referred here to the "ovipositor" by which a female insect places her eggs).

Lamarck narrates the whole life story of scale insects (similar to aphids and mealybugs), both female and male versions, which are starkly different, although they start out in the same way, the young ones running with abandon all over the leaves and stems of plants looking "almost like small whitish wood lice with only six legs." But after a while, "the female alone fixes herself to a place on the plant in which she lives." For the rest of her life, she'll remain perfectly motionless, sucking food from that spot on the plant by means of a sucker on her beak. Her body slowly swells, growing to the size of a peppercorn or larger. As she lays her eggs, she passes them under her body and seems to incubate them. Her skin tightens, becomes smooth, dries out, fades. "In a word, the animal generally loses the shape and figure of an insect." Instead, she begins to look like "a miniature shield, or escutcheon." She ends her life in this transformed state. After she's laid her eggs, her dried body covers and protects them. Meanwhile, "the male of this singular female hardly resembles her" at all, except at the very beginning when they're all carefree larvae running about on the plant together. Once grown, he's "very small, equipped with two wings longer than his body, and six legs. His body is reddish, often covered with a white powder." The instant he achieves his adult state, "he uses his wings to fly toward the females. As they are much larger than him, he walks on them, and [in this way] succeeds in fertilizing them."

The extraordinarily diverse sex lives of invertebrates take up a lot of the *Natural History of Animals Without Vertebrae*. Mollusks, for instance, Lamarck explains, are sometimes either male or female—as is the case with cephalopods, including octopuses—and in these cases they can't mate. Instead, "the males spread a fertilizing liquor on the eggs already laid by the females." On the other hand, gastropods, such as snails, are hermaphroditic, having both male and female organs united in the same animal. Some of these hermaphrodites—again, snails are an example—"need reciprocal mating and are provided for this purpose with a singular organ, which is nothing but an exciter, but necessary to give rise to the act of fertilization." Lamarck offers a spirited description of the foreplay of snails and slugs, "a very singular and very curious exciting prelude": "Besides the well-known penis of these animals, . . . they possess a kind of elongated dart or sting which comes out through the same opening in the neck which gives issue to the penis; . . . when two individuals approach, the dart of one pricks the other, and falls to the ground or remains attached to the one which has been pricked." The animals both withdraw but soon approach each other again, and then their mating takes place. Other hermaphrodites, lacking such an exciter organ, don't mate with another animal "and appear to be sufficient to themselves."

Sex and feeling are essentially coextensive, Lamarck thinks: Those animals that can feel also reproduce sexually and vice versa. In fact, he uses the sexual response as an example of the power of the inner feeling of existence, that defining feature common to all beings with sensation. "Nothing," he says, is more striking, especially in men, than the body's response to particular sights and thoughts, which act upon their inner feeling and on their various organs. "What man is ignorant of the effects that the sight of a young and beautiful woman can produce on his person?" Even after she's no longer present, the mere thought of her in his imagination can have the same effects.

Sexual differences between male and female animals, and sexual feelings and relations throughout the animal world, were of great interest to Lamarck. He studied them closely and wrote a lot about them. Yet, remarkably, he drew no social implications from his observations in these areas whatsoever: nothing with regard to the social role or proper functions of human women. Again, he is remarkable for what he *doesn't* say, because the practice of prescribing the social role of women based on descriptions of female anatomy and physiology was utterly commonplace in Lamarck's world (as indeed it has been ever since).

Lacépède, for instance, the professor of reptiles and fish at the museum, wrote in his *Natural History of Man*—published posthumously in 1827 with a prefatory eulogy by Cuvier—that, for reasons of anatomy and physiology, the institution of marriage was vital to human society. Without marriage, says Lacépède, men become impetuously passionate and untamable, but the situation is far worse for women. They develop a "uterine fury. A real mania then troubles their mind; their imagination is especially kindled when it has been excited by obscene images and licentious remarks; their delusion even taking away all modesty, they abandon themselves not only to the most lascivious speeches, but also to the most indecent acts." From there the situation goes downhill fast. Lacépède warns ominously that "the consequences of excessive pleasures are much more terrible still: the strength weakens . . . the features become deformed, the hair falls out, the hearing becomes dull, the sight is extinguished, the memory is erased, the spirit disappears, and death ends all these miseries."

Lacépède's *Natural History of Man* had a counterpart in the two-volume *Natural History of Woman* by Jacques Louis Moreau de Sarthe, a doctor and anatomist. Moreau opens the work by observing that although people have written a lot about women, no one has yet conducted a thorough scientific treatment founded in female anatomy, so he's stepping in to fill the gap. Moreau then declares himself to be no Lamarckian. He's entirely different from those naturalists who attribute the physical organization of animals to their "habits and inclinations." Instead, like Cuvier, Moreau does just the opposite. He derives the habits, inclinations, and appetites of animals from their physical organization. "It is in this spirit and taking Buffon's beautiful discourse on the nature of birds as a model," writes Moreau, "that we will try to gather together some observations on the nature of woman."

Just what sort of model, you may be wondering, is Buffon's "Discourse on the Nature of Birds" for a treatise on women? Good question! Well, Buffon opens the essay by explaining that the word "nature" has two senses: an active and a passive. "Nature" in its active sense has imprinted a particular character upon birds, a set of instincts and habits, which is their "nature" in the passive sense. For instance, the necessity of incubating their eggs leads birds to form stable marriages with a conjugal fidelity never seen, according to Buffon, among quadrupeds. Their uxoriousness is a result not of their active decision but of their "nature," over which they have no control.

Women, like birds, have a passive nature imprinted upon them, according to Moreau; indeed, women resemble birds in various ways such as their mellifluous voices and dependence on the institution of marriage. Moreau promises to do the same sort of thing for women as Buffon has done for birds: to explain how their passive nature determines their instincts and habits. As it turns out, it's pretty simple: It all comes down to the uterus. The constant agitation, "upheaval and storm" of menstruation, Moreau explains, throws women into a perennial state of weakness, nervousness, and sensitivity. As a result, they're inclined to benevolence and pity, but also to whims and fancies. They have "more aptitude for emotion than for thought," and they're incapable of anything that "require[s] sustained attention, prolonged recollection and long meditation." The "uterine influence" is responsible for all physical and moral aspects of "woman": her "anxieties," her "extraordinary oddities and caprices," her occasional "great exaltation of sensibility, imagination and spirit." While man can transcend his sex, woman is tied unceasingly to hers by her uterus, which determines her being even more powerfully than "the trunk of the Elephant" or the "digestive organs of carnivores."

Lacépède's and Moreau's zoological extrapolations regarding the nature and social function of women were absolutely standard; Lamarck was the exception. The reduction of social behaviors to natural structures was completely alien to his way of thinking. To be clear, he was neither a feminist nor a militant for the rights of the enslaved, at least as far as the written record can attest, and it's not that there were no feminists or militants for the rights of the enslaved in his vicinity. In 1788, a journalist, Jacques-Pierre Brissot, and a financier, Étienne Clavière, founded the abolitionist Society of Friends of Blacks in Paris. They gathered more than a hundred members, met for several years, published abolitionist literature, and presented their case to successive Revolution-era governing bodies, disbanding about a year before February 4, 1794, when the National Convention passed a law abolishing slavery.

Also while Lamarck was beginning his career in botany, then moving into zoology and surviving the Revolution, the playwright and political philosopher Olympe de Gouges was writing abolitionist plays such as *The Enslavement of Blacks*, performed briefly by the Comédie Française in 1785 before being sabotaged by pro-slave-trade hecklers, and composing the *Declaration of the Rights of Women and of the Female Citizen* to propose to the National Assembly in response to the *Declaration of the Rights of Man and of the Citizen* in 1791. Two years later, de Gouges was guillotined for her opposition to Robespierre and the Committee of Public Safety.

Lamarck, all this time, was strenuously avoiding politics: just trying to remain employed, to keep the garden open, and to continue his scientific work. He was avoiding politics, that is, except for the fact that, to his astonishment and dismay, his scientific work kept turning out to be deeply political. This was partly because of his dismissal of God and all religious explanations for natural phenomena. But it was also because of his insistence on the agency of living beings, and on their irreducibility to passive parts. De Gouges might have found some use in Lamarck's kind of science, if only he'd been a few years earlier with it. On the eve of the Revolution, in her 1788 abolitionist pamphlet "Reflections on Negro Men," she wrote that although the defenders of slavery claimed the enslaved were brutes by nature, she had come to see that enslavement was the work of force, prejudice, and "the unjust and powerful interest of the Whites." Nature, she wrote, "had no part in it." She leveled a similar charge against those who mounted arguments in defense of the subjugation of women. "Bizarre, blind, bloated with science" was her pithy characterization. Bloated with science they certainly were, and they remained so long after de Gouges had lost her head for speaking her mind. But it wasn't Lamarck's kind of science, which never reduced people or other animals to parts.

. . . .

As a final example of the contrast between Lamarck's kind of science and the kind de Gouge's foes were bloated with, consider the kangaroo; or rather, consider European naturalists' earliest writings on the kangaroo.[1] Moreau, in his *Natural History of Woman*, observes that the pouch of the female kangaroo is the physical cause of her maternal love. It's Moreau's womb fixation again. The womb in kangaroos, he says, is so small that the young must be born prematurely and held in the pouch, "pressed against the breasts." The result of this arrangement is that mother kangaroos form a close bond with their young and become the most tender and solicitous of mothers.

Lamarck too wrote about the kangaroo, and he also took a particular interest in the pouch, which is indeed a striking feature of kangaroos. But to Lamarck, the pouch didn't cause any specifically feminine or maternal behaviors in female kangaroos. Rather, it played a crucial role in his understanding of how all kangaroos had shaped themselves into their distinctive form of tiny forelegs, great and powerful hindlegs, and a thick, strong tail. For the sake of the baby in her pouch, says Lamarck, the kangaroo has adopted the habit of standing upright, resting on her

Kangaroo from Buffon's *Histoire naturelle*
new edition by C. S. Sonnini, an 8
(1799–1800)

hind legs and tail, and "moving only by means of a succession of leaps, during which she maintains her erect attitude in order not to disturb her young." The result is that her forelegs, which she uses very little, have become short and weak, while the hind legs and tail have grown large and powerful, and the tail especially thick and strong at its base. It's not that her body determined her behavior, but that her behavior shaped her body.

Lamarck knew perfectly well that only female kangaroos have pouches to carry their offspring. For one thing, it says so in the first edition of Déterville's *Dictionary of Natural History*, in the article "Kanguroo or Kangurou," written by Cuvier's student and disciple the then-nineteen-year-old zoologist Anselme Gaëtan Desmarest. Early English writers on kangaroos had also described the pouch as a distinctive feature of females.[2] Even more important, Lamarck had had that pair of kangaroos to observe for two years before having to relinquish them to the empress. He unhesitatingly used this uniquely female feature not to argue that female animals' maternal parts determined their behaviors but just the opposite: to explain how female kangaroos had transformed the shape of all kangaroos by their habits and willful actions.

TWENTY-FOUR

Darkness

I n the summer of 1818, darkness fell: Lamarck abruptly lost his eyesight. He had been suffering from intermittent, intense pains in his eyes for almost a decade. Those around him attributed his eye trouble to strain caused by his unremitting use of magnifying glasses and microscopes to scrutinize the tiny structures of invertebrates. Still, he persevered, refusing to be deterred by the pains. But then, all at once, he was blind. René Desfontaines, the botanist who decades earlier had resented Lamarck's encroachment on his territory in the garden, was now a solicitous visitor. On August 3, the minutes from the weekly meeting of the Academy of Sciences record that "Mr. Desfontaines gives news of the unfortunate state of Mr. de Lamarck's eyesight." The following week, a hopeful update: "Mr. Desfontaines saw Mr. Lamarck today, whose ophthalmia has almost completely cleared up." But his vision was gone; already in June, Lamarck had requested at a meeting of the museum professors that Latreille partially replace him in his teaching and other duties.

Within a couple years, Latreille became the sole instructor of Lamarck's course in invertebrate zoology. But Latreille's health was also fragile, and he soon began receiving assistance from another of Lamarck's most devoted students, Jean-Victor Audouin, who had enrolled in Lamarck's course for the first time in 1813, at age sixteen. Audouin had evidently found the experience so inspirational that he had decided to become an entomologist and reenrolled in the course for five years running. (In 1823, he would be appointed an assistant librarian at the museum, and in 1833, after Latreille's death, Audouin would succeed him in the chair of crustaceans, arachnids, and insects.) Audouin's magnum opus, *History of Insects Harmful to Vines and Particularly the Pyralid Moth . . . with an Indication of the Means Which Must Be Used to Combat It* (1842), shows the influence of his teacher Lamarck in its emphasis on the moths' "tastes," "ways of life," "habits," "customs," and cunning "tricks."

At around the same time that Lamarck lost his vision, in July 1818, his eldest son and second child, André—the one who was always running away to sea, and who became an ensign and then lieutenant in the marines aboard the vessel the *Salamander*—died at the age of thirty-seven in the Antilles, from yellow fever.[1] Characteristically, Lamarck carried on working despite both blindness and bereavement. With Latreille's help, he taught his course for the last time in the spring of 1819. One of his students that year was a young medical student named Philippe Buchez, who would later become a socialist politician and president of the National Assembly. As with René Villermé, Lamarck's student from his very first year of teaching, who would later lead a public health movement, Buchez incorporated central Lamarckian principles into his politics. "We must remember," he admonished, "that in natural history, race is understood to mean a variety within the species . . . resulting from the influence of habits and the environment." To represent races as "invariable and incommunicable type[s]" like species was profoundly wrong. According to Buchez, such a doctrine existed purely to justify the abomination of slavery by claiming a spurious basis for it in natural history. No real naturalist or doctor could accept such a claim, Buchez wrote, which must revolt any person of feeling.

A fellow student of Buchez's in Lamarck's last class in 1819 was Isidore Bourdon, another young doctor who had just finished his studies at the Paris Faculty of Medicine. Bourdon marveled at the "infinite skill" with which the blind Lamarck "described and demonstrated shells whose shape he could trace by touch alone." Despite the half-century age difference, Bourdon became friends with his teacher. Decades later, he would write a biographical sketch of Lamarck drawing partly on conversations they'd had, and not just about invertebrates; for instance, Bourdon is the source of the story about Buffonet mischievously splattering ink on Lamarck's clothes: "He was seventy-five years old when he told me this sad story, already in his memory for half a century." While taking Lamarck's course, Bourdon was finishing a memoir on vomiting, arguing what might seem the fairly obvious point that the stomach plays an active role in the process. He later published a manual of health and hygiene with recommendations about diet, dress, housing, exercise, and lifestyle, in which he makes the Lamarckian observation that moral harm can suffice to turn good health to illness.

From 1820, as mentioned above, Latreille took over the famous course on invertebrate zoology. But with Rosalie's help, Lamarck carried on with his research. The "Homer of naturalists," as Latreille called him, had endured the worst blow a naturalist could suffer, but he was lucky in his eldest daughter, who never left her father's side even "for a single instant." Rosalie devoted herself entirely to zoological research, to completing the *Natural History of Animals Without Vertebrae*, and to writing the *Analytical System of Human Knowledge* "under dictation" from her father. (Given Lamarck's frailty, it seems likely she acted at least to a degree as an unacknowledged coauthor; the book also incorporates most of the writing Lamarck had done for Déterville's dictionary on human nature and psychology.)

Rosalie accompanied her father everywhere he went for as long as he could still go out. According to Cuvier, her sacrifices "went beyond anything that could be expressed." (Cuvier's own daughter Clémentine had died at age eighteen in 1827, and his other three children had died in early childhood, but he had a devoted stepdaughter, Sophie Duvaucel, who managed his scientific research activities and hosted a salon. Her brother Alfred Duvaucel had collected animals for the museum in India and Southeast Asia but had died in 1824 after being mauled by a rhinoceros in Bihar.) Once her father was confined to his room, Rosalie apparently never left the house; on her first time outside after his death, she received a shock from the fresh air, to which she'd grown unaccustomed. Cornélie's devotion too became renowned. In fact, Latreille might have compared Lamarck to a different blind poet than Homer: John Milton dictated *Paradise Lost* to his daughters, according to a legend favored by Romantic writers and artists, echoed in the images of Rosalie and Cornélie by their father's side throughout his final, dark decade.[2]

Just as his last course was coming to an end, on August 9, 1819, Lamarck attended a weekly meeting of the Academy of Sciences, presumably with Rosalie. The minutes for that day record that "Mr. de Lamarck presents the 6th volume, 1st part, of his Animals without vertebrae." A few weeks later, on August 27, Julie Mallet, Lamarck's wife of twenty-one years, died at the age of fifty-one. Whatever else he might have felt, he must have sensed that time was closing in; he kept working. The *Analytical System of Human Knowledge* was published at the end of February 1820. The minutes of the Academy of Sciences meeting on March 6,

1819, record its reception of the book on that day; perhaps Lamarck came with Rosalie to attend the meeting and present it.

The *Analytical System* was Lamarck's last book, apart from the final volume of the *Natural History of Animals Without Vertebrae*, and in it he carries his theory to its ultimate extent. The book contains the full flowering of his "very branchy" tree of animal life, growing not upward toward a unique state of perfection but outward in every direction, so that animals can't be ranked in a hierarchy as better or worse, more or less perfect. The animals on their different branches are just . . . different. Lamarck presents humans as natural beings like other animals. Everything about animals including humans is "an organic phenomenon," even the loftiest product of the human mind, an idea. This is the result of impressions upon the organ of intelligence, those wrinkled, pulpy hemispheres common to birds and mammals. Lamarck also takes on his colleagues in zoology for one last time. They're prone to mistakenly ascribing intentions to nature, he says. To be sure, we see animals everywhere beautifully suited to their environments in a way that might look to us like the fulfillment of a purposeful plan, but this harmony doesn't follow from any preexisting design. It's the result of the animals' own actions.

In August 1822, Lamarck published the final volume of the *Natural History of Animals Without Vertebrae*, just two months before another death. At the end of October, Eugénie, Lamarck's youngest child, died in the garden apartment at age twenty-five, barely a year older than her mother, Charlotte Reverdy, had been at her death. Lamarck had now outlived three wives and four of his eight children. He carried on attending meetings of the museum professors and the Academy of Sciences, but he finally stopped writing. Three of his surviving children remained at home: hard-of-hearing Antoine, working when he could as a painter; and

Rosalie and Cornélie, their father's closest companions and collaborators. The family was apparently so strapped for money that in 1824, Lamarck sold his precious herbarium to a German botanist. Begun decades earlier when he was a young soldier in Provence, it now left the garden for the University of Rostock, in the north of Germany near the Baltic Sea. (Happily, it returned home several decades later when the museum purchased it back again.)

Of the four surviving children, Auguste was the most independent one, and the source of hope and posterity. A civil engineer, he had his diploma from the École Polytechnique, the engineering and military academy founded during the Revolution, which remains one of the ultra-selective and prestigious *grandes écoles* today. In July 1825, at age thirty-four, Auguste married a woman named Mélanie Nicolle. The following year, their son, Eugène, was born, perhaps named for his late aunt, and the year after that came their daughter, Louise.[3] After all the premature deaths and unremitting illness, the births of two grandchildren perhaps affirmed their ailing grandfather's theory that the "power of life" carried on its eternal struggle against death, as living beings developed themselves and their progeny in ever new directions.

By the autumn following Auguste's wedding, however, Lamarck's attendance at the weekly meetings of the Academy of Sciences had at last begun to drop off. The minutes for October 3, 1825, record a decision to count him as present whether or not he actually made it to the meeting so that he could continue to receive his attendance fees: "The Academy, given the age and state of blindness of Mr. de Lamarck, unanimously decides that he will be considered present at all its Sessions." Still, on rare occasions he went to meetings of the museum professors, attending for the last time in July 1828. The following October, he took to his bed for good; a note from Geoffroy to Latreille suggests it was because of a broken femur. Rosalie and Cornélie took turns reading him novels; he especially loved the *Waverley* novels by Sir Walter Scott, historical adventure stories set in the political turmoil of mid-eighteenth-century Britain, around the time of Lamarck's birth. The stories must have been thematically familiar to Lamarck at the end of a long life lived in political turmoil, but sufficiently remote in time and place to transport an old, blind, bedridden sufferer.

. . . .

Jean-Baptiste Lamarck died on December 18, 1829, at the age of eighty-five. His capital, all that remained to support his three dependent children, amounted to 1,000 francs, about an average year's earnings for a carpenter in Paris in 1820.[4] Cuvier nastily insinuated that Lamarck had been the "credulous" dupe of speculators, drawn into bad investments, but his salary had also been quite small relative to those of his colleagues and to the size of his family. Auguste, who seems to have done nothing to help his poorer siblings, had a different criticism, recounting somewhat bitterly to his own son, Eugène, that his father's conduct hadn't been blameless: "It is doubtless beautiful to devote oneself to science without any view of ambition or fortune, but only on condition that the family's interests will not suffer."

The museum and institute paid for Lamarck's funeral, which took place two days after his death, on December 20, 1829, in the fifteenth-century church of Saint-Médard in the rue Mouffetard, about five minutes' walk from the garden, where Aristide and Eugénie had been baptized. But they apparently did it on the cheap: Lamarck was buried in a collective, unmarked grave in the Montparnasse Cemetery. Moreover, it was only a temporary arrangement. The grave's contents were emptied into the Paris catacombs every five years to make way for new occupants. Cuvier was charged by the National Institute to write Lamarck's eulogy on behalf of the Academy of Sciences. He solicited information from Auguste but apparently refused Auguste's repeated attempts to come see him. In the end, Auguste resorted to writing two letters about his father's life and career.[5] Cuvier's draft was so inflammatory, however, that the institute's committee refused to authorize its reading and demanded changes, which Cuvier refused to make. These would be made literally over his dead body: An amended, though still exceedingly non-eulogistic, version of Cuvier's eulogy was ultimately read at the Academy of Sciences only in November 1832, several months after Cuvier himself had died, and published in the academy's journal in 1835.

In the event, Latreille delivered the eulogy at the funeral on behalf of the Academy of Sciences, and Geoffroy Saint-Hilaire the one for the museum. Latreille's was highly emotional from start to finish. "Crushed by the weight of a double pain," he said he felt the general sorrow at losing a great naturalist and also a personal grief for one who had been like a father to him. "Nestor of Zoologists," he exhorted, "receive our last farewells . . . in particular those of your adoptive son." He promised to remember the "tenderest and best of fathers" until reuniting with him, not in heaven, but in "this restful earth." The immortality he hoped for, for both of them, was simply that they might live on "in the memory of the friends of science, and even more in that of good people!" Geoffroy, more poised and formal, nevertheless described a sense of "vast emptiness" and "eternal regret." He gave an appropriately eulogistic summary of Lamarck's career, but he also had a practical purpose in mind on behalf of Rosalie, Cornélie, and Antoine, who seemed on the verge of destitution: Geoffroy reflected in closing that Lamarck had surely secured for his children "the recognition of society" and "legitimate rights to government benefits." Later, he would write, "Lamarck lived for a long time poor, blind, and abandoned, not by me; I loved him and venerated him always."

Rosalie, aged fifty-one at her father's death, did receive his small pension. The assembly of professors at the museum petitioned the Interior Ministry to grant her this, and they also hired Cornélie, aged thirty-seven, to work in the museum's botanical laboratory, arranging and mounting dried plants. "I myself saw in 1832 Mlle. Cornélie de Lamarck attaching for a small salary on sheets of white paper the plants of the herbarium of the Museum where her father had been a professor," reported Martins, the Lamarckian doctor and naturalist whom we encountered in the Interlude, who in the middle decades of the nineteenth century restarted a meteorological correspondence and yearbook and produced a new edition of the *Zoological Philosophy*. Martins lamented that had Cornélie and Rosalie been daughters of a minister or general, they would have received better treatment, but, alas, "their father was only a great naturalist."[6] To eke out a living, the children also sold their father's books and furniture, including several hundred unsold copies of the *Zoological Philosophy* and the *Analytical System of Human Knowledge*.

The professorship in insects and worms, which Lamarck had transformed into a position in invertebrate zoology, was now split between Blainville and Latreille. The number of invertebrate animals had grown so much "thanks to the incredible activity of the [chair's] first occupant," as Martins noted, that it required two people to do the job. Poor Latreille, sixty-eight years old and in poor health, had been working at the garden for more than thirty years and would live for only another three. Upon being appointed to the position of professor of crustaceans, arachnids, and insects, he remarked, "They waited to give me a piece of bread until I had no more teeth."

. . . .

I n his infamous non-eulogy, ultimately delivered only after the death of its author as well as its subject, Cuvier scornfully describes Lamarck's biology, with its central idea that "desires" and "efforts" can shape bodily organs, as suited only to "amuse the imagination of a poet." This remark reflected a world in which science and poetry were rapidly becoming antithetical, but Lamarck's mode of science was different in this respect as well. Imagination, he writes in the *Zoological Philosophy*, is crucial to literature and poetry, and is also essential—in close partnership with reason—in science. Without imagination, he says, there can be no genius, and without genius there can be no chance of making discoveries other than isolated and inconsequential facts.

In some private notes, Lamarck unleashed his frustration about his contemporaries' methodological strictures, condemning any large view of nature, as though one could understand the geography of the earth's surface by contemplating each spot on the ground with a magnifying glass. To insist that science consists just of gathering facts, Lamarck protested, is like being an architect who does nothing but cut stones and prepare mortar, wood, and ironwork but would never dare to construct a building. "We deceive ourselves and others by repeating every day that we reason only from the facts," he argued. Between fact and explanation, "there is always an interposed hypothesis."

During the centuries since Lamarck's death, he has served time and again as an example of how *not* to do science: no imaginative interpretation, no ascriptions of agency to living beings, for God's sake, no poetry!

Some, though, continued to defend and to practice modes of science that were, if not poetry, also very importantly not *anti*-poetry: science that owned its imaginative, interpretive, and even literary nature, and described the natural world as filled with many forms of living agency. We'll follow Lamarck's legacy now through the nineteenth and twentieth centuries, and through . . . let's call it the science and poetry wars.

PART FOUR

Science and Poetry

In which we behold a Battle over Mollusks and their Meaning with regard to Science, Religion, and the Forms of Life. We then Witness the Failure of Beings to grow in Boiled Broth and the Excision of Tails from Mice as Spurious Disproofs of our Hero's Idea of the Self-Making Agency of Living Beings. Lamarck's Theory, say the Broth-Boiler and the Tail-Cutter, is Not Science, setting the tone for more than 150 years of Disdain. We follow the Fortunes of Lamarck's Not-Science to England, Germany, America, and the Soviet Union and through the Twentieth Century, as it serves as a Perennial Example in the Battle over Science— what it Is, what it Isn't, how it Differs from Poetry, and how it should Explain the World (non-Poetically).

TWENTY-FIVE

The Battle of the Mollusks

N ow, what do you think of this great event?" exclaimed the German writer and philosopher Johann Wolfgang von Goethe to a friend stopping by his house in Weimar—a Genevan scholar named Frédéric Soret—on the afternoon of August 2, 1830. "The volcano has finally erupted, everything is in flames, and we no longer have a transaction behind closed doors!" Soret naturally assumed that Goethe was referring to the same thing everyone else was talking about that day: the events taking place in France known as the July Revolution. Over the previous month, France had once again erupted into political revolt, this time against the reactionary policies of King Charles X. On the very day of Soret's visit to Goethe, the king had been forced to abdicate the throne and would soon be replaced by a distant relative, Louis-Philippe d'Orléans, who would rule under the constraints of a new constitutional order known as the July Monarchy (after the revolution it followed).

But no! This wasn't at all what Goethe had in mind. "We do not appear to understand each other, my good friend," he protested when Soret referred to the royal expulsion. ". . . I am speaking of the contest, so important for science, between Cuvier and Geoffroy Saint-Hilaire, which has come to an open rupture in the academy." So unexpected was this retort that Soret was struck dumb with surprise "and for some minutes felt my thoughts perfectly at a standstill."

Indeed, Geoffroy and Cuvier had been duking it out before an expanding, increasingly riveted public at the weekly Monday meetings of the Academy of Sciences. Their old, close friendship had soured dramatically, and in fact their battle was perhaps not entirely unrelated to the July Revolution, as we'll see.

First, some important background. Before the Battle of the Mollusks of 1830, there had been an earlier skirmish in 1825: the Controversy of the Crocodile of Caen. The crocodile in question was not in fact a whole crocodile, but just part of a fossilized skull that had been sent to Cuvier in 1823 by Félix Lamouroux, the professor of natural history up in Caen whom we met when he sent notes on polyps, sea stars, and jellies that Lamarck used in his *Natural History of Animals Without Vertebrae*. Soon after receiving the crocodile fossil, Cuvier had published a brief description of it, concluding that it belonged to some extinct species. Geoffroy then essentially hijacked the fossil. He published a memoir thoroughly analyzing it, showing it didn't belong to any known species of crocodile living or extinct, and naming it "Teleosaurus," from Greek roots for "perfect reptile," because he thought its respiratory canal resembled those of mammals, suggesting to him that it was a higher form of crocodile than previously known. He also hypothesized that current crocodiles had descended from the Teleosaurus. As the crocodile's environment had changed, its bodily organs had transformed too, "precisely in keeping," Geoffroy wrote, with Lamarck's theory in his *Zoological Philosophy*.[1]

In a footnote, Geoffroy warmly recommended Lamarck's theory to all "young people," though he said he differed with Lamarck on some specific details of fact and causation, and he carefully affirmed Newton's view that the order exhibited in the living world indicated "the wisdom and intelligence of a powerful Agent." Despite these caveats, Geoffroy had now used Cuvier's crocodile fossil as an occasion to declare in print his opposition to Cuvier's firm position regarding the fundamental question of whether living organisms changed over generations. Geoffroy's notion of a mammal-like aspect to the Teleosaurus's respiratory canal also violated Cuvier's division of the animal world into four distinct "embranchments": vertebrates, mollusks, articulated animals (annelids, insects, spiders, crabs, lobsters), and zoophytes, or radiates (sea stars, jellies, corals, sea anemones). Cuvier insisted on the absolute distinctness of these four large groups, rejecting any idea of unifying patterns throughout all animals, which might indicate development over time from a common origin.

"Our learned and venerable colleague," wrote Geoffroy the year before Lamarck's death, had always known his ideas were ahead of their time and would be rejected as dangerous, but he courageously published them anyway. Geoffroy accused Cuvier of dogmatism. The only alternative to development over generations was God's "work of six days," repeated anew after each extinction—the view Cuvier clearly preferred—and this, asserted Geoffroy, was contrary to the evidence and repugnant to reason.

. . . .

This was the tense state of affairs in 1830 when the Battle of the Mollusks broke out. The drama began when two unsuspecting young anatomists named Laurencet and Meyranx submitted to the academy a treatise with the modest title "Some Considerations on the Organization of Mollusks." Geoffroy and Latreille had the task of reviewing it and preparing a report. The treatise suggested a connection bridging the gap between invertebrates and vertebrates. The authors proposed that if you imagine bending a vertebrate animal backward into a circle, attaching the top and bottom of its spine, its internal organs would be arranged similarly to those of a cuttlefish, a kind of cephalopod mollusk. Geoffroy seized upon the authors' findings as evidence of a fundamental unity of composition among all animals, a kind of overarching "Animality," which was important to him because he believed it supported his conviction, in keeping with Lamarck's theory, that all animals had developed generation by generation from a common origin.

Worse yet, Geoffroy cited Cuvier's 1817 "Memoir on the Cephalopods and on Their Anatomy" to epitomize the wrong approach to zoology. Cuvier insisted in this treatise that cephalopods were utterly distinct in structure. They were "not on the way from anything; they did not result from the development of other animals, and their own development has produced nothing superior to them." Anyone claiming the contrary, according to Cuvier, was just a fabricator of "vain systems." This was the very passage Geoffroy invoked to exemplify the old, benighted zoological method. When he read his report to the assembled colleagues at the Monday meeting of the Academy of Sciences on February 15, 1830, an outraged Cuvier rose to his feet in protest, insisting that the authors of the treatise on mollusks were "completely mistaken" in their observations and promising to prove it at the next meeting of the academy. He also demanded that for the written proceedings of the day's meeting Geoffroy remove from his report the offensive section in which he disparaged Cuvier's earlier treatise on cephalopods.

Poor Laurencet and Meyranx were aghast at the trouble they'd caused; no doubt they feared the consequences of Cuvier's wrath, as did

Latreille, who was just in the process of jockeying with Blainville and Audouin to fill Lamarck's position at the museum. Latreille wrote to Cuvier disclaiming any attempt to establish structural similarities between invertebrates and vertebrates. And Meyranx told Cuvier, "I cannot find words to express how devastated I am that our memoir has given rise to disputes." He and his coauthor had never dreamed anyone could draw such extravagant implications from "a single, simple consideration on the organization of mollusks."

At the next Monday meeting on February 22, Cuvier laid down the law: The common plan among vertebrates *did not appear in mollusks.* For goodness' sake, hadn't he made himself abundantly clear on this point? "I expressed my sentiment on this subject, to be sure with the moderate tone that the sciences demand, and with the politeness of any well-bred man," was his aggrieved observation, "but still in a clear and positive enough manner that no one could have misunderstood me." Cuvier cataloged the "enormous differences" between cephalopods and vertebrates and demanded how anyone could claim to identify a unity of composition "without deforming the words of the language."

What had begun with an unassuming treatise on mollusks and then been subsumed into the ongoing battle over the theory that living organisms transformed in time now turned into a struggle over the fundamental methods of zoology. For two millennia, ever since Aristotle, Cuvier scolded, zoologists had wisely stuck to describing the fitness of animal structures to their functions. This was the only legitimate foundation of zoology. Geoffroy had gone off the deep end by introducing a new idea of unity of composition among animals. What did this "unity of composition" even mean precisely? Cuvier accused Geoffroy of using impressionistic language that lacked scientific rigor, a dangerous practice in which "the mind quickly goes astray." Once one starts speaking in "metaphors and figures of rhetoric," Cuvier warned, one "quickly sinks into a labyrinth without an exit." The only legitimate considerations for a properly scientific zoology were the suitability of animal parts "for the role that the animal must play in nature."

"I do not know of any animal that MUST play a role in nature," was Geoffroy's heated retort. Rather, animals create their own roles, behaviors, and functions. Geoffroy, as Lamarck had done, accused Cuvier of reversing causes and effects. For instance, when considering a bird that flies high in the upper atmosphere, Cuvier might conclude that it has been accorded a body suitable to this region and admire the lightness of its hollow bones and fine feathers. Such a mode of reasoning, argued Geoffroy, was like saying that a man who uses crutches was destined to have one of his legs amputated. Of course, the accident and amputation preceded the crutches, not the reverse. The experience and behavior of the animal precedes and brings about the form of the body. (The analogy is a bit odd, since he seems to consider the crutches an extension of the body, but we can still grasp his point.)

An excited throng gathered the next week, on March 1, to witness Geoffroy's full rebuttal to Cuvier's presentation of February 22. The debate was growing more boisterous week by week, covered in the popular press in serial, the seats in the hall filling and overflowing. To the spectators' delight, things now got even more explosive. Geoffroy rejected Cuvier's demands for cleaving to ancient tradition and maintaining a rigorous scientific language. "In my opinion," Geoffroy proclaimed, we should not restrict but rather "increasingly expand the scope of philosophical considerations."

As an example of new ways to approach zoology, Geoffroy proposed a method of "analogues." You look for analogous structures in different animals and then seek explanations for their differences. He used the hyoid bone as an example. In humans, the hyoid sits in the middle of the neck, in front of the epiglottis and pharynx, connected above to the muscles of the mouth's floor and below to the larynx. It plays a central role in swallowing and in movements of the tongue. Using his method of analogues, Geoffroy compared its structure in humans and in other mammals and concluded that the upright posture of humans had gradually produced a transformation of the hyoid. Much to the disappointment of the fashionable Parisians who were now eagerly awaiting each new installment of the Battle of the Mollusks, there followed a three-week hiatus. On March 8, Geoffroy said he was unwell, and on March 15 it was Cuvier who announced he was unable to attend the meeting.[2]

The audience was at last satisfied on March 22, when Cuvier returned with a counterattack minutely documenting the myriad differences among hyoid bones in animals of different kinds and pointing out that there are even animals, in fact a veritable "crowd of animals," with no hyoid whatsoever. In dramatic exasperation, Cuvier demanded, "Who would dare to say that the jellyfish and the elephant, the giraffe and the starfish, result from an assemblage of organic parts that repeat themselves uniformly?" Finally, Cuvier also made explicit what had been implicit from the very start, that he regarded Geoffroy's philosophical innovations as irreligious: "Your pretended identities, your pretended analogues, *if they had the least reality,* would reduce *Nature* to a kind of slavery, to which happily its author is far from being chained."

Geoffroy dug in and insisted on his model of science in his rebuttal of March 29. He said that he and his former friend differed not over facts but over their interpretation. Despite Cuvier's love of repeating the phrase "positive facts," great science must be more than that; Geoffroy invoked Lamarck as one whose strength of mind allowed him to think beyond the facts currently established. As for Cuvier's admonitions against an overly imaginative and poetic science, Geoffroy declared that "power and elevation of thought are necessarily imbued with imagination and poetry." On April 5, Cuvier issued a dire warning against what he characterized as a lapse back into ancient pantheism: usurping the role of God and ascribing it to nature. "I was counting on arguments from one naturalist to another," retorted Geoffroy. "The argument became theological."

At this climax of fireworks, Geoffroy called a halt to the live spectacle, saying he wouldn't respond to Cuvier's arguments in the academy meetings any longer, the scene had become too rowdy—instead he would take them up in the quieter and more dignified medium of writing. The result was a book with the resonantly Lamarckian title *Principles of Zoological Philosophy*. It received a glowing review from none other than Bory de Saint-Vincent, the teenager we met in chapter 15 when he rescued Latreille from prison and certain death. Bory had become a military officer under Napoleon and participated in several scientific expeditions around Africa and the Pacific. Along the way, he'd found what he thought was an unanswerable argument for transformism: the extinct dodo, a "monstrous bird" of "ridiculous appearance" that had existed only in Mauritius. It strained credulity to suppose the preposterous creature had been divinely created just for Mauritius. Yet how could a flightless bird have arrived there from elsewhere? Transformation over time was the only reasonable explanation.

And so, the battle continued in print beyond the academy debate. In the popular press, during the lead-up to the July Revolution, the progressive journalists who supported the revolutionaries, such as Jules Guérin, a good friend of Geoffroy's and editor of the *Medical Gazette of Paris*, took Geoffroy's side. Meanwhile, the conservatives, such as the Catholic, royalist *Journal of Debate*, supported Cuvier. That journal's science editor, a doctor named Alfred Donné, was a disciple of Cuvier's and adopted Cuvier's rhetorical strategy. He wrote that while Cuvier presented "positive facts," Geoffroy's theory was a "poetic idea," a "philosophical principle that one must believe in out of sentiment, like a revealed truth," and an "article of faith that he defends with all the zeal of a believer upholding his dogmas."[3] In this way, Cuvier and his supporters neatly turned the tables. Although Cuvier's was the theory predicated on the existence of an omnipotent, supernatural God, they cast Geoffroy's Lamarckian science as the one driven by unscientifically poetic and metaphysical sensibilities.

Geoffroy's defenders, meanwhile, continued championing their cause in the progressive publications. An elderly but still vehement Goethe, aged eighty-two, wrote two articles about the affair that appeared in translation in French journals; the second of these was Goethe's last piece of writing, published in March 1832, the month he died. In these reflections, he eagerly embraced Geoffroy's theory as kindred to his own view of a dynamic unity embracing all living organisms. While Cuvier minutely analyzed each individual material part, said Goethe, Geoffroy perceived the whole; while Cuvier described facts, Geoffroy developed ideas.

Goethe's irrepressible excitement at the beginning of August 1830, which so bewildered his friend Soret, was inspired by a new twist that had just taken place in Paris. At the July 12 meeting of the Academy of Sciences, Cuvier had read a paper on the dodo, which Bory had invoked as evidence of transformism, and which Cuvier sought to claim for his side. By now, the July Revolution was fully under way. At the next meeting, on July 19, Arago, whom we first met standing near Lamarck at a reception with Napoleon in 1809, and who was now the new permanent secretary for mathematical sciences (while Cuvier was the permanent secretary for physical sciences), did something revolutionary. Instead of just reading the title of Cuvier's paper on dodos, as was customary, he also read an excerpt from it to be included in the minutes. This might seem a minor and innocuous departure from tradition, but it was in fact momentous, because it rendered the content of the paper public in a newly official way. Arago was responding to the great popular interest in academy meetings that the Battle of the Mollusks had engendered and to the anti-elitest sentiment of the moment, with a general call for openness and transparency in official bodies. A kerfuffle ensued, with Cuvier vigorously objecting to Arago's summary departure from the established practice of a century and a half, while Geoffroy—surprise!—warmly approved of Arago's change.

Arago's revolutionary move delighted Goethe too. He saw it as bringing science—and in particular the struggle between Cuvier's "analytical" and Geoffroy's "synthetic" approaches—into public view. "I see how great must be the interest of the French scientific world in this affair," he remarked to Soret, "because, notwithstanding the terrible political commotion, the sitting of the nineteenth meeting was very fully attended." The best of it was that the "synthetic manner of treating nature," as Goethe called it, would no longer be "referred to secret committees and dealt with and got rid of, smothered behind closed doors."

Goethe was right: Throughout the 1830s and 1840s, Geoffroy vigorously publicized the theory that current organisms had developed from earlier forms of life. In 1833, he published a memoir following up on his earlier one about the Teleosaurus, in which he developed his own view that the "ambient world" acted upon the developing fetus, bringing about tiny changes that, when inherited and accumulated over many generations, added up to transformations big enough to bring about new species. Geoffroy once again invoked "our profound physiologist Lamarck," approving of Lamarck's basic principle of change over time, although Geoffroy now questioned Lamarck's evidence that the actions and habits of animals led to modifications in their organization. To Geoffroy, the crucial point was that living forms changed in keeping with a changing environment. Cuvier this time offered no rebuttal: He had died the previous year.

Also as Goethe had predicted, Geoffroy drew prominent followers to the theory that organisms emerged over time rather than being created all at once by a supernatural God. He held Sunday evening salons at his residence in the Garden of Plants, where his illustrious guest list included the novelists Honoré de Balzac and George Sand; the republican historian, Romantic poet, and political revolutionary Edgar Quinet; and François-Vincent Raspail, a doctor, chemist, and naturalist who did important work in microscopy, also a political revolutionary who was imprisoned after his participation in the July Revolution, and later a republican politician. Raspail was a partisan for Geoffroy against Cuvier, who, he wrote dismissively, produced "descriptions and nothing but descriptions," while Geoffroy's theory of analogues was "a grand law" that would soon engulf all the categories and classifications of the old zoology. As for Quinet, he judged that Geoffroy had brought to light the era's founding idea, the concept of a "vast unity" joining the parts of the natural world; now it remained to extend this unity to the "civil, political and religious world."

The regulars at Geoffroy's Sunday evening salon formed a flamboyant circle of radicals and carried Geoffroy's Lamarckian theme outward into their works and lives. Balzac dedicated his novel *Father Goriot*, which appeared in serial in 1834 and 1835, "to the great and illustrious Geoffroy Saint-Hilaire, as a testimony of admiration for his works and his genius." *Father Goriot* was the pivotal novel in *The Human Comedy*,

Balzac's vast collection of linked novels and stories, and in the preface to the whole compendium, written in 1841, Balzac announces that his central idea is the "comparison of Humanity and Animality." Geoffroy has provided his inspiration, he says, by establishing that "there is only one animal," which takes on various forms according to the different environments in which it develops, producing "Zoological Species." Announcing and establishing this idea "will be the eternal honor of Geoffroi Saint-Hilaire, the conqueror of Cuvier," proclaims Balzac grandiloquently; moreover, his "triumph was hailed by the last article written by the great Goethe."

Society works in the same way as Nature, Balzac goes on to say. People start off fundamentally the same but develop into "Social Species" according to their environments. He explains that *The Human Comedy* represents the social world the way Buffon's *Natural History* represents "the whole of zoology." In another of the *Human Comedy* novels—*Louis Lambert*, published in 1832—Balzac sends his eponymous hero, accompanied by his friend "Meyranx" (the very name of the young naturalist who, along with his coauthor Laurencet, inadvertently set off the Battle of the Mollusks) to a lecture at the Museum of Natural History clearly by Geoffroy, though he's not named, on the "unity" of the natural world.

Sand, for her part, was twenty-six at the time of the Battle of the Mollusks and still went by her original name (Amantine Lucile Aurore Dupin de Francueil, so "George Sand" must have had the advantage of convenience as well as masculinity), and she would soon embark on a career that would make her one of the most prolific and popular writers of the Romantic period. She took her male pen name a year later and around the same time separated from her husband and acquired a permit from the city of Paris to dress in men's clothing. An ardent advocate of women's rights, she publicly criticized the institution of marriage and had lots of love affairs with other writers, politicians, and artists, famously including Frédéric Chopin. She was also close friends with her fellow novelist Sainte-Beuve, whom we met when he attended Lamarck's lectures in the Garden of Plants and later sent his fictionalized young self to them in the character of Amaury, the protagonist of his novel *Volupté: The Sensual Man*. In one letter to Sainte-Beuve, Sand invokes Geoffroy on the influence of the "ambient world" to explain her own "life of presumptuous carelessness and brazen heroism." She, along with Balzac, Quinet, Raspail, and the others, became partisans of Geoffroy in the struggle over the nature of life and of science.

After Geoffroy's death in 1844, however, there was a resurgence of the Cuvierian party. In 1861, two years after the publication of Darwin's *Origin of Species*, Pierre Flourens—Cuvier's disciple whom we met when he participated in the "examination" of Saartjie Baartman, and who was now permanent secretary of the Academy of Sciences—lamented that it was astonishing to see "in our century, men of genius" lending credence to the "ridiculous" ideas of Lamarck: "The truth is that no species has ever changed. Since Aristotle, the animal kingdom has remained the same. The fixity of species is, in all of natural history, the most important and most completely demonstrated fact."

The official definitions of the two sides established during the Battle of the Mollusks long outlived that early skirmish. Those who believed in direct creation by an omnipotent God were, in the upside-down funhouse mirror of reputation, the conservative empiricists, adherents to sober "positive facts"; while those who described living organisms as creating and transforming themselves from generation to generation were the romantics, the revolutionaries, the literary types, the people with "ideas" that went beyond "positive facts": in a word, the poets.

TWENTY-SIX

The Battle of the
Boiled Broth

I n February 1859, a second controversy began brewing in the Academy of Sciences, which would erupt by the end of the year, and which was in certain ways a continuation of the Battle of the Mollusks, or anyway a further skirmish in the same larger war. This new battle, which I'll call the Battle of the Boiled Broth for reasons that will soon become apparent, once again turned upon a central principle of Lamarck's theory of life. This time, it was the idea of spontaneous generation. This year of 1859 was a watershed year in the history of evolutionary theory. Exactly half a century after the publication of Lamarck's *Zoological Philosophy*, Charles Darwin published *On the Origin of Species*, drawing importantly on Lamarck's work (more on this connection between Darwin and Lamarck in chapter 27). The appearance of Darwin's book and its translation into French three years later also figured in the Battle of the Boiled Broth.

Occupying Cuvier's role in the new drama, representing the scientific establishment against dangerously poetic, radical thinking, was Louis Pasteur, then an ambitious thirty-six-year-old chemist from the Franche-Comté region in the Jura Mountains on France's eastern border. Pasteur's career was on the rise; after a succession of increasingly important positions, he had recently been appointed director of scientific studies at the ultra-prestigious École Normale in Paris, one of the *grandes écoles* established during the Revolution. Pasteur's research focused on fermentation, on diseases afflicting wine grapes, and on infectious diseases more generally. His development of vaccines against anthrax and rabies, which would make him into a national hero and even a kind of civic saint, were still more than two decades away, in the 1880s.

Pasteur's opponent was a fifty-nine-year-old naturalist named Félix-Archimède Pouchet, director of the Museum of Natural History of Rouen in Normandy. Pouchet's main claim to fame was that he had shown that ovulation occurs in female animals "spontaneously" and not, as had been widely believed, as a result of contact with sperm. Further pursuing this idea of the spontaneous appearance of eggs, Pouchet developed a theory that eggs could be spontaneously generated not only in the ovaries of female animals but in the organic debris of animals and plants. In 1859 he published *Heterogenesis, or, Treatise on Spontaneous Generation*, making his case for the spontaneous generation of eggs. Pouchet was careful to distance himself from the atheistic, materialist sort of believers in spontaneous generation: God, he said, was certainly behind it all. But in the political moment, this was still daringly radical.

Since the July Revolution and the installation of Louis Philippe on the throne, there had been yet another revolution in 1848, in the name of democratic rule and social welfare; this had spawned popular rebellions across central Europe and Italy. On February 24, 1848, Louis-Philippe had abdicated the throne, and Alphonse Lamartine, a poet, playwright, historian, and revolutionary, had declared the Second Republic. This republic had lasted just under four years, during which time it had instituted universal male suffrage, re-abolished slavery in the

French colonies (which had been abolished by the revolutionary government in 1794 and reinstituted by Napoleon in 1802), and designated Algeria—which had been the object of colonial warfare and conquest since Charles X—a "department" of France (the difference was essentially orthographic, since Algeria remained a colony, subject to imperial violence and oppression, until it won its independence in 1962).

The democratic Second Republic had soon been its own undoing: Male voters had used their universal suffrage to elect Louis-Napoléon Bonaparte—Napoleon's nephew—as the republic's first and only president. He had already twice attempted coups d'état, been imprisoned, and escaped. Now he promised to make France great again: religion, family, empire, work for the workers, more wealth for the rich. The peasants, who had just gained the right to vote, overwhelmingly supported Bonaparte. He seemed likely to win the election, but contemporaries were surprised by the extent of his electoral victory: 74 percent of the vote and almost every department. At the end of his term in 1851, he predictably declined to step down and instead staged another coup, this one successful, declaring himself the emperor Napoleon III (the disputed "rule" of Napoleon II, the son of Napoleon I, had lasted just a couple weeks following the end of Napoleon's Hundred Days back in power in 1815, at which time the ostensible emperor was four years old and living in Vienna).

The Battle of the Boiled Broth therefore took place in the reactionary world of the Second Empire. The imperial state and the Catholic Church were together reasserting their authority against republicans, liberals, freethinkers, and atheists. Even before his coup d'état, while he was still President Bonaparte, Napoléon III had signed a law permitting Catholic religious instruction in public schools, which had been essentially secular since the first French Revolution. Many French conservatives during this time regarded philosophical materialism and spontaneous generation as threats to the political and religious order.

The publication of Darwin's *Origin of Species* in 1859, an immediate bestseller in England that presented a theory of life in keeping with Lamarckism, and then its French translation in 1862, intensified the political and cultural antipathy for any scientific idea connected with theories of the transformation of living organisms over generations. Making matters worse, Darwin's irrepressible French translator, Clémence Royer, added an unauthorized preface to her translation representing Darwinism as antithetical to Christian doctrine and attacking the Catholic Church (while Darwin himself earnestly tried to persuade people that evolution and religious faith were compatible and that they needn't choose between God and Darwin).[1] In the midst of these struggles over science, religion, and politics, Pasteur was a Second Empire establishmentarian: A devout believer in a divine creator, he scorned materialists, and he was a patriotic conservative and a Bonapartist, like Cuvier before him.

These convictions drove Pasteur's campaign against spontaneous generation as they had inspired Cuvier's against Lamarckism and the transformation of living forms. The English comparative anatomist Richard Owen, who was a young man of twenty-six during the first debate and a middle-aged fifty-five-year-old during the second, later reflected that both Cuvier and Pasteur had had the advantage not only of rhetorical skill but of serving the established political and religious order, and it was mutual: Their careers thrived on their utility to the reigning powers. Following Louis Napoleon's 1851 coup d'état, Pasteur warmly supported the self-proclaimed emperor, to whom he sent a copy of his work on spontaneous generation. And decades later, at his inaugural address in April 1882 at the French Academy, a body so distinguished its members are known still as "the immortals," he boasted that his research, by demonstrating that life had never been shown to be a product of material forces, had bolstered the "spiritualist doctrine" against materialism.

The Battle of the Boiled Broth began at a low simmer on February 28, 1859, when Pasteur responded to a letter that Pouchet had written to him. Pouchet's letter had been provoked by a presentation Pasteur had made two weeks earlier, on February 14, at the weekly Monday meeting of the Academy of Sciences. Pasteur had described some experiments on lactic acid fermentation and closed by remarking that the lactic yeast in these experiments had always originated in the surrounding air. When the air had been subjected to intense heat first, no yeast had grown. Pasteur had concluded insinuatingly that "the question of spontaneous generation has made progress." His unstated implication was that the living organism of yeast originated not in spontaneous generation but in some sorts of seeds or germs in the atmospheric air, which were killed by heat.

Pouchet, who was about to publish *Heterogenesis*, wrote to Pasteur about a paper of his own presenting experimental evidence for spontaneous generation: Microorganisms had appeared in infusions of hay that Pouchet had kept under mercury to preclude organic contamination, then exposed to pure oxygen.[2] Pasteur responded confidently that Pouchet had somehow "unwittingly introduced common air" into his infusions, and therefore "the conclusions you arrived at are not based on facts of irreproachable accuracy."

Next Pasteur—who, remember, was more than a generation younger than Pouchet—went on to lecture his senior colleague about scientific attitudes and practices: "I think therefore, Sir, that you are wrong, not to believe in spontaneous generation, because it is difficult in such a question not to have a preconceived idea, but to affirm spontaneous generation. In experimental sciences one is always wrong not to doubt when the facts do not command affirmation." Finally, adopting the stance of the objective and disciplined experimentalist, Pasteur acknowledged judiciously that his own experiments didn't positively establish that there were germs in the air responsible for the organisms in the infusions. "In my opinion, Sir, the question is wholly and completely devoid of decisive proof. What is there in the air which causes the organisms? Are they germs? Is it a solid body? Is it a gas? Is it a fluid? Is it a principle like ozone? All this is unknown and invites experimentation."

Later that year, undeterred by his upstart junior colleague, Pouchet published *Heterogenesis*, presenting his case for spontaneous generation. He began by citing past authorities, especially featuring Lamarck, thereby aligning himself with the dangerous Lamarckian tradition that described living organisms as their own creators. Unfortunately, Pouchet wrote, ever since Lamarck there had been nothing but vagueness and equivocation on the subject of spontaneous generation; it was time for a clear statement of the case.

Meanwhile, Pasteur set out in the opposite direction, on a systematic campaign of experimentation to disprove spontaneous generation. His first step was to demonstrate that there were organisms floating in the atmosphere. He collected "dust" from common air by using a vacuum aspirator to suck the air through a wad of cotton wool. Then he soaked the cotton wool in a tube filled with an ether-alcohol solution, which caused the collected particles to fall to the bottom. When he examined these particles under a microscope, he found that there were plenty of "corpuscles whose form and structure indicate that they are organized" (that is, organic). Were these the germs of the mold and infusoria that appeared in solutions such as his and Pouchet's? Pasteur tested this question using a broth of sugared yeast water, which he boiled in a flask, then he fed sterilized air into the flask and sealed it. A flask prepared in this way, he wrote, would sit indefinitely in a warm oven without anything ever growing in it. After a few weeks, Pasteur introduced into the flask a wad of cotton wool filled with particles of organic "dust" from the air. Within a day or two, the broth was teeming with microorganisms.

Finally, in what became his most famous experiments for their clever minimalism, he altered some flasks by drawing out and curving their tops so that they were "swan-necked." Into these altered flasks went the same sugared yeast water concoction as before. Pasteur boiled most of them but left three or four unboiled and then left them all, unsealed, to sit. After a day or two, the liquid in the unboiled flasks was cloudy with mold, while the boiled flasks remained perfectly clear "not only for a few days, but for entire months." This indicated that boiling had killed whatever preexisting germs must have given rise to the growth in the unboiled flasks. As for the boiled flasks, Pasteur surmised that their "sinuous" necks must have trapped all the organic dust particles from the air. When he broke off the necks and dipped each into the liquid in its flask, within a day or two he reported that these flasks too were full of mold and bacteria.

Wait, though—if each drop of air carried germs to generate any sort of microorganism, retorted Pouchet and his partisans, surely there would be a fog in the air as dense as iron! Ah, responded Pasteur, but you make a false assumption that any drop of air anywhere on Earth carries germs to form any microorganism. In some places there are lots of germs in the air; in others there are few or none. Logically, the farther you go from an inhabited area, the fewer germs of living organisms float in the air. To demonstrate this, Pasteur prepared flasks with a boiled solution and sealed them, then opened them in a variety of settings, exposing them to the ambient air: "in an apartment, in a laboratory, in a garden; in the Jura mountains at an elevation of over eight hundred meters," and with dramatic flair, Pasteur even carried some out onto the Mer de Glace (Sea of Ice), a glacier in the French Alps on the northern slopes of the Mont

Blanc range above Chamonix. He confirmed that the number of flasks in which microorganisms appeared varied according to the density of habitation where they were opened. Only five out of the twenty in the Jura Mountains, and only one of the twenty on the Mer de Glace, showed any growth.

Finally, to counter the argument that the heat he had used to sterilize the broth might have destroyed its capacity to generate life, Pasteur replaced the broth with blood and urine that he drew directly from live, healthy (but very unfortunate) animals, with no possibility of contamination, and not heated. When he exposed these to sterile air, nothing grew in them.

Of course, Pouchet and his allies maintained that their mercury wasn't contaminated, and that Pasteur was the one whose experiments were flawed. In 1862, the Academy of Sciences decided to settle the matter in its customary way, by naming a prize competition: 2,500 francs to one who "by well-conducted experiments, throws new light on the question of so-called spontaneous generations." The composition of the prize committee reveals the abiding influence of Cuvier in French science several decades after his death. It included Flourens; Adolphe-Théodore Brongniart, another Cuvier disciple who did foundational work in paleobotany; and Henri Milne-Edwards, a zoologist and former student of Cuvier's. The other members were the physiologist Claude Bernard, who had been Pasteur's teacher, was his close friend and collaborator, and had already criticized Pouchet publicly; and a marine biologist named Victor Coste, who was close to the emperor and would later be appointed president of the Academy of Sciences. Pouchet submitted an entry with two coauthors, but they withdrew it when some of the committee members announced their decision for Pasteur before even considering the entries.

Undaunted, Pouchet and his collaborators hiked up into the Pyrenees to conduct their own version of Pasteur's Mer de Glace experiment, except they used their hay-water infusion instead of Pasteur's sugared yeast water. All eight of their flasks, they reported, had grown organisms. So, in 1864 the Academy of Sciences named a new committee to reexamine the question; however, three of the members were holdovers from the old committee, and the other two were close friends and supporters of Pasteur's. At this point, Pouchet gave up. The Cuvier–Pasteur scientific establishment, conjoined with the imperial state and the Church, constituted an inexorable force. The same year, Flourens published a book criticizing Darwin's *Origin of Species*, arguing (incorrectly) that Darwin's theory relied upon the existence of spontaneous generation, which had been disproven by Pasteur; ergo, Darwin's theory too was disproven.

To this day, Pasteur's experiments are cited in introductory biology courses as definitive disproof of spontaneous generation. Yet the prevailing view among scientists now is that "abiogenesis"—the emergence of living organisms from inanimate matter, or spontaneous generation— has happened at least once: around 3.8 billion years ago.[3] To be sure, it wasn't happening in Pouchet's hay infusions. Pasteur was right about that . . . but for the wrong reason. Pouchet might well have been correct when he said that organisms grew in his boiled hay infusions without any contamination from the air, though he too drew the wrong conclusion. Pasteur never repeated Pouchet's experiments using boiled hay infusions; if he had, he might have disliked the results. A few years later, during the 1870s, English and German experimenters confirmed that microbes often did grow in boiled hay infusions and found that the life cycle of the hay bacillus, a bacterium commonly found in soil, includes a highly heat-resistant phase during which it can survive boiling.

. . . .

L et us now travel in our mind's eye to a Thursday evening in the
spring of 1864, the evening of April 7, to a "scientific soirée" hosted
by the Paris Academy of Sciences. Pasteur is the evening's entertainment.
We find ourselves at an elegant affair. All around us are the leading lights
of the city, once again including George Sand, now fifty-nine years old
and the author of more than fifty novels, plays, and works of criticism;
she had also been a member of the revolutionary republican government
in 1848.

Glancing around the room we glimpse Sand's good friend the novelist Alexandre Dumas, just back in Paris after several years in Italy, where
he's been hanging out with Giuseppe Garibaldi, the revolutionary republican leader of the Risorgimento. With the proceeds from *The Three
Musketeers* and *The Count of Monte Cristo*, among his other bestsellers,
Dumas has even been supplying arms to Garibaldi's campaign to liberate
the states of the Italian peninsula from Austrian rule and unify them
into a single sovereign nation. Continuing our survey of the room, we
recognize Princess Mathilde Bonaparte, cousin of Napoleon III, and
Victor Duruy, a historian of antiquity currently serving as minister of
education.

Before this prestigious, clever, and influential crowd, Pasteur begins
to present his case against spontaneous generation. He starts by laying
out the implications of his topic for the great problems of the day, the
questions that "keep all minds awake: the unity or multiplicity of human
races; the creation of man over a few thousand years or a few thousand
centuries; the fixity of species or the slow and progressive transformation
of one species into another; matter eternal, beyond it nothingness; the
idea of God useless." Public opinion is perennially divided, he tells us,

between two ancient ideas, "materialism and spiritualism. What a conquest, gentlemen"—apparently, he's indifferent to the presence of George Sand, Princess Mathilde, and no doubt other women in our midst—"What a conquest, gentlemen, for materialism" if it can show "that it is based upon . . . proven fact."

The stakes are terribly high. If materialism shows itself to be factual, "what is the point of the idea of a creator God?" Pasteur assures us he won't try to solve any of these conundrums, but he will take up a question connected with all of them and "accessible to experiment": "Can matter organize itself? In other words, can beings come into the world without parents, without ancestors?" Strategically allowing his stance of positivist neutrality to slip, he proclaims, "It is this belief I come to combat."

First, Pasteur describes for us Pouchet's experiments, promising to show that their results are "completely illusory" because Pouchet's mercury was certainly contaminated by dust particles in the air. "There is no one among you, sirs, who does not know that there are always dust particles suspended in the air." In fact, Pasteur says, there are particles in the air all around us right now in this very room! We don't see them for the same reason we don't see the stars during the daytime: There's too much ambient light. He now theatrically darkens the room and projects a single beam of light across it, and in its glow we perceive illuminated specks of dust dancing about. Next, in semidarkness, he shows us a basin of mercury, sprinkles its surface with dust, then presses a glass wand into its surface. We see the dust particles descend into the hole produced by the wand, leaving a clear, shining surface. QED: Mercury is inadmissible in any experiment on spontaneous generation, because there is no way to prevent dust particles from entering it and hiding invisibly inside.

Now that he has us in his thrall, Pasteur recounts his own series of experiments, including his journey out onto the Mer de Glace. He confides that the sunlight reflecting off the ice created such a glare he couldn't see the alcohol jet he was trying to use to re-close his flasks. "You will say: you could have shaded the lamp with your clothes. Yes, but the clothes would have been a source of dust!" Heroic experimenter that he is, he went so far as to spend an extra night at the little inn at Montanvert and repeat the experiment the next day before sunrise. Meanwhile, he hadn't thought to seal the bottles he'd brought back to the inn with him until the next day, so they had been exposed to the dust of the inn overnight, whereupon they obligingly confirmed Pasteur's hypothesis: Ten of the thirteen exhibited a growth of organisms, indicating the greater amount of organic dust in inhabited places.

The experiment with the flasks of boiled broth, Pasteur tells us, has settled the matter once and for all. He has removed from the flasks the one thing "that it has not been given to man to produce": life itself, "because life is the germ and the germ is life." He has demonstrated that without the God-given germ of life, no new life appears. "Never," he exults, "will the doctrine of spontaneous generation recover from the mortal blow that this simple experiment deals it." Spontaneous generation "is a chimera." The audience rises to its feet in a standing ovation.

. . . .

With consummate skill, Pasteur has delivered the rigorous science of objective facts to the reigning powers of church and state. Even while laying out the theological stakes of the question of spontaneous generation and promising to "combat" materialism, even while declining to repeat Pouchet's experiments, whose results seem to contradict his own, he has insisted that this is a purely experimental matter, "not a question of religion, philosophy, or any system whatsoever," that the sole considerations are experimental objectivity and rigor. "It is a question of fact," he repeatedly assures us. And "gentlemen," he declares, "there has been enough poetry . . . ; it is time for science."

Pasteur has accomplished an extraordinary feat. He has rendered God a rigorously, objectively, scientifically, experimentally demonstrated *fact*. As for the emergence and development of life from matter by natural processes alone, well, that's clearly poetry. We can see that now as clearly as we see the shimmering particles of dust in the glowing beam of light through the darkened lecture hall: each one irrefutable, experimental evidence of the presence of God.

Darwin's Grandfather Problems

Spontaneous generation was a sore topic for Charles Darwin. His grandfather Erasmus Darwin had believed in it and had even conducted experiments to demonstrate it. The Romantic elder Darwin, who had died seven years before the younger one's birth, was a source of both embarrassment and pride for his Victorian grandson. Erasmus Darwin had been a doctor, naturalist, poet, libertine, abolitionist, republican, and admirer of the revolutionaries in France, for whom he composed some heroic verses in 1791. In addition to believing in spontaneous generation, he had described living organisms as continually transforming, with species emerging gradually one from another over time. Often, he presented his ideas about natural history in verse and in the footnotes to long poems. For instance, he reported his spontaneous generation experiments in a poem titled *The Temple of Nature*: "Next to our wondering eyes the focus brings / Self-moving lines, and animated rings." The note clarifies that he's describing the experience of looking through a microscope at a paste of flour and water that he had left to putrefy in a closed container, yielding microscopic "animalcules called eels, vibrio anguillula" that seemingly materialized and then moved about with "wonderful strength and activity."

Spontaneous generation was in ill odor in the decades after Pasteur's campaign, and the younger Darwin struggled to put it in its place. "It is curious," he wrote in 1861, "how largely my grandfather, Dr. Erasmus Darwin, anticipated the erroneous grounds of opinion, and the views of Lamarck," that simple forms of life are continually being spontaneously generated. Moreover, "there is no doubt that Goethe was an extreme partisan of similar views." The idea had of course become quite out of the question: "I need hardly say that Science in her present state does not countenance the belief that living creatures are now ever produced from inorganic matter." But note the word "now": Darwin leaves open the possibility that living creatures might at some point in the past have been produced from inorganic matter.

A decade later, he was even less sure what to think. His friend Alfred Russel Wallace recommended that he read *The Beginnings of Life*, a book by a doctor and neurologist named Henry Charlton Bastian who had defied the prevailing headwinds to make a case for spontaneous generation, which Bastian called "archebiosis" from the Greek roots *arche* (beginning) and *bios* (life). Darwin wrote to tell Wallace that he had "at last finished the gigantic job of reading" the book and found himself "deeply interested," also "bewildered & astonished," but still "not convinced, tho' on the whole it seems to me probable that Archebiosis is true." He even hoped so: "I shd. like to live to see Archebiosis proved true, for it wd. be a discovery of transcendent importance."[1]

Spontaneous generation in short was a quandary: Darwin liked the idea, but he couldn't quite believe it or shake off his consciousness of its poor reputation. Another somewhat embarrassing thing about Grandfather Erasmus was that he rhapsodized in verse about the loves and sex lives of plants, such as the female of the aquatic plant Vallisneria yearning for her male. In a footnote, the poet explains that the female of this "extraordinary plant" sends her flowers to float on the surface of the water by means of elastic, spiral stalks that expand as the water level rises and contract as it falls. The male's flowers are underwater, but when their farina or "dust" particles are mature, they rise to the surface and drift to the female flowers. The mutual yearning and ultimate union of the Vallisneria lovers inspired Erasmus Darwin to suggest the possibility that the first insects had actually been the anthers and stigmas of flowers—the detachable parts of the stamens and pistils, the plants' reproductive organs—that had broken loose from their parent plant as in the Vallisneria. In this way, he supposed, plant life had perhaps given rise to animal life.

Evolution is an idea that arose in poetry and science together, during a time when the two forms of thought were not distinct but all blended together, and Erasmus Darwin personified their union. The idea of evolution represented some favorite themes of Romantic-era writers and artists: the oneness of all living things, and of humans with the rest of living nature; the perennial struggle of life against death. "With finer links the vital chain extends," wrote Erasmus Darwin, "And the long line of Being never ends"; also "The births and deaths contend with equal strife, / And every pore of Nature teems with Life." He was Lamarck's contemporary, but there's no indication they knew of each other, which is a shame, since they clearly would have had a lot to talk about.

Charles Darwin did not love his grandfather's poetry, though he made allowances for changes of taste. "I have myself," he acknowledged, "met with old men who spoke with a degree of enthusiasm about his poetry, quite incomprehensible at the present day. . . . [N]o one of the present generation reads, as it appears, a single line of it." The elder Darwin's fusion of poetry, natural philosophy, physiology, and medicine was badly out of fashion, tainted by its association with revolutionary radicalism and by the influential efforts of pillars of the scientific establishment such as Cuvier and Pasteur, who vanquished distasteful ideas like spontaneous generation and the transformation of living organisms by disparaging and dismissing them as poetry. Yet the younger Darwin was in sympathy with his grandfather's ideas about natural history, including the idea of a continuity between plants and animals.

"It has often been vaguely asserted that plants are distinguished from animals by not having the power of movement," wrote Charles Darwin, but this common view was mistaken. In fact, plants did move when they needed to. They just didn't often see the need, since "food is brought to them by the air and rain." A tendril-bearing plant, Darwin observed, displays a remarkable ability to move itself in several ways. "It first places its tendrils ready for action"; then the tendrils spontaneously revolve; when they strike an object, they can quickly curl around it and grasp it. They can also move toward or away from light, or disregard light altogether, as they see fit. Darwin named this phenomenon—the spontaneous rotational motion of the growing parts of plants—"circumnutation," from the Latin roots *circus* for "circle" and *nutare* for "nod" or "sway." In *The Power of Movement in Plants*, he explained that the radicle, or embryonic root of the plant, has a major part in this motion. It "acts like the brain" of a simple animal, directing the movements of the parts in response to the environment.

When Darwin began a private notebook on the transmutation of species in the summer of 1837, at the age of twenty-eight and recently back from his around-the-world journey aboard the HMS *Beagle*, he titled it *Zoonomia*, a reference to his grandfather's work of that title—this one in prose—which attributed to animals a "power" of "animality": of "acquiring new parts, attended with new propensities, directed by irritations, sensations, volitions and associations; and thus possessing the faculty of continuing to improve by [an organism's] own inherent activity."

Notebook B, as Darwin's transmutation notebook of 1837 is known (Darwin lettered his notebooks on the front cover), also has several references to Lamarck's "monads," the spontaneously generated infusoria, each an "animate point" that, through its upward-striving power of life, leads to all higher animals according to Lamarck's theory. Together with Erasmus Darwin's power of animality, Lamarck's rising monads constituted the context in which Darwin began to think his way into the possibility of species change. "Each species changes," he wrote in the notebook. "does it progress? Man gains ideas. the simplest cannot help—becoming more complicated; & if we look to first origin there must be progress. If we suppose monads are constantly formed, would they not be pretty similar over whole world under similar climates & as far as world has been uniform at former epoch." A bit later he added, "Man is derived from Monad each fresh—" but he crossed that out and tore out the page. Later yet we see him still struggling: "Lamarck's 'willing' doctrine absurd (as equally are arguments against it—namely how did otter live before being modern otter—why to be sure there are a thousand intermediate forms)." He couldn't endorse the idea that animals could transform themselves by acts of will, and yet he couldn't quite reject it either.

Lamarck, a kind of intellectual grandfather to Charles Darwin, and sharing Erasmus Darwin's reputation as a poetic thinker, caused Darwin as much trouble as his actual grandfather. Darwin's close friend and intellectual role model, the Scottish geologist Charles Lyell, early on shaped his attitude toward Lamarck's theory. The first volume of Lyell's monumental *Principles of Geology* had recently been published when Darwin, at the age of twenty-two, set out on his five-year voyage aboard the *Beagle* as naturalist without portfolio and traveling companion to Robert Fitz-roy, the ship's captain. Fitzroy gave Lyell's new book to Darwin as a gift; Darwin received the second volume—whose opening chapter contained a meticulous and scathing refutation of Lamarck, at Montevideo in 1832, and the third before reaching Valparaiso in 1834.

Lyell's book imported Cuvier's rendition of Lamarck to England. In it, Darwin read that Lamarck showed an "unpardonable" disregard for evidence, that his "efforts of internal sentiment" and "acts of organiza-tion" were no better than so many "fictions . . . and other phantoms of the middle ages," that his theory was "defective in evidence," "fallacious in reasoning," and lacked even "the semblance of a foundation." As for Lamarck's argument that habits and behaviors gave rise to bodily organs rather than vice versa, Lyell calls it "staggering and absurd." He dis-misses as groundless Lamarck's suggestion that dogs have descended from wolves and, citing Cuvier, points out that there's no difference be-tween the mummified animals and people of ancient Egypt and their living equivalents. Rather than entertaining all this nonsense about ani-mals transforming themselves, chides Lyell, "we must suppose, that when the Author of Nature creates an animal or plant, all the possible circum-stances in which its descendants are destined to live are foreseen, and that an organization is conferred upon it which will enable the species to perpetuate itself and survive under all the varying circumstances to which it must be inevitably exposed." There: good, sober, evidence-based science.

After Lyell's treatment, anyone wanting to defend the idea that liv-ing organisms transformed from generation to generation must have felt they'd better distinguish their own theory from Lamarck's, and that's just what the author of the next bestseller in natural history did, or any-way he tried. The book, titled *Vestiges of the Natural History of Creation*, caused a sensation when it appeared in 1844, thanks to its winning combination of scandalous content and cloak-and-dagger presentation. The author remained stubbornly anonymous despite a storm of attempts to discover his identity, which was revealed only posthumously in the twelfth edition of the book in 1884. Throughout the first eleven editions, which enjoyed great notoriety and generated lots of controversy, people referred simply to "the author of *Vestiges*." He was in fact Robert Chambers, a Scottish geologist, writer, and publisher, the last of these callings perhaps explaining his knack for generating publicity and sales. Only four people were in on his secret: his wife, Anne Kirkwood Chambers, who copied out the whole manuscript and all the correspondence so that his handwriting wouldn't be recognized; his brother William, and two friends, Robert Cox and Alexander Ireland, who transmitted all the publication arrangements and correspondence.

Vestiges offers a naturalistic account of the cosmos, beginning with the formation of the stars and planets, and includes a chapter on living things. It says that life originated in a *"chemico-electric operation"* that first produced the simplest organisms. Then came a gradual progress *"from the simplest forms of being, to the next more complicated."* Each being *"gave birth to the type next above it,"* producing a gradual ascent to the *"very highest"* beings. You're no doubt thinking it could be tricky to distinguish this theory from Lamarck's, and so it was, but Chambers did his equivocal best. "Early in this century, M. Lamarck, a naturalist of the highest character, suggested an hypothesis of organic progress which deservedly incurred much ridicule," he wrote, "although it contained a glimmer of the truth." And on he went, first condemning and then, sort of, endorsing. For instance, regarding Lamarck's scandalous idea that humans had descended from other mammals, Chambers said, true, humans have no tail, "but the notion of a much ridiculed philosopher of the last century is not altogether, as it happens, without foundation, for the bones of a caudal extremity exist in an undeveloped state in the *os coccygis* of the human subject."

Chambers mischaracterized Lamarck as having ascribed *all* transformations to the behaviors of organisms in willful pursuit of their needs and desires, whereas, as we know, Lamarck ascribed habits and acts of will only to birds and mammals; and he also described other important forces of transformation, including the complexifying power of life and the more rudimentary responses of simpler organisms. While Chambers allowed it was possible that "wants and the exercise of faculties" had played some role, it was "certainly not in the way suggested by Lamarck, whose whole notion is obviously so inadequate to account for the rise of the organic kingdoms, that we only can place it with pity among the follies of the wise." His own theory was superior, he said, because he had founded it upon "the laws of organic development. . . . I also go beyond the French philosopher to a very important point, the original Divine conception of all the forms of being which these natural laws were only instruments in working out and realizing."

Here was the state of affairs when Darwin started his Notebook B by writing down his first thoughts on the transmutation of species. Cuvier and Lyell were the authoritative voices on Lamarck. A few years later, even the anonymous author of *Vestiges*, who seemed to endorse Lamarck's main ideas, described him as ridiculous, pitiable, foolish, and atheistic. And yet Darwin was interested in developing these very ideas himself. Little wonder that his private written musings show signs of ambivalence and struggle, as do Darwin's letters. "Heaven forfend me from Lamarck nonsense of a 'tendency to progression' 'adaptations from the slow willing of animals' &c," he wrote to his closest friend, the botanist Joseph Hooker, in 1844, "but the conclusions I am led to are not widely different from his." Darwin hastens to explain that his idea of "the means of change" is "wholly" different from Lamarck's. "I think I have found out (here's presumption!) the simple way by which species become exquisitely adapted to various ends.—You will now groan, & think to yourself 'on what a man have I been wasting my time in writing to.'"

What Darwin meant by a wholly different mechanism of change was his new idea of natural selection. Organisms that were well adapted to their environment would enjoy a reproductive advantage, producing lots of offspring who would inherit their parents' advantageous traits; meanwhile, less well adapted organisms would have fewer offspring, so their disadvantageous traits would gradually disappear. In other words, natural selection, as Darwin explained it, worked by accumulating favorable variations and eliminating unfavorable ones. But where did the variations come from in the first place? Throughout his life, Darwin remained unsure about the "complex and little known" laws of variation, but with regard to one particular cause he was confident: "use and disuse." By using a part or organ, animals strengthen and develop it, Darwin believed, while by not using it, they weaken and diminish it, and then they pass on these changes to their offspring. Darwin included use and disuse as an important cause of variation in every edition of the *Origin of Species*, emphasizing that "changed habits produce an inherited effect." Despite Darwin's protestations in his letter to Hooker, in other words, the mechanism of change in his theory doesn't really seem *wholly* different from the one in Lamarck's theory.

Lyell certainly had trouble seeing it as wholly different. A surprising thing happened after Darwin sent him the proofs of the *Origin of Species* in the fall of 1859. Having read Darwin's book, Lyell began having second thoughts about Lamarck. Writing with warm praise for *Origin*, Lyell congratulated himself on having urged Darwin to publish his majestic idea. But he had a few gentle queries and corrections. First, Darwin had written that "the most eminent naturalists have rejected the view of the mutability of species." Surely, wrote Lyell, you don't mean to ignore Geoffroy Saint-Hilaire and Lamarck? Darwin omitted the sentence in question from the published version. Lyell also pointed out that "Lamarck's monads coming daily into being supply a perpetual crop of the simplest forms." In Darwin's system, where did novelty and variety come from? What produced "new powers, attributes & forces"? And was there no "principle of improvement" at work, driving organisms toward ever better, more complex and perfect versions?

Darwin responded testily with regard to "the continued Creation of Monads"—spontaneous generation happening in the present—"This doctrine is superfluous (& groundless) on the theory of Natural Selection." There was no need for the supply of simplest organisms to be continually replenished because Darwin's theory didn't assume that the monads were necessarily progressing upward in complexity, leaving vacancies at the lower levels. As far as he was concerned, a monad could stay the same for age upon age as long as it underwent no beneficial (or, presumably, harmful) changes. Moreover, he wrote, "I entirely reject as in my judgement quite unnecessary any subsequent addition 'of new powers, & attributes & forces'; or of any 'principle of improvement.'" If the theory of natural selection required such forces and powers to account for complex forms of life, "I would reject it as rubbish. . . . I would give absolutely nothing for theory of nat. selection, if it require miraculous additions at any one stage of descent."

In this very grumpy response to Lyell, Darwin distinguished his own theory from Lamarck's by the fact that he, Darwin, assumed no intrinsic tendency toward ever greater complexity and perfection, an assumption, he said, that would encumber the theory with "miraculous additions" and render it total rubbish. It's a bit unfair, since Darwin owned the first volume of Lamarck's *Natural History of Animals Without Vertebrae*, in which Lamarck explicitly rejects the idea that organisms reveal a single line of progression toward perfection. Not only did Darwin own the volume, but he read it: It's listed as read in August 1842 in his "Books Read/Books to Be Read" notebook. (Darwin apparently didn't own Lamarck's *Analytical System of Human Knowledge*, containing the full flowering of his "very branchy" tree.) True, the "power of life" still played an important role in Lamarck's theory, driving organic matter to transform and complexify, but Lyell pointed out that Darwin's "variety-making power" was perhaps "only a change of name."[2]

To Darwin's frustration, Lyell kept on rethinking Lamarck. Darwin had converted Lyell from a creationist to an evolutionist, or what people were then calling a "transformist," and therefore he found he had to reconsider Lamarck. In 1863, he wrote a sort of apology to Lamarck's memory. "It is now thirty years since I gave an analysis . . . of the views which had been put forth by Lamarck, in the beginning of the century . . . and what Lamarck then foretold has come to pass." The more people discovered new forms, the less they could perceive sharp distinctions between species, suggesting that the species were in flux and blending into one another. Lyell credited Lamarck with having been the first to "introduce the element of time into the definition of a species," and indeed with having had a clearer appreciation than Lyell himself for the great depth of deep time, the slowness of change, and the "insignificance of thirty or forty centuries." Lyell therefore saw Darwin's *Origin of Species* as a development of Lamarck's theory.[3]

This was too much for Darwin! "You refer repeatedly to my view as a modification of Lamarcks doctrine of development & progression," he wrote; "if this is your deliberate opinion there is nothing to be said." But he did not see that *Origin* had anything more in common with La-marck's "wretched book," from which he had "gained nothing," than with "Plato, Buffon or my grandfather," all of whom had "propounded the *obvious* view that if species were not created separately, they must have descended from other species." Darwin understood that "Lamarck" was a dirty word. To associate his theory with Lamarck's, he wrote, would be "very injurious to its acceptance." Even Henrietta, Darwin's then twenty-year-old daughter, "who is a first rate critic & to whom I had *not said a word* about Lamarck, last night said, 'Is it fair that Sir C. Lyell always calls your theory a modification of Lamarcks? Why is it more a modification of his, than of any one's else?'"

But Lyell held his ground even against the innocent judgment of Henrietta, writing back, "When I came to the conclusion that after all Lamarck was going to be shown to be right, that we must 'go the whole orang,' I re-read his book, and remembering when it was written, I felt I had done him an injustice." Many of Lamarck's ideas now seemed to him "very Darwinian."

Despite Darwin's defensiveness at Lyell's suggestion that his theory was derivative, he was too conscientious to leave Lamarck unacknowl-edged. In the third edition of the *Origin of Species* he featured Lamarck in a historical section in which he credited "this justly-celebrated natu-ralist" as the originator of "the doctrine that all species, including man, are descended from other species. He first did the eminent service of arousing attention to the probability of all change in the organic as well as in the inorganic world being the result of law, and not of miraculous interposition." For the moment, Darwin still rejected Lamarck's idea of ongoing spontaneous generation in the present, but by the sixth edition he was open to that too, replacing his earlier categorical denial with the observation that science had "not as yet proved the truth of this belief, whatever the future may reveal."

As Darwin continued to develop his theory after the *Origin of Species*, he moved ever closer to Lamarck. For instance, he developed a tentative theory of inheritance that he called "pangenesis," which explained how organisms could pass on to their offspring the changes they underwent during their lifetimes. Darwin hypothesized that each part of an organism threw off particles he called "gemmules," which served as the mechanism of inheritance. During the next few decades, this idea would give rise to other theories of particles of inheritance, leading to the idea of genes in the early twentieth century. But Darwin's gemmules were quite different from genes. He imagined that the gemmules were modified by the organism's use or disuse of the corresponding parts, and that they shared a "mutual affinity" for one another, leading them to gather in the reproductive organs, whence they transmitted qualities to the offspring. In this way, Darwin proposed a mechanism by which animals could exert evolutionary agency, shaping the direction of development by their behaviors and self-transformations.

Another example of Darwin's interest in the evolutionary agency of animals was his theory of sexual selection, which he mentions briefly in the *Origin of Species* but develops much further in a later work, *The Descent of Man, and Selection in Relation to Sex*. His idea was that animals choose mates based upon their desires and preferences, and these choices, accumulating over generations, produce striking features and capacities, such as the peacock's tail; the brilliant colors of birds, butterflies, fish, and snakes; the lovely songs of songbirds; the displaying and combating behaviors of all sorts of male animals; the extravagant structures of creatures from birds to reptiles to fish to insects, such as plumes, spurs, topknots, and frills. According to the theory of sexual selection, not only do living organisms transform themselves heritably, but they also direct evolutionary development in furtherance of their particular tastes and desires.

. . . .

Darwin never insisted on a distinction between science and poetry. As a young man, he loved to read all kinds of literature, though he lamented that later in life he seemed to lose his earlier ability to appreciate poetry, painting, and music (though he still enjoyed reading novels). Still, he never suggested that science and literature were antithetical—or even distinct—ways of thinking. Writing was the medium of his science. The music of the words and the resonances of their layered meanings didn't just convey his theory but helped constitute it. He famously ends the *Origin of Species* by contemplating an "entangled bank, clothed with many plants of many kinds, with birds singing on the bushes, with various insects flitting about, and with worms crawling through the damp earth." This image, worthy of Erasmus Darwin and his generation of Romantic poets, leads to Charles Darwin's concluding thought, that "whilst this planet has gone cycling on according to the fixed law of gravity, from so simple a beginning endless forms most beautiful and most wonderful have been, and are being, evolved." Science and poetry, poetry and science.

Pangenesis, too, with each being a throng of transforming gemmules, was an idea made of a kind of thinking to which Cuvier's and Pasteur's distinction between science and poetry was profoundly alien. "An organic being is a microcosm—a little universe," Darwin wrote, "formed of a host of self-propagating organisms, inconceivably minute and numerous as the stars in heaven." The same is true of sexual selection, which led

Darwin to the "remarkable conclusion" that animals can direct evolutionary change. Animal and human qualities, from courage to colorful ornamentation to a capacity for music, had originated in "the exertion of choice, the influence of love and jealousy, [and] the appreciation of the beautiful in sound, colour or form." Lamarck had called it "the influence of the moral on the physical." Feelings, tastes, desires, an appreciation of beauty—poetry and its makings—these were essential elements of Darwin's science as of Lamarck's, and were powerful forces shaping the natural world that their science described.

TWENTY-EIGHT

He Cut Off Their Tails

(NOT WITH A CARVING KNIFE)

Once upon a time in the western foothills of Germany's Black Forest, there were twelve little mice: seven girls and five boys. They all lived together in a laboratory in the Zoology Department at the University of Freiburg. One autumn day in 1887—it was October 17, to be precise—a middle-aged professor with a stern look and a bushy beard came into the lab and, just like that, cut off all their tails! Despite this painful insult, the little mice carried courageously on with their lives, and one month later there came to pass a happy event. On November 16, two litters were born, eighteen babies in all. Each baby mouse had an excellent tail measuring eleven to twelve millimeters in length. But not for long. Old Bushy Beard returned and lopped off fifteen of the babies' tails. Like their parents, they took it bravely in stride, going about their business, and soon grew up to bear many fine children of their own, all graced with beautiful tails. You can guess what comes next.

The stern-faced, bearded, knife-wielding zoologist was a man named August Weismann, and the little mice were collateral victims in his campaign to defeat Lamarckism. Since pregnancy in white mice lasts twenty-two to twenty-four days, Weismann pointed out that the first two litters of babies were all conceived by two tailless parents. He carried on with his experiment and over the course of the next fourteen months de-tailed five generations of mice and produced 901 typically tailed young. The Lamarckian theory, Weismann declared, *"assumes the transmission of acquired characters."* Lamarck had supposed that bodily changes were inherited, and Darwin, regrettably, had followed him. But if this assumption were shown to be false, "we must completely abandon the Lamarckian principle."

Lamarck might have objected, if he could have, that his theory didn't assume the transmission of *all* acquired characters, only those acquired through responsive behaviors, habits, and ways of life. He never claimed that amputations or other mutilations would be inherited. Of course not. By acting certain ways, Lamarck had thought, an organism gradually altered its organization from within, redirecting the fluids that maintained its bodily parts and organs. A brusque change imposed from outside was another matter entirely.

The poor little mice had suffering rodent counterparts in Paris: a set of guinea pigs who had their spinal cords partially severed by Charles-Édouard Brown-Séquard, a Franco-American neurophysiologist, inducing epilepsy. Brown-Séquard claimed that the offspring of the mutilated guinea pigs *did* in fact exhibit various defects including epilepsy, deformities of the eyes and ears, missing digits, and muscular atrophy, and that these constituted positive evidence of the inheritance of acquired characters. These experiments were of course as irrelevant to Lamarck's theory as Weismann's were, and for the same reason: Lamarck never claimed mutilations would be inherited. (Brown-Séquard's other claim to fame was his research on the rejuvenating properties of the testicles of virile young animals, which he ground up to make a liquid extract and

then injected into older, feebler animals. Having tried it first with rabbits, with no apparent ill effect, at the age of seventy-two, he injected
himself with an extract from the testicles of young guinea pigs and dogs
and reported remarkable rejuvenating effects including renewed vigor
and powers of concentration, and a much stronger jet of urine. These
observations, when published, set off both a storm of ridicule and a stam-
pede of those eager to try it themselves. And now you know the origin of
hormone replacement therapy.)

In addition to his experimental amputations of mouse tails, Weismann considered other evidence of the non-inheritance of mutilations.
For instance, certain breeds of dogs, cats, and sheep whose tails are traditionally docked persist in bearing offspring with normal tails, and the
same goes for traditional alterations of human body parts, such as cir-
cumcision and foot binding, neither of which has apparent hereditary
effects.

Weismann did acknowledge that amputations and other mutilations
constituted only one kind of bodily transformation and considered some
instances that more closely resembled Lamarck's examples of habits and
ways of life, such as the recurrence of musical talent in the Bach family
and mathematical genius among generations of the Bernoulli family in
Switzerland, with many members doing landmark work in probability
and mathematical physics. These cases might seem to suggest that the
exercise of musical or mathematical ability produces a heritable increase
in aptitude. By way of refutation, Weismann observed that the father of
the German mathematician and physicist Carl Friedrich Gauss was not a
mathematician, that the father of the composer George Frideric Handel
was a surgeon, and that the Renaissance Italian painter Titian was the
son and nephew of lawyers.

Well, obviously, there are plenty of anecdotal examples on each side, so they hardly settle the matter. Weismann concluded rather weakly that he mentioned these "to show that, in my opinion, talents do not appear to depend upon the improvement of any special mental quality by continued practice." The amputee mice were by far his strongest argument. What they lacked in relevance, they made up for in clarity, and they made a strong impression on Weismann's readers. The mouse-tail experiment is cited to this day in textbooks and popular science books as the decisive refutation of "Lamarckian" biology.[1]

Our hapless, tailless little mice were participating not only in the defeat of Lamarckism but in the founding moments of scientific research as practiced in modern universities. Their truncated, documented lives took place in one of the first modern research universities, which emerged in Germany during the mid- to late nineteenth century as the universities there developed from medieval theological seminaries to juggernauts of scientific and industrial research and key organs of the state. By January 1871—when the kingdom of Prussia and its allies had defeated Napoleon III, ending the Franco-Prussian War, and founded the imperial German state—the research university had taken on a pivotal importance in German society and politics. Shortly afterward, beginning in 1876 with the founding of Johns Hopkins on the template of the University of Berlin, the United States began to import the model, which subsequently became the primary locus of scientific research worldwide. (At the time of this writing, it looks like the reigning political regime may be bringing the age of the American research university to an abrupt and unforeseen end.)

In the new, modern universities in Germany, the science and theology faculties established a mode of harmonious coexistence. They stayed out of each other's way by sticking to their own domains. The scientists confined themselves to describing material mechanisms, leaving all questions of agency and meaning to the theologians. In this arrangement, biologists represented living beings as assemblages of moving parts, assuming that the source of their forms and structures lay not within themselves but with a divine engineer beyond the machinery of nature. Weismann helped constitute this model of living beings as the passive objects of external agencies by insisting that organisms varied only as the result of outside forces acting upon them, never through their own behaviors, and that no change taking place in the body of an individual organism could ever inscribe itself in the reproductive cells. This principle, which came to be known to biologists as the "Weismann barrier" separating body cells from reproductive cells, implied that no living being could possibly transform itself in any way that could be inherited.[2]

The "inheritance of acquired characters," in other words, was impossible according to Weismann. He attached this phrase indelibly to Lamarck, although Lamarck himself never used it. In fact, the expression became current only in the second half of the nineteenth century, several decades after Lamarck's death. It was Herbert Spencer, a philosopher and social theorist, who popularized it. Spencer had a gift for catchy phrases. For Darwin's theory, he coined "survival of the fittest," which first appeared only in the fifth (1869) edition of the *Origin of Species*, and "evolution" in its Darwinian meaning, which is absent from all but the very last, sixth edition in 1872.[3]

By the time Darwin incorporated these terms into his later editions, Spencer had already inaugurated them in bestsellers of social theory and popular science that he churned out and that Darwin described as "detestable," obscure, unedited, clever, but empty, a lot of "dreadful hypothetical rubbish," and a disappointing amalgam of "words & generalities." They inspired Darwin's friend Hooker to characterize Spencer as "all oil and no bone . . . a thinking pump." (If you're finding that image obscure, so did Hooker. "I can attach no meaning to the simile," he confessed, but "it ought to have one." Darwin was so pleased with it that he read it aloud to his family, by whom "it was unanimously voted first-rate, & not a bit the worse for being unintelligible.") But Spencer had such spectacular success at promoting his own version of Lamarckism–Darwinism that Darwin eventually came around and included Spencer's key words and phrases in his books. These became the slogans by which people grasped, and continue to grasp, Darwin's theory of what he had originally called "descent with modification."

Weismann eagerly took up Spencer's phrase "inheritance of acquired characters" and ran with it, although in the opposite direction. He made it his mission to eradicate every trace of this Lamarckian idea from Darwinian evolution. The distinctive, defining feature of Weismann's version of Darwin's theory was that organisms entirely lacked evolutionary agency. Darwin himself, like Spencer, had seen "use and disuse" as an essential feature of his theory. So did most of his followers, such as Weismann's close friend Ernst Haeckel, Darwin's other leading proselytizer in Germany, who wrote bestselling popularizations in which he represented Darwin's work as the culmination of a transformist theory of life inaugurated by Goethe and Lamarck. (Weismann and Haeckel's difference of opinion did not break their friendship but did strain it, as is clear in their letters to each other.)

With his little mice, Weismann was determined to change all that: to establish once and for all that whatever characteristics an organism might manage to acquire during its lifetime would disappear with its death. In Weismann's natural-selection-only version of Darwinism, the variations undergone by organisms might bestow an advantage on them, in which case they would be likelier to reproduce, or a disadvantage, in which case they would be less likely, but that was purely a matter of luck.

Living beings were entirely passive with regard to evolution, as powerless in their own creation as God's creatures had ever been.

It wasn't just a matter of correct science to reject stories of "Lamarckian" transformations, Weismann maintained, but a "duty" and an occasion for a kind of moral outrage and "contempt": "No one can be prevented from believing such things, but they have no right to be looked upon as scientific facts or even scientific questions." Weismann rejected all of Lamarckian biology categorically: "The whole principle of evolution as proposed by Lamarck, and accepted in some cases by Darwin, entirely collapses."

Like Cuvier and Pasteur, Weismann emphasized that his approach to science rested on the assumption of a divine presence behind the whole process. He assured his readers that his biology didn't just leave room for a "Universal Cause" but actually demanded one. He wrote of a *Final Cause* operating "behind" the mechanism of the universe. All agency lay with this remote, divine presence and none with the evolving, mortal beings themselves. Therefore, Weismann said his science was "absolutely opposed" to materialism.

In the decade following Darwin's death, Weismann and his followers conducted a successful campaign to establish their own version of Darwin's theory as "pure Darwinism" and to purge it of all traces of "Lamarckism." Things got particularly intense in the summer of 1888, with heated letters flung back and forth in the columns of *Nature*. The main partisans for the Weismannian position were two Oxford zoologists: Ray Lankester, who introduced the notion of ideological purity with regard to Darwinism, and Edward Poulton, Weismann's English translator, who adopted the fighting slogan "Lamarckism *versus* Darwinism," whereas previously there had been no "versus" about it. George Romanes, Darwin's youngest close friend and collaborator, struggled to make that point.

Romanes had worked with Darwin during his final months on some botanical experiments intended to confirm the theory of pangenesis. Now, in response to Lankester and Poulton, he protested that surely "pure Darwinism" should refer to Darwin's theory as Darwin himself presented it, which incorporated crucial elements of Lamarck's theory. Romanes therefore suggested a different name for Weismann and company. "The school of Weismann may properly be called Neo-Darwinian," he wrote, since "pure Darwinian it certainly is not."[4]

The name "Neo-Darwinian" caught on beautifully, but in other respects Romanes fought a losing battle. Lankester was particularly influential, founding what one neo-Darwinist referred to admiringly as "a school of selectionism at Oxford," meaning a Weismannian natural-selection-only interpretation of evolution. The Columbia University embryologist and early geneticist Thomas Hunt Morgan, pioneer of the experimental use of *Drosophila* (fruit flies) in genetics research, later wrote approvingly that the disfavor into which Lamarck's theory had fallen was all thanks to Weismann's ability to enlist "common sense." With just "a few experiments," said Morgan, Weismann had handily dispatched Lamarck's theory. Weismann's valiant, irrelevant little mice carried the point. But not for absolutely everyone.

TWENTY-NINE

Inside
Mammoth Cave

A lpheus Spring Packard was an entomologist and evolutionary biologist at Brown University, and an out-of-the-closet Lamarckian. In 1901, he wrote the first English-language biography of Lamarck, traveling to France and speaking with everyone he could find who had any connection to his subject, even visiting and photographing Lamarck's childhood home in Bazentin, Picardy.

Twenty-five years earlier, Packard had declared a "distinctively American school of evolutionists," which he had soon renamed—in answer to Weismann's neo-Darwinism—"neo-Lamarckianism."[1] You may be wondering why Packard thought Lamarckism was distinctively American, given the undeniable Frenchness of the theory's originator. Packard explained that Americans were in an especially good position to measure the creative responses of organisms to their environments because of the great extent and dramatically varied environments of North America.

In August 1871, in heroic pursuit of extreme environments and their adaptive inhabitants, Packard traveled to Mammoth Cave in central Kentucky, the world's longest known cave system, and intrepidly explored deep inside it to record the adaptations of the creatures there to profound darkness. His companion was Frederic Ward Putnam, an ornithologist for whom the journey would apparently change his course in life. Seeing traces of aboriginal human habitation deep inside the cave system would fill him with a fascination for archaeology, and he would soon become the director of the new Peabody Museum of Archaeology and Ethnology at Harvard. Packard and Putnam were friends from their student days, when they had both studied with the Harvard creationist ichthyologist and geologist Louis Agassiz but had rebelled against their mentor by embracing the theory of evolution.[2]

The two cave explorers traveled with a group of fellow naturalists aboard the Louisville and Nashville Railroad. Arriving at the mouth of the cave, they set forth with "our trusty guide, Frank." This was probably Frank DeMonbrun, a white guide who worked at the cave for several decades around mid-century, but the original guides and explorers of Mammoth Cave were three enslaved Black people named Stephen Bishop, Mat Bransford, and Nick Bransford, whose knowledge of the cave system and excellence as guides were legendary. Stephen Bishop was the first person to cross Bottomless Pit, a chasm more than a hundred feet deep, and so discover the whole region of the cave beyond it. He had died more than a decade earlier, but Mat Bransford and Nick Bransford, now free, were still working at the cave when Packard and Putnam visited it eight years after Emancipation.

Frank brought along oil lamps and a can of extra oil. Together, they stepped into a darkness so profound it "must be felt to be appreciated." All those "mortals who have never descended to [Mammoth's] cavernous depths, nor trod its gloomy corridors," recounted the voyagers, couldn't possibly imagine what it was like. The group had hiked for thirteen hours, penetrating ever deeper into the limestone interior, when Frank asked them if they'd like to see Mammoth Dome, a part of the cave system that was rarely visited because of its extreme remoteness and the treacherous route to get there. Feeling "footsore and weary," and no doubt a bit trepidatious, they hesitated, but a fellow visitor who had been there once before assured them it was so magnificent as to be worth the effort and danger, so they decided to go for it.

Carrying magnesium and calcium torches provided by Frank, they first "crawled and climbed" their way to the brink of a deep pit, then descended "a rickety ladder, slippery and dripping with water." They found themselves in a cathedral-like chamber, its ceiling towering to a height of more than 250 feet. The spray from a cascade falling from high above their heads into a chasm deep below their feet so filled the air with mist that the rock walls were dripping wet. Once everyone was in place, they lit their torches, and "the immense dome was revealed to us in all its majestic beauty." Arriving back up at the top of the terrifying ladder, they gave three elated and relieved cheers for Thomas Kite of Cincinnati, the fellow who had urged them to brave it.

In the cave's dry areas, meanwhile, the group marveled at the otherworldly ceilings, which were "covered with a white efflorescence that displays itself in all manner of beautiful shapes," looking just like wreaths and rosettes of flowers, especially lilies. This "beautiful stucco" was the result of crystallized gypsum forming on the surface of the limestone. In areas with somewhat more moisture, the gypsum instead formed rounded white blobs, giving the "Snow-ball room" its decorative motif and name.

Mammoth Cave was bustling with living beings. Packard and Putnam saw fish, insects, and crustaceans. So far no worms, Packard reported, but he was sure they were there somewhere and just needed finding. They observed many specimens of *Ceuthophilus stygius*, or cave camel crickets,

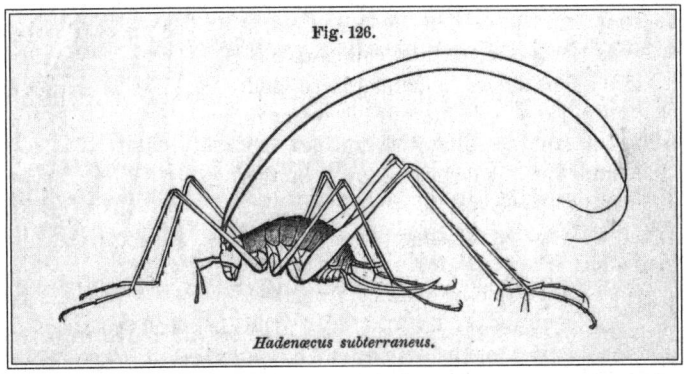

Fig. 126.

Hadenæcus subterraneus.

Cave cricket from Packard and Putnam, *The Mammoth Cave* (1879)

"jumping about with the greatest alacrity upon the walls" and scuttling across the ceilings. These creatures were extremely reactive, leaping away at the sound of approaching footsteps, then freezing when the noise stopped, "turning about and swaying their long antennae in a most ludicrous manner, in the direction whence the disturbance had proceeded." The slightest sound "would increase their tremulousness," although neither currents of air nor lamplight seemed to disturb them. Yet they had perfectly formed eyes and could obviously see, since they jumped away from a slowly approaching hand, making it necessary to seize them rapidly if you wanted to catch one.

Near Bottomless Pit there lived a brownish centipede-like animal, just over an inch long, that "moved off in a rapid zigzag motion" when approached and so proved impossible to capture. And in a shallow pool of water four or five miles from the mouth of the cave were tiny, blind crustaceans creeping across the sandy bottom, hanging out with a host of blind *Campodea*: little, white, hairy, be-tentacled arthropods. There were also many blind fish and blind crawfish. The blind crawfish were especially interesting because, while the eyes of the adults were tiny and rudimentary, the young animals had larger eyes. Packard took this as evidence for his Lamarckian view that the embryos of these blind arthropods developed similarly to sighted ones, and that their blindness resulted from a change that had first occurred in adults and then been transmitted to their offspring. Transmission of an acquired blindness would in turn support the idea that the differences distinguishing genera of animals arose from changes in adults who transformed in response to their "changing conditions of life," then passed these changes on to their young.

The cave creatures, Packard thought, must have arisen from outside-dwelling animals that had migrated into the cave and transformed in response to "the new and strange regions of total darkness." On a geological timescale, he imagined this transformation must have happened comparatively quickly. The caves of the region contained fossilized remains of animals only from the Eocene (which began about fifty-six million years ago) and later. Natural selection acting upon chance variations, Packard suggested, could not account for a rapid enough change in the cave-dwelling organisms. Rather, they must have acquired the features they needed to survive in total darkness by a Lamarckian process of changing their habits, providing beneficial variations for natural selection to work upon, gaining features they needed and losing ones that were now useless to them. Some became "blind, others very hairy, others with long appendages." The sheer variety in their forms showed that they must have descended from ancestors that had themselves had different features and habits. If they had all been specially created for subterranean existence in Mammoth Cave, Packard reasoned, they would have exhibited "a much greater uniformity in the organs adapting them to a cave life."

One of Alpheus Packard's closest neo-Lamarckian comrades was also named Alpheus. He was Alpheus Hyatt, a zoologist and pale-ontologist at MIT. In 1867, the two Alpheuses, along with Putnam and Edward Morse, a zoologist and archaeologist from Maine, founded a popular science journal, *The American Naturalist*, which is still pub-lished today, and in 1883 Packard established a society to go with it, the American Society of Naturalists, also still in business.[3] The cofounders had all met as students of Agassiz's, and all strayed from the fold by be-coming evolutionists. In the introduction to their journal's first issue, they struck an ecumenical stance, presuming their readers would be in-terested in facts about the origins and forms of life "whether by direct creation, or by secondary laws as claimed by the followers of Lamarck and Darwin."

Less than a decade later, the journal was overtly not just evolutionist but Lamarckian. An explicit endorsement of Lamarckism appeared in *The American Naturalist* in April 1876, three years after Agassiz's death. The article stated that the laws of evolution recently advocated by Darwin had "first been presented to the scientific world . . . by Lamarck, of Paris." It was only "owing to the adverse influence of Cuvier" that Lamarck's the-ory had lain dormant for half a century, until "Darwin resuscitated it." The author of this bold sentiment was Edward Drinker Cope, a principal figure in the establishment of American paleontology and, along with Packard and Hyatt, the third member of their neo-Lamarckian triumvi-rate. Hyatt and Cope were the leading advocates of the view, which Pack-ard endorsed, that natural selection acting upon chance variations would be too slow to account for the pace of evolutionary development, so there must be other forces influencing the supply of variations.

None of them questioned the role of natural selection in shaping the development of living forms, and—like Romanes—they didn't see Lamarckism as an alternative to Darwinism but rather saw the two theories as overlapping and complementary. Natural selection was all very well but could originate nothing. Darwin himself, they pointed out, had said as much. Natural selection could select only among existing forms. Where did the variations come from in the first place? That's where Lamarck's theory came in, offering answers to this question, including the transformative effects of organisms modifying their habits in response to a changing environment.

In 1885, Packard coined the term "neo-Lamarckianism" to designate the version of evolutionary theory developed by him, Hyatt, and Cope. He wrote that Cuvier, Agassiz, and "popular prejudice" had prevented Lamarck's theory from having the influence it deserved. He and his comrades considered that there must be not just the responsive behaviors and habits of organisms but a whole "series of evolutionary agencies or causes" supplying the forms of life for natural selection to select among. "In other words, we believe in a modified and greatly extended Lamarckianism, or what may be called neo-Lamarckianism." During the later 1880s and the 1890s, neo-Lamarckism gained some traction, especially in America, much to the disgust of Weismann and also of Poulton, his English translator, who called it "remarkable" that the movement included "many of the most distinguished American biologists."

Where before there had been two mutually consistent theories of evolution, there were now two opposing camps: neo-Darwinism and neo-Lamarckism. A colleague of Packard's at Brown University, a botanist, paleontologist, and sociologist named Lester Ward, conducted a thorough examination of the conflict between them, in which he pointed out that the original Darwinism and Lamarckism had not been in conflict at all. Ward's sympathies lay with Packard and the neo-Lamarckians. He remarked upon a similarity he noted between the neo-Darwinist movement and the program of Francis Galton, a mathematician, naturalist, and social theorist who was also Charles Darwin's first cousin. Inspired by his cousin's theory of evolution, Galton proposed a science devoted to "the cultivation of race": "improving stock" through "judicious mating" and other means of enhancing "the more suitable races or strains of blood." He called this new science "eugenics," from Greek roots meaning "good in stock."

Galton grumbled that he had "no patience" with the empty platitude that "babies are born pretty much alike." He rejected "pretensions of natural equality" as morality tales for children and declared that innate "mental capacity" was widely variable. Education could develop innate qualities to their fullest potential but could do nothing to enhance them; the way to improve the human race was therefore through breeding. In presenting eugenics to the public, Galton emphasized "the vastly preponderating effects of nature over nurture." The traits of humans and other animals, he said, were all essentially inbred, inherited from parents and passed along from generation to generation.

Ward objected vehemently to both Galton's eugenics and neo-Darwinism, which he took to be essentially the same theory, a theory of human passivity and helplessness. If everything humans did within their lifetimes died with them, he wondered, why do anything at all? "If, as Mr. Galton puts it, nurture is nothing and nature is everything, why not abandon nurture and leave the race wholly to nature? In fact the whole burden of the Neo-Darwinian song is: Cease to educate, it is mere temporizing with the deeper and unchangeable forces of nature."

The self-making agency of humans, as Ward saw it, resonated with his deeply held egalitarian belief in the unconstrained potential of all human individuals. True, like almost everyone around him, he ranked races on a spectrum from primitive to advanced. He also wrote that only Europeans were advanced enough to experience romantic love, and he described the "gynaecocratic" (women-governed) stage of civilization as more primitive than the "patriarchate" (men-governed). But he saw these distinctions as "purely sociological" and not biological. He thought all humans were equal in potential: in what they could make of themselves. He was accordingly a dedicated abolitionist, an advocate of women's suffrage, a proponent of social welfare and labor unions, and a fierce critic of organized religion.

But the American neo-Lamarckian movement, small though it was, also encompassed sentiments antithetical to these. "We all admit the existence of higher and lower races," opined Cope. These he said were defined by their greater or lesser proximity to apes, the most apelike being "the negro in his most typical form." Women, meanwhile, he described as more childlike and less developed than men. In "Two Perils of the Indo-European," Cope warned that the "American-African" must urgently be stopped from voting and promptly transported to Africa, and that "woman," because of the "sex character of her mind," must never be granted suffrage. A devout Quaker, he believed evolution was the expression of an invincible "Divine Spirit" that animated the changing behaviors and beliefs of human beings and was in this way inexorably bringing about "a perfect humanity."

. . . .

It was in the new neo-Lamarckian journal *The American Naturalist* that the Princeton psychology professor James Mark Baldwin published his landmark 1896 paper on what came to be known as the Baldwin effect, which he himself called "Organic Selection," and which was hotly contested for more than a century but has recently achieved widespread acceptance among biologists.

"Organic Selection" means selection that an organism enacts upon itself by behaving in certain ways. When you respond to your environment, you choose features of yourself to use. For instance, you might use your opposable thumbs to grasp things: climb trees, peel fruit, dig termites out of a hole with a stick. When used in these ways, your thumbs can be important to your survival, but otherwise they're just cumbersome appendages. By selecting your thumbs for use in certain ways, you make them into an advantage for survival and reproduction. Natural selection will favor organisms with thumbs as long as they employ them to good advantage. Organic selection can be a subconscious process: Baldwin writes that even plants, unicellular organisms, and "very young children" act spontaneously to "rise to the occasion" and make use of what they have. But organic selection can also be the result of "conscious agency": "intelligent" processes such as learning from your parents or from experience, imitating others, and "reasoning from means to ends."

Organic selection channels natural selection in certain directions, such as in the direction of favoring opposable thumbs, so organisms aren't just the passive objects of natural selection but its active conductors. Baldwin saw organic selection as a "direct substitute" for Lamarckian inheritance of the effects of use and disuse. In his view, there was no need to suppose that organisms directly inherit changes such as a strengthened limb. Rather, they bring about stronger limbs by directing natural selection to favor them. Baldwin's is therefore an indirect sort of Lamarckism, filtered through natural selection, but as in Lamarck's original theory the energetic activity and intelligence of organisms shape the course of evolutionary development.

Baldwin was not in fact a biologist but a psychologist and philosopher with a background in French and German literature. His interest in what he called "problems of genesis—origin, development, evolution" was inspired by a very intimate and emotional instance of genesis: the birth of his first child, Helen. From the moment of her arrival, as her doting father later recalled, she—and two years later, her sister, Elizabeth—became "the focus through which all the problems of general biology and psychology presented themselves." Baldwin described and analyzed the behaviors of his two little girls in his book *Mental Development in the Child and the Race*, which inspired the Swiss psychologist Jean Piaget—often regarded as the founder of developmental psychology—to embark on a methodologically and philosophically similar research program in child development.

Charmingly, Baldwin recounts the exploits of Helen (H.) and Elizabeth (E.), such as the day in early May 1894, when they were four and two years old, and they pretended to be "mama" and "baby." H., addressing E. as "my darling," instructed the impressively cooperative baby to fall asleep, wake up, get dressed, have breakfast, take a nap, get ready for a walk, and go greet her papa. Although the actual papa was sitting nearby pretending to read a newspaper, they assigned the role of "papa" to a column of the porch, which Helen then animated in a *"gruff, low voice"* that severely taxed the surreptitious observer's ability to contain his amusement. Baldwin's intimacy with the experimental subjects was, in his view, absolutely essential to his scientific observation of them. Because he recognized the origin of every phrase they used, and what it meant in its original context, he was able to understand how they had adapted it in their game. Therefore, he had a special word of advice for parents, "especially the fathers!": "You can be of no use whatever to psychologists . . . unless you know your babies through and through."

322

Helen and Elizabeth's game exemplified a special and "extremely important" form of "Organic Selection" that Baldwin called "Social Heredity": all the cooperative and social processes by which gregarious animals learn from one another, enabling "young creatures" to acquire the behaviors of their parents. Social heredity isn't physical heredity, Baldwin wrote, but it's still very real heredity, because it transmits adaptive physical functions from one generation to the next. It therefore influences physical heredity in the way that organic selection always does: by preserving certain physically inherited variations and directing the course of natural selection to preserve them.

Now let's just pause for a moment to consider Baldwin sitting on his porch on a fine spring day, chuckling behind his newspaper as his daughters play their make-believe game, paying attention to every word they utter, and building it into his theory of evolution. Baldwin blended biological and neurochemical factors with social, cultural, emotional, and psychological ones in his theory of organic selection. And his methodology was also a mixture. He acted as a father, a psychologist, and a storyteller all at once, the three roles combining and reinforcing one another.

This was science in a very different mode from the reductive, objective stance adopted by the neo-Darwinists (and Cuvier, Pasteur, and Weismann before them). Humanistic and scientific thinking were inseparable to Baldwin. In his 1909 book, *Darwin and the Humanities*, he wrote that the humanities and evolutionary science must be equal partners because the mind, province of the humanities, shaped the course of evolution as much as the brain shaped culture and society. The crucial point was that an organism isn't just an assemblage of parts. The fitness of any part depends on what the organism *does* with it. Once again, in Lamarck's terms, the influence of the "moral"—thoughts, feelings, desires—on the physical.

. . . .

When Packard died in 1905 at the age of sixty-five, his neo-Lamarckian movement was pretty much over, while neo-Darwinism's star was ascendant. The author of Packard's obituary for the National Academy of Sciences, an entomologist named Theodore Cockerell, wrote that "modern experimental methods" had defeated the observations of Packard and company. "It appears certain," he concluded, "that we shall never return to Lamarckism, new or old."[4]

Nevertheless, the first English translation of Lamarck's *Zoological Philosophy* appeared nine years later, in 1914. The translator, a young English popular science writer named Hugh Elliot, wrote in his introduction that Lamarck had originated the idea of evolution and defended it at a time when all the worldly powers were aligned against him and that for half a century his theory had been "almost the only public representation of a belief which no one now questions," the belief that life evolves. This, especially at a moment when Lamarck was "slowly entering upon the final stage of oblivion," made the translation important even though, Elliot said, Lamarck had been entirely mistaken about the inheritance of acquired characteristics and was a second-rate philosopher and zoologist prone to making unwarranted assumptions, to violating the rules of scientific method, and to outrageous, fantastic, and absurd speculations. Lamarck had been no genius, Elliot wrote, but a tireless worker who gave up everything in pursuit of truth: a martyr to science. In this strange way, the first English text of Lamarck's magnum opus made its appearance through a neo-Darwinist lens that the translator insistently held up to the reader's eye.

THIRTY

The Gospel According
to Saint William

On May 8, 1900, William Bateson, an English biologist and writer, was riding on a train from Cambridge to London to deliver a lecture when he received the word of a new Messiah. He would devote the rest of his life to interpreting and spreading this gospel Truth. We will come back to Saint William the Apostle presently, but for now we will leave him rumbling along, awestruck, on the train. We have some unfinished business with Weismann, whose other activities—beyond mice and their tails—are relevant to this part of our story.

One of Weismann's main contributions to the study of reproduction and inheritance had to do with cell division and chromosomes. During the 1880s, experimenters had begun to notice certain filaments in cell nuclei that, when exposed to a dye, readily took on its color, and had taken to referring to these as the cell's "chromatic elements," or "chromatin," which soon gave rise to the word "chromosome," from the Greek roots for "color" and "body." Weismann, along with others, believed that these parts of the nucleus contained the machinery of inheritance, and he gave an early description of the process that would later be called meiosis: the division of germ cells that halves their number of chromosomes in the production of egg and sperm cells so that when they come together at fertilization, their union contains the correct number. This process, Weismann said, provided "an inexhaustible supply of fresh *combinations* of individual variations." He wasn't sure exactly what the hereditary factors within the chromosomes were—the hypothetical little things that were continually being combined and recombined in sexual reproduction to create fresh combinations of variations—but he called them "determinants."[1]

Weismann's determinants were units of inheritance like Darwin's "gemmules," but they had a crucial difference. You might even say they represented the opposite idea. While Darwin's gemmules traveled from all throughout the body to the reproductive organs, bearing marks of any changes undergone by the various body parts, Weismann's determinants were absolutely separate from body cells and isolated from all bodily transformations. Weismann believed that the determinants could be changed only by *"the direct effect of external influences,"* in particular, fluctuations in nutrition. According to his theory, new variations of organisms arose by a combination of the environment acting directly upon the determinants and then sexual reproduction continually "intermingling" these in new combinations. Nowhere in this process could the behaviors of organisms make a difference.

While Weismann was messing with mice and developing his ideas about determinants, other people were also taking up Darwin's idea of physical units of inheritance. One of them was Hugo de Vries, a botanist at the University of Amsterdam. De Vries praised Weismann for having "shattered" the belief in inheritance of acquired characteristics, and he rejected Darwin's idea that the gemmules traveled from all over the body to the reproductive organs bearing marks of bodily changes. But he still thought Darwin had gotten something right in the notion that there must be material particles of inheritance that carried combinations of characteristics into organisms in a "mosaic-like" way. Instead of imagining that these particles traveled throughout the body, he proposed that copies of them existed in every cell, but each was only active in certain cells. To distinguish his version of pangenesis from Darwin's, de Vries called his own theory "intracellular pangenesis" and his particles of inheritance "pangenes" instead of gemmules.

To his likely dismay, in March 1900, as de Vries was developing his theory of pangenes, he received from a colleague a copy of a pamphlet titled *Experiments in Plant Hybridisation* by a Moravian monk named Gregor Mendel. Although this paper had been published in 1866, no one had paid much attention to it until now, when burgeoning research interest in the mechanism of inheritance brought several people to take an interest in the old report (poor Mendel had by then been dead for sixteen years).

The son of farmers, Mendel had lots of childhood gardening experience and some formal education in philosophy and physics from the University of Olmütz (now Olomouc). In the garden of the monastery where he lived, St. Thomas's Abbey in Brünn (now Brno, in what is now the Czech Republic), Mendel had experimented with crossbreeding pea plants to define a mathematical law governing their inheritance of traits. He had grown more than ten thousand plants, he explained, in garden beds and "a few also in pots," keeping them upright "by means of sticks, branches of trees, and strings stretched between." Mendel had studied the shape and color of the seeds and pods, the position and color of the flowers, and the length of the stem. For each of these characteristics, he had found that when he crossbred two true-breeding plants to create a "hybrid," one form turned out to be what he called "dominant," meaning that the hybrid offspring always resembled the parent with that form, and the other form he called "recessive," because it disappeared in the offspring. For instance, for the color of the pods, green was dominant over yellow. However, Mendel also found that the recessive trait reappeared in about 25 percent of the next generation. When he bred his hybrid green-podded plants, about one-quarter of their offspring had yellow pods.

Mendel explained this by surmising that there must be some kind of "factor" in each reproductive cell—egg or pollen—corresponding to either the dominant or the recessive form, and that each hybrid plant must possess both kinds of eggs or pollen cells. Breeding the hybrids would yield four possible combinations. For instance, when hybrid green-podded plants were bred, the four would be (1) two green-pod reproductive cells; (2) a green-pod egg cell and a yellow-pod pollen cell; (3) a yellow-pod egg cell and a green-pod pollen cell; (4) two yellow-pod reproductive cells. Mendel imagined that in order for the recessive form to appear in a plant, this plant must result from two reproductive cells both corresponding to that form. So, only the last of the four combinations would produce offspring with yellow pods. Assuming each possibility was equally likely, a quarter of the plants would have yellow pods.

"Vloek!" (or something equivalent) de Vries apparently thought to himself upon reading Mendel's old article. In order to develop his ideas about pangenes, he too had been experimenting with crossbreeding plants, in his case, poppies, and noticing a similar pattern. De Vries seems to have briefly considered pretending he'd never heard of Mendel. In a short note on inheritance in hybrids that he published in the proceedings of the French Academy of Sciences soon afterward, he incorporated Mendel's vocabulary and mathematical logic but omitted to mention his name. But de Vries then received a subtle admonishment from another person who had suffered the same blow, a German botanist named Carl Correns. "Verdammt!" (or something like it) Correns apparently thought to himself upon reading Mendel's old article, and then "verdammt nochmal!" when he read de Vries's note. Correns seems to have decided that if he wasn't going to get away with it, neither was de Vries. "The same thing happened to me which now seems to be happening to de Vries," he wrote. "*I thought that I had found something new.*" But then he realized the old monk had beaten him to it.

Both de Vries and Correns followed Mendel in assuming that in the hybrids the dominant and recessive reproductive cells had equal chances of participating in making offspring. The particular kind of egg or sperm cell that entered into each combination was therefore a random occurrence. Of course, none of them got perfect numbers, with exactly 75 percent dominant traits and 25 percent recessive, but they all considered that the fluctuations averaged out to a kind of mathematical perfection and that randomly chosen unknown factors in the egg and pollen cells determined the forms of the plants. This notion added a layer to Weismann's principle of the passivity of the evolving organism. Now the organism was subject not just to natural selection but to a kind of lottery taking place during conception. The randomness of the reproductive processes became a defining element of neo-Darwinist theory.

Mendel's idea of inheritance "factors" in the egg and pollen cells, factors that followed a simple mathematical rule of combination and determined the plants' characteristics, was of course extremely powerful. It inspired foundational research in what would become the field of genetics. But it doesn't really describe how inheritance actually works. First, most characteristics aren't binary either-or situations. A standard example of a trait that doesn't fit Mendel's model is human height. Humans aren't either tall or short; there's a range of possibilities. Second, most characteristics aren't determined by any single "factor" in the egg or pollen cell. Instead, a few genes, or even many genes, might have an effect on a given characteristic (I've just gone ahead and called Mendel's factors "genes"; more on that name change in a moment). Even the characteristics that Mendel studied turned out not to be determined by a single gene. Third, it's not just a question of the genes in the egg and pollen cells. DNA can't do anything on its own; it is an integral part of a host of developmental processes. These include many other causes and influences that contribute to creating an organism and influencing its characteristics.

So, Mendel's rule was not an accurate account even of how pea plants inherit traits, but it was an extremely effective tool to think with. And people began thinking furiously with it, once Mendel's old paper was back in circulation. They especially thought about how to use it to develop their ideas about particles of inheritance in a direction that carried them away from Darwin's gemmules and Lamarck's inherited effects of habit and behavior.

This was especially true of Bateson, whom we can now rejoin on his train journey. As the train carried him toward London, he read Mendel's idea, and the experience was immediately life changing, starting right then and there with Bateson rewriting the end of the lecture he was on his way to deliver at the Royal Botanic Society. From that moment on, he dedicated his professional life to establishing a new science inspired by Mendel. He even named his son Gregory after the gardening monk (Gregory Bateson, in adulthood a prominent anthropologist, was abidingly resentful of his name, which he saw as an attempt to predetermine his course in life). Bateson père was a curious person: An avowed atheist who read the Bible to his family every morning after breakfast, he often described Mendel's and his own roles in the emerging science of heredity in resonantly biblical language—and apparently only half jokingly thought of calling a book of his writings on Mendel "Scientific Calvinism," as a biological version of the Calvinist doctrine of predestination, which holds that people are predestined to be saved or damned.

In this regard, Bateson saw eye to eye with his good friend Francis Galton, the founder of eugenics and also a kind of scientific-Calvinist predestinarian. After Bateson stepped off the train with a new sense of direction and made his way to the Royal Botanic Society's lecture hall, he delivered a lecture that featured both Galton and—with its hastily composed new ending—Mendel. Bateson praised Galton for having been the first to enunciate a law of heredity. Galton had announced, after studying the coats of basset hounds over several generations, that the average "ancestral contribution" of the progenitors to the offspring halved with each successive generation, diminishing geometrically but never disappearing: The grandparents have half the contribution of the parents, the great-grandparents a quarter, the great-great-grandparents an eighth, and so on.

Bateson and others received Galton's law with enormous enthusiasm, seemingly undimmed by their equally great confusion. For one thing, Galton himself stated his law in two different ways that were mathematically inconsistent with each other (one in terms of the average contribution of each generation of ancestors to the offspring, the other in terms of the deviation of each generation from a mean). For another thing, what did Galton mean by "contribution"? How did one measure it from observations of the dogs? Still, Bateson celebrated Galton as a pioneer in the science of inheritance and Mendel as its Messiah.

As for himself, Bateson assumed the role of Saint Paul the Apostle. He wrote that the second coming of Mendel had been "a moment of rejoicing, and they who had heard the news hastened to spread them and take the instant way." Alas, not everyone was ready to receive the Truth: "Every gospel must be preached to all alike. It will be heard by the Scribes, by the Pharisees, by Demetrius the Silversmith." Demetrius, who wanted to keep selling his silver idols and worshipping Artemis, and so incited a riot against Paul, was in this case Raphael Weldon, an invertebrate zoologist at Oxford. Weldon had reservations about the teachings of Mendel as preached by Bateson. For instance, Weldon said Mendel's categories were ambiguous: "'Green' and 'yellow' are not quantitatively definite terms; each includes a considerable range of recognisably different colours." In general, Weldon argued that Mendel's paper didn't provide a reliable basis for generalizing about inheritance. He had his own, competing theory of inheritance based on biometrics—measurements of body parts, notably skulls—that was also closely connected with Galton's eugenics movement; Galton was Weldon's friend and collaborator. A bitter quarrel between Bateson and Weldon ended only when Weldon died suddenly in April 1906 at the age of forty-six from acute pneumonia.

With Weldon out of the way, three months later, at the Third International Conference on Plant Hybridization, Bateson proposed a new Mendelian science of heredity to be called "genetics," the term derived from Darwin's "pangenesis" and de Vries's "pangenes." The proposal was received with such enthusiasm that the conference immediately renamed itself; its published proceedings have the historically revisionist title of *Report of the Third International Conference 1906 on Genetics*.

The following year, Galton and a fellow eugenicist named Sybil Neville-Rolfe founded the Eugenics Society. Genetics and eugenics were born as conjoined twins. When Bateson delivered the annual Galton Lecture to the Eugenics Society in 1919—his lecture was titled "Common-Sense in Racial Problems"—he explained the connection. Anyone interested in genetics must necessarily be sympathetic to eugenics: The nature of the parents obviously determines the nature of their offspring, and by now surely no one "would venture to assert that men are born equal," although probably "few realise how unequal they are." Eugenicists were in possession of a truth that "a reluctant and unheeding world" was very slow, at its peril, to acknowledge: "The physiological fact of the diversity of mankind is of prime importance in every consideration of human affairs." To deny this was to act "in defiance of common-sense."

Among other things, Bateson recommended that the "feeble-minded" not be allowed to "interbreed," and he criticized promoters of mass education for regarding "mankind as a homogeneous plastic substance which can be modelled to taste." He said that "racial discrimination" was going on all the time (there at least we might agree with him), and he asked rhetorically, "Might we not gain by recognizing this fact and allowing it prominence in political philosophy?" He concluded on an ominous note. "Equality of political power has been bestowed on the lowest elements of our population," he warned, and he prophesied darkly that his and his audience's children and grandchildren would "learn something of the consequences of un-applied biology. . . . The truth has been recognised too late."

. . . .

Mendel's unknown "factors" acquired the name "genes" in 1909 from a Danish botanist, Wilhelm Johannsen. He coined "gene" to replace Darwin's "gemmules" and de Vries's "pangenes," both of which had predated Mendel, or at least Mendel's second coming. Johannsen didn't pretend to know just what a gene was, but he knew from Mendel that it must be something, and it corresponded nicely to Bateson's "genetics." The word "gene," Johannsen said, was neutral and would "prejudice nothing." It was simply "a very applicable little word, easily combined with others." But of course, "gene," like all terms, certainly did carry implications: Johannsen wrote that individual organisms could never influence their offspring by means of their "personal qualities" but only through their genes; that "Mendelism" was defeating "Lamarckism"; and that these facts were foundational to "the modern view of heredity."[2]

Genes remain foundational to the modern view of heredity, and yet they also remain as hypothetical as they were in 1909. When Francis Crick, Rosalind Franklin, James Watson, and Maurice Wilkins identified the DNA molecule in 1953 (more on that in a bit), it didn't help clarify what a gene was. On the contrary, their discovery rendered the question murkier yet. How do DNA molecules, made up of long chains of nucleotides, correspond to the idea of a "gene"? When we say "gene," do we mean a section of DNA that encodes the information for making a protein? Or do we mean a heritable factor that determines a specific characteristic? These are different things, yet various versions of each appear as standard definitions of "gene" in biology textbooks, often both within the same textbook, implying that they're the same. In fact, rather than converging, these two conceptions of "gene" have diverged over time. Research has shown that most traits correspond to more than one section of DNA, and even those that correspond to only a single section are not determined by it but depend on other developmental processes in the organism.

In short, the word "gene" continues to imply something fictional: that there are atoms of inheritance tightly correlated with phenotypic traits. The fact that we still use the word represents a mismatch between the dominant way of thinking about inheritance even among scientists and the much more complicated physiological situation. This presiding fiction has its roots in Saint William's Calvinist science of genetics. The eradication of Lamarckian inheritance and the idea that living beings were the passive objects of their "determinants" were founding principles of neo-Darwinism, eugenics, and genetics alike. And as a result of these conjoined developments, by 1920 the word "Lamarckism," according to the Harvard entomologist William Morton Wheeler, designated "the ninth mortal sin" in biology.

Harvard in fact became a center of neo-Darwinist biology, due in part to Ernst Mayr, one of the most renowned evolutionary biologists of the twentieth century. Mayr learned Weismannism at his mentor's knee during the 1920s, when he was a graduate student at the University of Berlin, working under the guidance of the Weismannian ornithologist Erwin Stresemann. "An animal does not act for itself," Stresemann observed, "but under a higher commission." So strongly did he feel about this that he put it in Latin: "Animal non agit, sed agitur" (an animal does not act but is acted upon).

Mayr imbibed these principles as an impressionable young student and carried them with him to America and throughout his career at Harvard as an ornithologist and zoologist. He praised Lamarck for his courage in breaking with a dogmatic, religious view of the direct divine creation of living beings and liked to startle people by telling them he'd been a "Lamarckian" as a young man. But he said Lamarck's theory wasn't "testable" and therefore wasn't really science but "philosophy." It was Weismann, according to Mayr, who had established biology on the right footing. In fact, he went even further than Weismann, representing animals as lacking agency not only with regard to evolution but even over their own actions. Late in his career, when he was in his eighties,

Mayr wrote that "a bird that starts its migration, an insect that selects its host plant, an animal that avoids a predator, a male that displays to a female—they all act purposefully because they have been programmed to do so." These actions were "no more nor less purposive than the actions of a computer." In fact, he said, "subhuman forms of life" never acted truly purposefully.

It seems an extraordinary point of view for a man who spent his professional life observing birds. Here was someone who devoted much of the twentieth century to watching birds as they built their nests, communicated with one another, raised their young, navigated great distances . . . and concluded that they were all just a bunch of little programmed machines? Moreover, this was a century that included extensive research on birds learning from experience, exhibiting spatial and temporal intelligence, constructing and using tools, communicating through song, and understanding abstract concepts such as "same" and "different."[3]

Even more striking is that Mayr himself argued influentially for the importance of animal behavior in evolution. New habits and behaviors, he wrote, were "very often as important in evolution as are new structures," and he urged biologists to take seriously "the importance of behavior in initiating new evolutionary events." Clearly, Mayr was deeply ambivalent on the question of animal agency. Despite his view of the importance of habits and behaviors, he was sure that Lamarck had gotten it all wrong, and Darwin too with regard to use and disuse, and that Weismann's elimination of their mistakes had been essential to the progress of evolutionary theory.

In his last book, which he wrote in 2004 at age one hundred, Mayr said he had been mistaken about nonhuman animals: Research had shown purposive behavior to be widespread among them. Even then, however, he added that "the goal is already coded in the program that directs these activities."

Truly, the dogma of the passivity of living organisms held a viselike grip on biology in the twentieth century, and even into the twenty-first. This central dogma of neo-Darwinism developed in conjunction with the major political struggles of the twentieth century. We will begin at the beginning, on the eve of World War I.

THIRTY-ONE

The Scandal of the
Midwife Toad

W hile Bateson and company were founding the new science of genetics, an Austrian Lamarckian biologist named Paul Kammerer claimed that he had succeeded in producing experimental evidence of the inheritance of acquired characteristics in various animals, including one instance in midwife toads. The name of these interesting creatures comes from the fact that when they mate, the female expels her eggs in necklace-like strands, which the male fertilizes and then twists around his legs to protect them from predators. He faithfully carries them around for about three weeks, keeping them safe and moist until they're ready to hatch. Usually midwife toads mate on land, but Kammerer obliged his toads to mate in the water by increasing the air temperature in their tank so they had to remain submerged to keep cool. He claimed that in these conditions the males developed rough black pads on their feet to give them more traction in clinging to the females, a "nuptial pad" structure that is found in some frogs and salamanders. He also claimed that these pads were inherited by the male offspring.

Kammerer was by all accounts a complicated person. Although he worked as a zoologist, he loved the arts and had studied musical composition at the Vienna Academy of Music. He had a flair for performance and self-promotion and met with success as a popular lecturer. A romantic, he passionately courted the widowed Alma Mahler, who recorded in her memoirs that after a "reluctantly granted kiss" he had threatened to shoot himself on Gustav Mahler's grave unless she agreed to marry him. In a professional context, too, Kammerer tended toward extravagance of expression; he was also irascible and combative. After some photographs in one of his articles received criticism, he blamed the journal's editor, a geneticist and botanist named Erwin Baur who became a vehement enemy. Kammerer seems to have been a bit sloppy with evidence of all kinds. His photographs were indistinct, his specimens and cross sections unsystematically kept, his drawings unskilled.

The midwife toad experiments took place at the Institute for Experimental Biology, an independent research institute nicknamed the Vivarium because it was housed in what had been an exotic animal exhibit in the Prater, the big public park in Vienna. The founder and director of the Vivarium was a man named Hans Przibram who suffered from the professional handicap of being Jewish in the antisemitic world of early twentieth-century Austria. He solved the problem by buying the Vivarium, turning it into a zoological laboratory, and hiring other researchers who were also mostly Jewish, including Kammerer on his mother's side. From the beginning, the Vivarium had a reputation as a "Semitic" place. One zoologist named Ludwig Plate issued dark warnings about the "unscrupulousness of the Jews": "All these reports by Kammerer the Jew must be taken with extreme skepticism." And Weismann wrote in his notebook, "Kammerer (Vienna) is a little, miserable, sticky Jew, who has proven himself on earlier occasions to be a quite unreliable worker."

The Vivarium was well equipped in certain ways but not in terms of photographic equipment. Most of Kammerer's photographs were made in studios whose usual clientele tended not to be toads. The portraits turned out blurry and inadequate as scientific illustrations. They were also generally retouched, but that was a usual practice in scientific publications. All of these factors—Kammerer's style, method, institutional affiliation, equipment, and Jewishness in an antisemitic world—added up to create a dubious reputation. At the same time, he was a compelling and successful speaker with a powerful message, and he had the support of colleagues who knew his work and the field intimately, such as Przibram.

Defending the inheritance of acquired characteristics against the "geneticians" was Kammerer's great mission, in pursuit of which he did other things in addition to meddling with the love lives of toads. For instance, he conducted experiments on salamanders, creating what he said were "graft hybrids." The graft hybrid was an idea that had been introduced by Darwin. He had suggested that if it were possible to create a hybrid plant by grafting two individual plants of different species, and if this offspring presented an intermediate form, blending the features of the parents, in the same way as a sexually produced hybrid, it should be called a "graft-hybrid." There were various stories and reports of graft hybrids but nothing conclusive. Darwin was interested in the possibility in relation to his theory of pangenesis. He reasoned that since the parts used in the graft didn't include the reproductive organs, a graft hybrid would confirm that the elements responsible for shaping the next generation were spread throughout the tissues of each plant, just as his theory proposed. These dispersed reproductive elements might then carry the effects of changes in the parents' bodies into their offspring.

Graft hybrids were in fact the subject of Darwin's final research project, the one he'd been working on with his young friend Romanes during the last months of his life. Romanes had been grafting root vegetables such as potatoes, carrots, and beets to try to produce hybrids, without much luck. Then, on New Year's Day 1882, Darwin wrote to Romanes of news that had arrived the previous evening at Down House, Darwin's home in the Kentish village of Downe. The French botanist Auguste François Marie Glaziou, director of parks and gardens in Rio de Janeiro, reported that he had created intermediate new varieties of sugarcane using grafts. Romanes offered to write a paper presenting the case of sugarcane in support of pangenesis, and Darwin provided a detailed outline. Their paper, "On New Varieties of the Sugar-Cane Produced by Planting in Apposition," was read at the Linnaean Society a fortnight after Darwin's death.[1]

Kammerer liked Darwin's idea of graft hybrids and saw no reason why it should be limited to plants. Why not salamanders? He used some that were naturally black with yellow spots, but he induced a number of them to change their markings and become striped by keeping them on a yellow background. Then he transplanted the ovaries from a regular spotted salamander into an "artificially" transformed striped salamander and bred her with a spotted male. The result, he reported, was an intermediate form of salamander with spots arranged in stripes. Since the sperm and the egg both came from spotted salamanders, Kammerer interpreted this as evidence that the striped surrogate mother's bodily change had exerted an influence on the offspring.

The amorous midwife toads and stripe-spotted salamanders generated a swirl of controversy. Claiming to have demonstrated the inheritance of acquired characteristics was like waving a red flag at the neo-Darwinists, geneticists, and eugenicists. Add to that the ambient antisemitism. And then there was Kammerer's flamboyant style and flaky evidence. His leading critic was none other than Bateson, the scientific Calvinist who believed that the fates of all living beings were predestined by being in-scribed in their genes. In September 1910, after writing to request more information and evidence, to no avail, Bateson showed up at the Vivar-ium asking to see Kammerer's specimens. But Kammerer apparently de-murred; he had only live toads at the time, and they showed the nuptial pads only in spring.

Then came the war to end all wars. The Vivarium fell on hard times, operating with a much-reduced staff under the auspices of the Academy of Sciences. Kammerer was exempted from active military duty because of a heart condition but was assigned to work for the military censorship department censoring letters from prisoners of war (who weren't al-lowed, for instance, to use the word "hunger"). His experimental ani-mals all died, and almost all of his evidence disappeared, leaving just a few specimens, and only one midwife toad showing a single remaining nuptial pad.

After the war ended, Kammerer did no new research, but he wrote up some old results, including experiments on the midwife toad. This caught Bateson's eye, and he wrote a letter to *Nature* warning ominously that Kammerer's results were beginning to pass into textbooks while his claims remained unproven. Even if Kammerer could show a midwife toad with "incontrovertible" nuptial pads, Bateson insisted, it would still be a matter of interpretation to know where they came from. Kammerer had surely made "mistakes of observation or of interpretation." The belief in inher-itance of acquired characteristics, Bateson observed, was "still almost universal among the laity" despite Weismann's definitive debunking; he thereby placed the idea among folk beliefs and old wives' tales.

In 1923, Kammerer traveled to England. His two main ambitions, he told his young friend and host Ivor Montagu, were "to eat a kipper and to meet Bernard Shaw." The second is perhaps self-explanatory, but also, Shaw had recently written a kind of humorous-serious tirade on the subject of neo-Darwinism and neo-Lamarckism, in which he declared himself, not entirely in jest, a neo-Lamarckian. The work included a "Lamarcko-Shavian Invective": "When I think of these poor little dull-ards, with their precarious hold of just that corner of evolution that a blackbeetle can understand—with their retinue of twopenny-halfpenny Torquemadas wallowing in the infamies of the vivisector's laboratory, and solemnly offering us as epoch-making discoveries their demonstra-tions that dogs get weaker and die if you give them no food; that intense pain makes mice sweat; and that if you cut off a dog's leg the three-legged dog will have a four-legged puppy, I ask myself what spell has fallen on intelligent and humane men that they allow themselves to be imposed on by this rabble of dolts, blackguards, impostors, quacks, liars, and, worst of all, credulous conscientious fools."

You can probably see how this was up Kammerer's alley. As for the kipper, he had promised an elderly English lady trapped in Vienna dur-ing the war that he would eat a kipper for her if he ever got to England. Montagu, who knew Shaw, obligingly arranged both experiences. Dur-ing the trip, Kammerer also went to Cambridge and tried to make peace with Bateson by showing him his one remaining midwife toad specimen, but Bateson was unpersuaded.

It didn't help that Kammerer was overtly ideological in his advocacy of the inheritance of acquired characteristics. He described himself as a socialist, though with reservations about political labels. Of course, his opponents were every bit as ideological as he was—Bateson regarded Mendelian genetics and eugenics as the foundations of a new social order—but they adopted a demeanor of scientific neutrality, whereas Kammerer did the exact opposite. In a manifesto titled *The Inheritance of Acquired Characteristics* published the year after his visit to England, he accused his opponents of "rigid dogmatism" and attributed their attitude to the current political situation. He'd been over at Einstein's house for dinner, he said, and they'd discussed it. (This is very possible: Einstein sympathized with Kammerer as an "internationalist and pacifist" and apparently liked him and his work, even once privately recommending him for a job in Switzerland.) Kammerer said the "Great War" had produced "an intensification of nationalistic and racial consciousness," and that this moment of bigotry was "not at all propitious" for the theory of inheritance of acquired characteristics, which taught that environment and education eliminated racial boundaries and national differences.

Above all, Kammerer assailed "that curious offspring of science, the race theory." Its advocates claimed to believe in the theory of evolution! Yet they maintained that there were "irreconcilable differences between races, nations, and classes," and that these were inherited and rigidly unchangeable. "Abusing the name of Darwin, the race theorists" perverted his theory into a "tool for creating national and class hatreds." In fact, "pure races" were "extremely rare," said Kammerer, especially on "Europe's embattled soil where repeated intermixtures of the various people have made the distinguishing marks of race so confusing that they are almost illegible." He especially detested the "miserable debacle of craniology," an attempt by the "race theorists" to "fix permanent racial characteristics by skull measurement." Their arguments were entirely spurious, he judged, and "race differentiations, based on mental and moral differences, stand the acid test of science even less, on account of being still more vacillatory than purely physical differentiations."

Kammerer hated what was being done to and in the name of Darwinian evolutionary theory. The "modern and fashionable mania of selectionism" was a perversion. "Genuine Darwinism" was more than just natural selection. "The unadulterated Darwinism," Kammerer proclaimed, "teaches of the transmutations of the species, the metamorphosis and interminglings of races and classes," and not only the power of the struggle for existence but also the "even higher power of help to existence." This was why real Darwinism, in Kammerer's view, shared a deep connection with socialism: "a doctrine of natural, world-embracing humaneness."

In place of selectionist eugenics, Kammerer proposed a "productive eugenics." Selective practices are purely negative, he explained, and merely change the composition of the population, not its elements themselves. He used height as an example. If you prevent shorter people from breeding, you might create a race of taller people, but they still won't be any taller than the tallest people were before. "Nothing essentially new would be created." To create something new, one needs positive measures such as education, nourishment, and exercise. For instance, Kammerer recommended making sure that babies never began their lives in prison by postponing punishment for pregnant women. "There exists hardly a problem of general and human biology," he lamented, "whose solution is not hampered by the dogma that environment and inheritance have nothing to do with each other." He called for a new ethics of mutual help and social welfare, devoted to "race hygiene" but "never in that antiquated form of a *single* race striving to forge aggressively ahead at the expense of other races."

. . . .

Not long after the publication of Kammerer's book, the controversy that had been simmering for almost two decades finally boiled over. Early in 1926, the curator of reptiles at the American Museum of

Natural History, a Weismannian herpetologist named G. Kingsley Noble, traveled to Vienna and demanded to see the toad specimen. Przibram arranged for him to view it, and Noble discovered that India ink had been injected at the spot of the alleged nuptial pad. Przibram agreed that the specimen was altered, but he maintained that Kammerer hadn't done it and that the nuptial pads had been real. He offered other evidence such as microtome cross sections, photographs, and eyewitness testimonials from Cambridge, from Kammerer's 1923 visit bringing his last specimen. Noble would have none of it; the pads, he said, were nonexistent. In a note in *Nature* the following August, he heavily implied that they had been fraudulently fabricated.

Przibram wrote an accompanying note acknowledging that the specimen contained ink, and therefore did not constitute valid evidence, but maintaining that the nuptial pads had earlier been present both in this and in other specimens. The only thing he could imagine, he said, was that someone had tried to preserve and enhance the visibility of the pads, which tended to fade from exposure to light, by injecting them with ink. "Kammerer himself was greatly astonished at the result of the chemical tests," wrote Przibram, "and it ought to be stated that he had been asked and had given his consent to the chemical investigations." Kammerer told Przibram he would have been inclined to think that someone had injected the ink in order to frame him, but he recalled seeing the same black color in the living toad. "Do you think I'm a *Dummkopf*?" he demanded when his friend Hugo Iltis, a plant geneticist, asked him about it. "Because that's what I would have to be if I left a forgery with ink standing around openly in the laboratory, where so many of my enemies have entry." Iltis found the argument persuasive and was among those who unwaveringly maintained Kammerer's innocence.

As this drama was unfolding, Kammerer made a trip to Moscow, where the Communist Academy offered him a position directing a new laboratory. He accepted the job and began organizing the shipping of equipment, books, and supplies. Then, on September 22, he took a train up into the Alps, to Puchberg am Schneeberg, a small town nestled into the mountain, and spent the night at a country inn called the Rose. The next morning, he walked up a footpath into the Schneeberg Forest, sat down leaning against the hillside near a rocky ledge known as Theresienfelsen (Theresa Rock), and shot himself in the head.

He had sent a letter the day before to the Communist Academy in Moscow saying that he had found Noble's claim about the ink to be accurate. Although he had "no part in these falsifications," his life's work was destroyed and his situation unendurable. "I hope that I shall gather together enough courage and strength to put an end of my wrecked life to-morrow," he wrote. To the Communist Academy he bequeathed his library, which was already en route, to "compensate it for all the efforts it has wasted upon me." Przibram, who continued to maintain Kammerer's innocence and the genuineness of his results, died in the Theresienstadt concentration camp in 1944.

. . . .

What can we make of this tragic tale? Regarding the ultimately unknowable truth of the matter, people have made various suggestions. One possibility, of course, is that Kammerer committed a brazen fraud. Another is that he did achieve toads with inherited nuptial-pad-like structures. Some biologists have argued that he might well have done this, though it wouldn't necessarily have confirmed the inheritance of acquired characteristics. Nuptial pads existed as a genetic atavism in midwife toads, an ancestral trait for which the genetic potential remained, and it would have been possible by selective breeding to re-"fix" it in the population.

A more recent proposal is that Kammerer might have been an unwitting pioneer in the discovery of epigenetic inheritance: the inheritance of structures outside the genome that influence gene expression. A Chilean researcher, Alexander O. Vargas, notes that Kammerer described changes in other traits besides the nuptial pads, such as a lengthening of the tadpoles' gills. Kammerer reported that when he crossed regular toads with ones he'd altered by keeping them in the water, either the "normal" traits or the "changed" traits were dominant, but it depended upon which parent was which. The dominant trait seemed to "follow" the father, so if the father had normal traits, those were dominant, and if he had changed traits, these became dominant instead. Vargas explains that this phenomenon is now well known in epigenetics research as the "parent-of-origin effect." An allele (a DNA sequence inherited from one parent or the other) can be influenced by the cellular environment of the male or female germ line. "It seems," he concludes, "that Kammerer had the misfortune of stumbling upon non-Mendelian inheritance at a time in which Mendelian genetics itself was just becoming well accepted."

Even if Kammerer did, one way or another, achieve toads with inherited nuptial-pad-like structures, there's still the question of how the India ink came into the situation. One possibility is that one of his many ill-wishers framed him by injecting ink into the last remaining specimen before Noble's visit. Przibram ultimately came to this opinion and even frequently lamented to his brother Karl that he thought he knew who had done it but lacked conclusive evidence. A person upon whom suspicion has fallen is a paleontologist named Othenio Abel, one of Kammerer's worst enemies, the organizer of a secret antisemitic club and participant in antisemitic riots at the University of Vienna, who shadowed Noble throughout his visit. Another possibility is that Kammerer himself, or someone working with him, injected the ink, but not with fraudulent intent. He routinely touched up photographs, and he or some associate might also have tried to render the structure on the toad more visible in preparation for a photography session. This would have taken place more than a decade before Noble's visit, and before the war, when Kammerer was still doing active research, perhaps while the toad was still alive.

Here, anyway, are four things I think we can be sure of: First, neo-Darwinists exploited the Midwife Toad Affair to solidify the reputation of Lamarckism as backward and fraudulent; second, Kammerer's antagonists were every bit as ideologically motivated as he was, although they assumed a stance of scientific neutrality; third, every theory of the nature of living beings and their development is necessarily invested with interpretive assumptions, intuitions, feelings, and values; and fourth, it's better—both more honest and scientifically better—not to speak ex cathedra in the name of Modern Science but to openly acknowledge the assumptions, feelings, and values one brings to scientific research. Whatever you think about Kammerer and his toads, he undeniably did that.

·····························

THIRTY-TWO

The Biologist Who Just Couldn't Seem to Come In from the Cold

L amarck's theory of life, born amid political turmoil, never could seem to outrun it; the turmoil always kept pace. He endured the successive revolutionary regimes, Napoleon, and two Bourbon Restorations during his lifetime, then became a lightning rod for reactionary politics in the later nineteenth century and a symbol of Jewish socialist thought in the early twentieth. He even, more than a century after his death, got caught up in the Cold War.

On July 31, 1948, a Soviet agronomist named Trofim Lysenko stood before the Lenin Academy of Agricultural Sciences in Moscow and announced that the "Lamarckian" principle of the heredity of acquired characteristics was "quite true and scientific," while Weismann was a reactionary, a mystic, and an enemy of science. Generations of historians have taken Lysenko at his word and described him as a "Lamarckian." But in fact, Lysenko's version of "Lamarckism" was not really Lamarckian. Like Weismann and Brown-Séquard, Lysenko conflated changes imposed brusquely from outside with changes enacted gradually from within. He did not describe organisms as transforming themselves heritably by means of their own actions, habits, or responsive behaviors. Instead, he advocated a theory more like the one Weismann refuted with his long-suffering mice: that transformations enacted upon organisms by outside forces would be inherited. Lysenko proclaimed: "We must not wait for favours from Nature; our task is to wrest them from her."

Specifically, Lysenko claimed that he could change the germination season of wheat from winter to spring by subjecting the seeds to cold and moisture (he could, but this was an established practice among farmers) and that the change was inherited in the next generation (it was not). Lysenko did not claim that the wheat was gradually adapting itself to a new environment, transforming its organization from within; rather, the cold forced it into germination. He also promoted a related anathema of bourgeois capitalist science: graft hybridization. This idea had become taboo under the reign of neo-Darwinism, because of its association with Darwin's idea of pangenesis and the supposition that the particles of heredity traveled throughout the body and carried bodily changes to the offspring. Lysenko used grafts to produce new varieties of plants, apparently with some success. (The phenomenon of graft hybridization, like the idea of inheritance of acquired characteristics, has recently been making a comeback and is now generally understood as an instance of "horizontal gene transfer," the movement of DNA from one organism to another outside the reproductive process.)

Following Lysenko's presentation to the Lenin Academy of Agricultural Sciences, Stalin endorsed his theory, and it became Soviet orthodoxy. Over the previous decade and a half, there had been a terrible famine, with millions of deaths, resulting from a combination of war, drought, and disastrous government policies. Lysenko, who came from a peasant family, made for good propaganda about recovery through the

practical science of the peasantry. After Stalin's endorsement of Lysenkoism, there followed a Soviet repression of genetics research as "bourgeois
science" that was reminiscent of the Nazi repression of relativity theory
and quantum mechanics as "Jewish science." At least one scientist who
resisted Lysenkoism, Nikolai Vavilov, was arrested and died in prison.

Despite Lysenko's tenuous connection with Lamarck's actual theory,
he provided too good an opportunity to pass up for those bent on eradi-
cating Lamarck's ideas from evolutionary biology. One such person was
Julian Huxley, a British biologist. A few years earlier, in 1942, in the
midst of World War II, Huxley had published a kind of constitution for
evolutionary biology, containing a declaration of what he catchily named
the "Modern Synthesis": a marriage of evolutionary theory with genetics
that became the central paradigm of evolutionary biology. For Huxley's
modern synthesis movement, Lysenko was the perfect villain.

. . . .

We'll come back to Lysenko and his excellence as an antihero, but
first let's spend a moment with Huxley and his modern synthesis movement. Julian Huxley's grandfather Thomas Henry Huxley had
been a good friend of Charles Darwin's. The elder Huxley had been a
naturalist and polemical Darwinist who referred to himself as "Darwin's
bulldog." Upon first reading the *Origin of Species*, he later said that he
had exclaimed to himself, "How extremely stupid of me not to have
thought of that!" and had written to Darwin professing himself ready to
defend his doctrine and even "prepared to go to the Stake" for it. His enthusiasm apparently overflowed his metaphors as he also said, "I am
sharpening up my claws & beak in readiness."

Darwin and T. H. Huxley didn't agree on everything, though. Darwin had exactly the opposite reaction from Huxley's when he read an essay by Huxley titled "On the Hypothesis That Animals Are Automata," in which Huxley argued that all organic beings, including people, were essentially automatic machines, their consciousness a mere "collateral product" of the mechanism. "If I were as well armed as Huxley," Darwin remarked wryly to a mutual friend, "I would challenge him to a duel on this subject." To Darwin, as to Lamarck, living beings were not passive machines but full of creative agency.

Julian Huxley seems to have inherited his grandfather's automaton-like model of organisms. Describing the "machinery of heredity," the younger Huxley wrote that many people disliked the idea of a "hereditary force pushing men blindly along predestined roads" and so they reacted by "belittling and discounting" the power of biological inheritance. But science, he said, didn't give a fig about people's "likes or dislikes." Sure, Lamarckism might seem a nicer idea—you might like to hand on the results of your life's work to the next generation—but "facts take no account of what we human beings think or desire."

Also from his grandfather, Huxley inherited the mantle of protector and interpreter of the Darwinian faith. From the start, he said, Darwinism had suffered from uncertainties about what caused organisms to change, producing new variations. Even Darwin himself had been "inclined to allow some weight to Lamarckian principles." This confusion had led to a period of "eclipse" and a "reaction against Darwinism," according to Huxley, at the end of the nineteenth century and beginning of the twentieth. But happily, the theory had been rescued by Weismann, Mendel, and Bateson. Weismann had brought about "a great clarification of the position" by drawing a sharp distinction between body cells and germ cells, precluding the Lamarckian inheritance of acquired characteristics. Then Mendel, as interpreted and generalized by Bateson, had provided a "particulate" mechanism of heredity, which had allowed evolutionary biology to unify around a methodology of combinatorial calculations. The "rebirth of Darwinism" on this new, mathematical foundation enabled biology to rival physics in its methodological unity.

Huxley's "reborn Darwinism, this mutated phoenix risen from the ashes," was founded upon the eradication of Lamarckism and of the idea that organisms played any role in influencing the course of evolution. Modern biology, with its basis in genetics, Huxley said, "repudiates Lamarckism." In the modern age, theories such as Lamarck's could appeal only to "literary men," such as Shaw and the French philosopher Henri Bergson, who had written approvingly about Lamarck's theory in his book *Creative Evolution* (1907). Huxley scornfully dismissed such views as based not upon "scientific fact and method, but upon wish-fulfilment." His verdict: Bergson was "a bad scientist" but "a good poet." In this way, Huxley declared a united front around the eradication of Lamarckism (wishful thinking, poetry) from biology.

Lamarck played a crucial role in the modern synthesis, the role of public enemy No. 1. This was all the more important since the founders of neo-Darwinism actually disagreed on some fundamental matters. Bateson never quite accepted the "chromosome theory," the idea that the hereditary matter of the genes was contained within the chromosomes, which Weismann and de Vries both believed. Meanwhile, Weismann assumed, with Darwin, that variation was gradual and that evolution was a slow accumulation of tiny changes, while de Vries and Bateson introduced the idea of big, sudden jumps: discontinuous variations, or "mutations," an idea that Thomas Hunt Morgan and Huxley adopted. But whatever their position on chromosomes or the pace of evolutionary change, the members of the modern synthesis could all agree on what became their founding principle: *There can be no inheritance of acquired characteristics.*

In the popularization of evolutionary biology, too, this principle became the unifying message beginning with Huxley and remaining so throughout the twentieth century. Richard Dawkins, the bestselling British biologist and writer, and Daniel Dennett, a philosopher of biology, were leading popularizers of the doctrine. Dawkins wrote that a revival of Lamarck's ideas would "devastate" his worldview, according to which living beings, including people, were the passive "vehicles" of their genes. Dennett called Lamarck a "loser" and Lamarckism a "heresy" whose acceptance would be "fatal" to Darwinian evolution.

The slogan "no inheritance of acquired characteristics" not only brought the members of the modern synthesis together but allowed them to add new elements to their doctrine. For instance, with the addition of mutations as a crucial element of the theory came a heightened importance for randomness: Not only were genes assigned to offspring randomly in sexual reproduction, but mutations, according to those who believed in them, also occurred entirely at random. Some people, such as Mayr, rejected the idea of sudden, discontinuous mutations, but still retained the principle that genetic variation was purely random. This randomness reinforced the passivity of organisms as the objects of biological fortune.

The passivity of the organism provided the rallying point for the modern synthesis. Organisms could never vary as a result of their own actions but were entirely subject to outside forces. Apparently failing to see a contradiction, the modern synthesis founders rejoiced in the idea that they themselves could *become* these outside forces. "It is as if man had been suddenly appointed managing director of the biggest business of all," Huxley wrote, "the business of evolution." Or, as he put it in a little book titled *If I Were Dictator*, "man can become the conscious trustee of Evolution." He even coined a new term, "transhumanism," to describe the human control of human evolution: "The human species can, if it wishes, transcend itself. . . . We need a name for this new belief. Perhaps *transhumanism* will serve." Humanity, he said, was "on the threshold of a new kind of existence."

Another word for becoming the managing director of evolution was "eugenics." Like Bateson, Huxley celebrated the intimate relations of neo-Darwinism and genetics with eugenics. As for how to evaluate eugenic fitness, he considered salary a "rough and ready" measure—the higher the salary, the more eugenically fit the person—and he therefore worried that the lower-paid laboring class reproduced at a greater rate than the higher-paid professional class. (You might think it would have occurred to Huxley that in strictly Darwinian terms the laborers' higher reproductive rate meant *they* were the more fit, since for Darwin "fitness" simply meant reproductive success. But apparently not.)

The same problem preoccupied Ronald Fisher, an English mathematician and geneticist whose work was foundational to the modern synthesis. Fisher affirmed that inheritance was strictly "particulate," that mutations were effectively random, having no shaping forces or causes, and that biologists should face the fact that natural selection was the only mechanism directing evolution. Yet natural selection didn't seem to be doing a great job of it. Like Huxley, Fisher assumed that richer people were more eugenically fit. But poor people plainly had more children, a conundrum he called the "inverted birth-rate."

Fisher was a devout Anglican, but he doesn't seem to have trusted his God to cope with the situation either. He meant to do something about it himself. In addition to being head of the Department of Eugenics and Galton Professor of Eugenics at University College London, Fisher edited the journal *Annals of Eugenics* and was the inaugural chairman of the University of Cambridge Eugenics Society. As a solution to the "inverted birth-rate" problem, he proposed that the state make family allowances proportional to salaries—the higher the salary, the higher the allowance— to encourage the rich to breed more and the poor to breed less.

Huxley, for his part, was a member of the British Eugenics Society, and later its president from 1959 to 1962, and sat on the society's Committee for Legalizing Eugenic Sterilization. He was also the inaugural director general of UNESCO, the United Nations Educational, Scientific, and Cultural Organization, where his project was to harness the modern synthesis to UNESCO's mission of human unity. Before World War II, he had written a vehement debunking of the idea of "race" as a biological category. Human races were not biologically distinct, he wrote, and "race" was therefore "a pseudo-scientific rather than a scientific term," and a myth that dangerously fueled prejudice. When he assumed his position at UNESCO, the experiences of World War II had cataclysmically confirmed his warning.

In June 1951, after lengthy negotiations, UNESCO issued a "Statement on the Nature of Race and Race Differences," asserting that "Scientists are generally agreed that all men living today belong to a single species, Homo sapiens, and are derived from a common stock." The statement said that races were always mixing and changing, that the differences within races were at least as big as the differences between them, and that there was no scientific evidence to support the idea that races differed "in their innate capacity for intellectual and emotional development." Twelve geneticists and anthropologists consulted in the preparation of the statement, and then Huxley composed its final wording together with a fellow founder of the modern synthesis, the Russian-born American geneticist Theodosius Dobzhansky. (Fisher, however, opposed it, saying he thought human groups had important, genetically caused differences in their innate intellectual and emotional capacities.)[1]

But although Huxley believed in human unity, he staunchly opposed what he called the "myth" of human equality. "Human beings are not born equal in gifts or potentialities," he asserted, "and human progress stems largely from the very fact of their inequality. 'Free but unequal' should be our motto."[2] In order to bring about an international betterment of human beings, Huxley urged that UNESCO work toward "a truly scientific eugenics" founded on the recognition of human inequality. "Biological inequality," he declared, "is, of course, the bedrock fact on which all of eugenics is predicated."

Dobzhansky agreed. He regretted that the Nazis had perverted the idea of eugenics, whereas "the eugenical idea has a sound core." Human welfare relied upon "the health of the genetic endowment of human populations." To further human genetic health, Dobzhansky served on the board of directors of the American Eugenics Society. Mayr too warned that it "does not require complex mathematics to figure out what would happen if the improvident moron or low I.Q. regularly had a dozen children and the prudent, superior citizen only two. . . . One can not afford to ignore this Achilles heel of natural selection." And in an essay titled "The Biology of Race and the Concept of Equality," he urged that it was important to understand the "underlying biological" factors of inequality. As the founder of the Society for the Study of Evolution in 1946 and its journal, *Evolution*, the following year, Mayr was a dominant figure in the field for many decades.

His younger colleague at Harvard E. O. Wilson recited the same catechism: Lamarckism is all wrong, no inheritance of acquired characteristics, science must study the genetic basis of human inequality. In 1975, he announced a new science with a catchy name, sociobiology, devoted to "monitor[ing] the genetic basis of social behavior." Wilson proposed that "ethics . . . be removed temporarily from the hands of the philosophers and biologicized." Human populations, he thought, might "diverge genetically . . . in ethical behavior," in which case ethics would require an "evolutionary approach" and "no single set of moral standards . . . [could] be applied to all human populations, let alone all sex-age classes." The book set off a storm of controversy. Looking back on it in 2004, Wilson attributed the response to the radical politics of the mid-1970s. He wrote with evident bitterness that his ideas had been perceived as a "threat to Marxist ideology" and that in "academia's now necktie-free zone, race was a radioactive issue, deadly to any who touched it without extreme caution. Talk of the inheritance of IQ and human behavior were punishable offenses. Anyone who dared mention these subjects in any manner other than formulaic condemnation was at risk of being called a racist." Necktie or no necktie, it's hard to imagine what else you would call someone who promotes the idea of genetically based "human racial variation in behavior" and intelligence as Wilson did; it could almost be a dictionary definition.[3]

Of course, outspoken advocates of eugenics were a dime a dozen throughout much of the twentieth century, and inevitably some of them invoked Lamarck. An example is Ernest William MacBride, a British marine biologist who defended Lamarck's theory and cited both Lamarck and Darwin in support of his views. MacBride wrote that the three human races populating the British Isles had each transformed in response to their original environments: The Nordic race was courageous, the Alpine race industrious, and the Mediterranean race mercurial and vengeful, with a downward tendency into the slums. Since evolutionary transformation took place very slowly, he reasoned, these characteristics were as good as fixed and couldn't be remedied by education. Therefore, social reformers should just allow natural selection to weed out the dissolute Mediterraneans or at least ensure that they didn't breed.

In France, meanwhile, eugenicists invoking Lamarck called for public health measures such as improvements to diet and living conditions by arguing that these measures would have heritable effects. In 1931, the French National Social Hygiene Office, established under the influence of the French Eugenics Society, conducted studies of cancer, tuberculosis, alcoholism, venereal disease, mental health, and housing for the poor. They also campaigned for a law requiring a premarital health exam but were unsuccessful, although the Vichy government, which was established in collaboration with Nazi Germany after France's surrender, did institute such a law. French Lamarckian eugenics was distinct, however, in its emphasis on improving the health of the current population in order to enhance the quality of future generations.

So not all eugenicists were members of the modern synthesis. But the converse is roughly true: The founding members of the modern synthesis were mostly eugenicists. Human inequality, along with no inheritance of acquired characteristics, was one of their main themes. *The Inequality of Man* is the title of a 1932 book of essays by J. B. S. Haldane, another founder of the movement, a British geneticist, physiologist, and science popularizer. Haldane elsewhere expressed skepticism about some eugenic programs, saying these used an anti-Darwinian notion of fitness and could be vehicles of racial and class prejudice. He also questioned the use of salary as a measure of fitness and pointed out that it was contradictory to believe in both the fixity of innate racial characteristics and the threat of race degeneration or possibility of race purification.[4] But in *The Inequality of Man*, Haldane endorsed Bateson's "Scientific Calvinism"—a person's destiny is written in their genes—and derived a lesson regarding inequality, or, as he called it, "diversity": "If innate human diversity is an ineradicable fact, the ideal society is one in which as many types as possible can develop in accordance with their possibilities." It sounds perhaps innocuous, when phrased that way, but consider what it meant in practice.

As an example of how to socially accommodate "human diversity," Haldane considered "the problem of the American Negro." Many Americans, he wrote, "hold that the negro is definitely inferior to the white man, and should, as far as possible, be segregated from him. Others believe that he should enjoy the same rights. The biologist cannot decide between them. He can point out that the negro's skull is more ape-like than the white's, but his hairless skin less so, and so forth." What the scientist could do, however, was to reveal that in the countryside, "the birth-rate of the negro population exceeds the death-rate," while it was the opposite in towns. "So if you keep the negro out of cars, factories, and so forth, or frighten him away from contact with whites by an occasional lynching, you drive him back to the cotton fields where he lives healthily and breeds rapidly, thus creating a negro problem for future generations." On the other hand, if "tomorrow the coloured population of the Southern States, but not the white, were given free access to cheap whisky and methods of birth control, the number of negroes would probably begin to fall off!"

. . . .

T he core principles of the modern synthesis have been genetic determinism, the passivity of living organisms—no "inheritance of acquired characteristics"—and human inequality. The idea of innate inequalities among groups of humans—races, social classes, sexes—has been commonplace among the major figures in genetics and evolutionary biology from the late nineteenth century into the early twenty-first. Even more commonplace has been the idea that there is "significant genetic variation in human mental abilities," as Dawkins expresses it. He writes that there may or may not be such variation, but it "cannot be denied" that humans must necessarily have been genetically unequal in mental abilities in the past, since he assumes we've gotten smarter over time simply by natural selection favoring the individuals in the population whose genes rendered them smartest. This reasoning, he says, shows "the inadvisability of dogmatic and hysterical opposition to the very possibility of genetic variation in human mental abilities."

Agreed, being dogmatic and hysterical is always a bad idea. But here's a calm and thoughtful objection. The modern synthesis has foundered in each of its basic tenets. There have not turned out to be "particulate" genes or, for the most part, straightforward correspondences between individual segments of DNA and bodily traits, let alone cognitive or behavioral ones. Research over the past several decades has demonstrated ever more ways in which living beings have evolutionary agency, playing various roles in shaping the course of evolution. Really when you think about it, how could they not? (More on this in the epilogue.) In light of these points, the claim that there might be "significant genetic variation in human mental abilities" can't even be right or wrong. The assumptions on which it's predicated are false.

. . . .

B iological inequality as a core principle of the modern synthesis brings us back at last to Lysenko. The importance of inequality was at the heart of the devastating connection that Huxley, using Lysenko, established between Lamarck, totalitarian Communism, and the death of science. In *Soviet Genetics and World Science: Lysenko and the Meaning of Heredity* (1949), Huxley wrote that Lamarckism was congenial to Stalin and the Soviet dogmatic egalitarians because they refused to accept the ugly but unavoidable implication of a strictly genetic theory of inheritance: that people are genetically unequal as a result of the random shuffling of genes. Rather than face the scientific fact of human inequality, the Communists clung to "Lamarckian" ideas that were "superstitions instead of legitimate scientific hypotheses."[5]

Although Lamarck never explicitly addressed the question of human equality, his materialism and overt irreligion, which had led people to connect his theory with the radical egalitarian extreme of the French Revolution during the early nineteenth century, now helped them to associate it with Communism in the mid-twentieth. Maybe this wasn't actually such a stretch, though. You might see a kind of equality nested inside the evolutionary agency that Lamarck attributed to living beings: not human equality, but an equal claim on the part of all organisms to act and not just be acted upon. Also, the Lamarckian view that living beings aren't reducible to physical parts but are defined by their creative activity in the world, and his corollary for humans, the important influence of the "moral" on the "physical"—these ideas undermine the notion of any fixed, unitary measure of fitness or worth. So perhaps it's understandable that the members of the modern synthesis associated Lamarck's theory with a belief in equality and despised it for that.

The Soviet endorsement of inheritance of acquired characteristics was ideologically driven, and so was the anti-Soviet repudiation of that idea. It became difficult, though not impossible, to conduct genetics research in the Soviet Union during the 1950s; at the same time, it became correspondingly difficult in Europe and the United States to pursue research on factors other than genes that might play a part in inheritance and in transmitting acquired characteristics from one generation to the next. The anti-Lamarckian position was as political as the "Lamarckian" one. Huxley's motto, "free but unequal," beautifully encapsulates the Cold War ideology of the modern synthesis.

Of course, the modern synthesis also had strong critics, such as Stephen Jay Gould and his colleague at Harvard the evolutionary biologist and population geneticist Richard Lewontin, both of whom objected to the idea that genes determine behavior and that natural selection is the only force shaping evolution. Still, with regard to Lamarck and his theory, they agreed with the others: Lamarck was wrong. To think otherwise had become unthinkable.

And yet Lewontin was one of the principal architects of the theory of "niche construction," that organisms aren't just shaped by their environments but also, reciprocally, are constantly building and transforming their ecological niches. This was Lamarck's fundamental idea: Living beings create themselves and the world around them, even as that world exerts its pressures upon them.

The person who coined the term "niche construction" to describe the role of behavior in evolution is an emeritus biological anthropologist at Oxford named John Odling-Smee. In the 1988 article in which he introduced the phrase, he traced the concept to a surprising source: not a biologist or ecologist but Erwin Schrödinger, the Viennese physicist who was one of the principal authors of quantum mechanics. Schrödinger wrote a short and momentous treatise in 1944 titled *What Is Life?* that became one of the founding texts of molecular biology. He proposed that the essence of life was a form of activity. "What is the characteristic feature of life?" he asked. "When is a piece of matter said to be alive?

When it goes on 'doing something,' moving, exchanging material with its environment and so forth."

Over the following two decades, as the modern synthesis dominated the scene in evolutionary biology, Schrödinger, an outsider to that scene, continued developing his idea. The behavior of an individual being, he suggested, plays not only a part but "the most relevant part in evolution." He argued that it made no sense for biologists to distinguish between a body part and "the urge to use it and to increase its skill by practice," because natural selection could select only a *"used organ."* Furthermore, by using an organ in new ways, an individual being must change its environment, so that the selective pressures exerted by the environment on the organism have also in some way been prepared by the organism itself.

It was, wrote Schrödinger, "as if Lamarck were right." Indeed.

The Life-Made World

In which we consider Lamarck's Warnings about the Human Tendency to Destroy the Planet and how Terrifyingly Apt they were; his Detractors' Worldview has prevailed during Two Centuries of Industrial and Imperial Devastation. Yet we also Glimpse the Return of Lamarck's Living World, so long Languishing in the Shadows, toward the Sunlight. Researchers studying Epigenetic Inheritance collect Instances of Living Beings Transforming and passing along the Changes to their Offspring. Epigenetic Inheritance is Widespread and is Inseparable from Genetic Inheritance, since Natural Selection can only Act upon Genes as they are Expressed in Organisms. Epigenetics is only one Area of Research connecting Individual Organisms with Evolutionary Development. Biologists are examining the Many Ways in which Living Organisms Influence the Course of Evolution: Through their Behaviors and Cultures, by Communicating, Learning, and Responding, they Shape the Ever-Changing Living World.

R ecently I was walking with my family in the hills near our house in Berkeley, California, on a chilly day of mixed sun and clouds. We came upon an extraordinary spot where it seemed as though an invisible line divided two different days from each other. On one side, it was cool, bright, and sunny; on the other, it was gray and drizzling. Looking around in mystification, we soon found the cause.

Eucalyptus trees on either side of the path had created their own micro-climate by trapping the moisture in the air, at once drinking and creating the dense mist. The transition was so abrupt that for a while we had fun standing with one foot in the sunny day and the other in the rainy day. The famous California coastal redwoods and California coastal fog bear a similar relationship. They create each other, the trees drinking the fog and also producing it by trapping water droplets in the air and exhaling moisture back out into it.

The environment is not a box or container with organisms inside. It is itself alive, dynamic, engaging, and transforming. It is the mutual interactions among all living beings and the changing landscape and climate. Organisms *are* the environment, and they are also always making and remaking the environment even as it is making and remaking them.

. . . .

Denying the agency of living beings has not been a purely intellectual mistake. It has informed two centuries of environmental destruction, allowing people to regard the living world as so much raw material to exploit for economic, industrial, and imperial gain. Lamarck, who witnessed the Industrial Revolution firsthand, warned of this. In his last work, the *Analytical System of Human Knowledge*, written with Rosalie in 1820 when Lamarck was infirm, blind, and dismissed and ridiculed by most of the powerful figures in science and politics, he lamented in a fit of despondency that people were selfish and shortsighted, careless for the future, wantonly annihilating their own means of conservation, and working toward the destruction of their own species and of all living things.

Perhaps he was thinking of the mines and quarries he had toured with Buffonet and, no doubt, of their ever-accelerating growth in the intervening decades. He pointed out deforestation and its consequences. Destroying the trees and large plants that protected the soil to "satisfy the greed of the moment, [man] quickly brings sterility to this soil . . . causes the springs to dry up, drives away the animals who found their subsistence there, and causes great parts of the globe, once very fertile and highly populated, to now be bare, barren, uninhabited and deserted." Human beings' passions led them to ignore experience, to carry on waging wars and plundering nature for riches. "It looks," Lamarck concluded despairingly, "as if man were destined to exterminate himself after making the globe uninhabitable."[1] Alas, it seems he was right about much of this, but let's hope not all of it.

Lamarck's central idea that living beings play an active role in their own transformation has recently taken on a new life in biology. For instance, over the last couple decades, some biologists have been working to launch a new approach to evolution and inheritance, an "extended evolutionary synthesis." This approach incorporates into the evolutionary picture an organism's growth, development, behavior, and engagement with the environment. Genetics are still fundamental to the story but are no longer the whole story. They share the stage with developmental biology and ecology. The central insights orienting these scientists' work are two points that, in retrospect, seem obvious: (1) The experiences and behaviors of living organisms are intrinsic parts of the evolutionary process; (2) DNA, while undeniably a crucial aspect of inheritance, works in tandem with other, equally essential elements.

To understand just what these other elements are, these extra-genetic forms of inheritance, the authors of a recent book about the extended evolutionary synthesis—Kevin Lala, Tobias Uller, Nathalie Feiner, Marcus Feldman, and Scott Gilbert—have suggested that it can be helpful to think about them in four overlapping categories: "the inherited microbiome," "parental effects," "epigenetic inheritance," and "animal culture." (These categories are overlapping because certain kinds of inheritance belong in more than one of them; for instance, "parental effects" can include members of any of the other three categories.) Here I adopt these categories and add two others, also overlapping but important enough to single out: "animal behaviors" and "niche construction."

THE INHERITED MICROBIOME

Let's begin with the "inherited microbiome." This includes all kinds of tiny, important beings such as symbiotic bacteria, fungi, viruses, and also protists and archaea, different sorts of single-celled organisms. These can travel from parents to offspring in both asexual and sexual reproduction in lots of ways. For instance, they can pass into eggs, seeds, or developing embryos, or they can be transferred in the birth canal, through the mother's milk, or in regurgitated food. Like the other categories of extragenetic inheritance, these microbiota are vitally important, making the difference between life and death.

A striking example can be found in the Mojave Desert, one of the most extreme environments on the earth. It is a landscape of ferocious, aloof magnificence with sweeping, rippling dunes and jagged granite peaks. During the winter, the temperature often drops below freezing, but in the summer months it is regularly in the 100s Fahrenheit and has reached into the 130s. It is the driest desert in North America. In other words, there are good reasons why the lowest area in the Mojave Desert is called Death Valley. But the Mojave is actually, improbably, thronging with living things. One of these is the unassuming desert woodrat, an extremely fetching little creature with soft gray and brown fur, round ears, and big black eyes. This resourceful animal lives partly by eating the seeds of the creosote bush, or *Larrea tridentata*, a yellow-flowering shrub that, especially after a rain, gives off a surprising stink of coal tar, hence the nickname. The plant is resourceful too, coating itself in a toxic resin that wards off most herbivores. But not the wood rat, who is able to feast on the creosote bush thanks to detoxifying bacteria in its gut; without these, it cannot tolerate the creosote bush diet. No genetic difference separates a wood rat who can thrive and reproduce from one who would languish from hunger in the punishing desert. The difference lies only in its inherited microbiota.

PARENTAL EFFECTS

"Parental effects" designates all the things that offspring receive from their parents apart from DNA. Parents give their children a vast number of things other than genes. (I'm sure any parent reading this will not be surprised to hear it.) In addition to cotton onesies, five-point-harness car seats, and saxophone lessons, these include hormones, symbiotic microorganisms, nutrients, antibodies, epigenetic changes, behaviors, and ecological resources like your beautifully appointed beaver dam. Perhaps these might sound incidental or secondary, but no, they are often absolutely essential, making the difference between life and death. They contribute to shaping the course of natural selection and the future of the genome.

It's hard to choose a single example to represent the whole universe of "parental effects," but since most of us focus more on imparting decent table manners than healthy gut bacteria, let's take an example that has to do with the education of young ones. Consider what happened when some meddlesome ornithologists decided to take eggs from the nests of blue tits and put them into the nests of great tits, and vice versa.

These two species of bird generally live peacefully side by side, but they do things quite differently. Blue tits like to eat twigs and buds high up in the trees, whereas great tits prefer to eat their meals down on the ground or on the lower, thicker branches. Both kinds of tit hunt for prey, including insect larvae, flies, and spiders, and bring some back to their hungry brood, but great tits tend to fetch back much larger prey than blue tits. They also sing different songs and fall in love with different kinds of tits—blue tits with other blue tits and great tits with other great tits. However, as you may be anticipating, the swapped baby great tits learned from their blue tit foster parents how to forage, hunt, sing, and even choose a mate, all like a blue tit. Likewise, the baby blue tits did everything just like a great tit. So much inheritance and none of it genetically "programmed."

EPIGENETIC INHERITANCE

A third category of extra-genetic inheritance, "epigenetic inheritance," refers to how the DNA molecule is packaged, since it works differently with different wrappings. Think of one of those Transformer toys that can be, say, a robot or a gorilla depending on how it's configured. Or think of a caterpillar and the butterfly it becomes: The DNA is identical. Clearly, the same DNA can be present in radically different bodies. Before I briefly describe how epigenetic changes work, remember that DNA is a long molecule with the famous double-helix structure that looks like a twisted rope ladder. Its function in the body begins with its being "transcribed" into RNA. To imagine what this transcription looks like, think of cutting the rope ladder in half vertically so that you have one side-rope with half of each rung attached to it. This half copy, called messenger RNA, then travels from the cell's nucleus into the cytoplasm and takes part in the synthesis of a protein.

Some epigenetic factors influence the transcription process, enhancing or decreasing it, or even turning it on or off. One of these is called methylation. When a methyl chemical group (a carbon atom and three hydrogen atoms) attaches to the DNA molecule at a certain spot, it can block its transcription. Another is called histone modifications. This refers to the fact that the long DNA molecule is spooled around proteins called histones and has different effects depending on how tightly wound it is and which parts are available for transcription. Another kind of epigenetic factor consists of certain kinds of RNA that can intervene after transcription has taken place, in the synthesis of proteins. Overall, epigenetic factors influence gene expression—that is, how a segment of DNA expresses itself, or doesn't, in the development of the organism. Crucial to an understanding of the evolutionary process, some epigenetic factors can be inherited over generations.

One experiment has found epigenetic changes in the finches of the Galápagos Islands, the birds who provided one of Darwin's earliest examples of variation and adaptation. He noted the "perfect gradation in the size of the beaks" across twenty-five different species of bird and mused that "one might really fancy that . . . one species had been taken and modified for different ends." The Galápagos finches grew famous as a result of Darwin's attention to them, and they have been much studied in the almost two centuries since he first encountered them. Recently researchers have noted striking differences in their colors, beak shape, and size depending upon whether they live in towns or more rural settings.

To investigate what might cause these differences, between 2008 and 2017, a team of researchers captured a thousand finches of two different species, *G. fortis* and *G. fuliginosa*, each having members of the species living in urban and in rural environments and showing the distinctive traits. The team measured the birds' beaks, took blood and sperm samples from them, then banded and released them. Next, they performed an analysis to look for any genetic or epigenetic changes that might correspond to the differences in beaks between the urban and the rural finches. While they found no genetic differences between the urban and the rural birds of either species, they did find many epigenetic differences involving the methylation of certain segments of their DNA. Same DNA, different expressions. The researchers don't yet have an explanation of how environmental factors or behaviors might have caused these differences. But since the towns in the Galápagos Islands have developed quite recently, during the last sixty years or so, the results suggest that epigenetic changes might have played a role in rapid adaptation to a new environment, perhaps leading the way for genetic differentiation.

A more sobering example of inherited epigenetic changes involves a fear that was transmitted across generations of mice in a laboratory. These mice didn't lose their tails, but they did receive mild electric shocks to their feet at the same time that they smelled a certain odor (acetophenone, a clear, sweet-smelling liquid that people experience variously as smelling like almonds, oranges, or cherries). After a few experiences like this, the mice unsurprisingly began to be startled whenever they smelled the odor. The extraordinary thing, though, is that when they later had children, the offspring too were startled by the odor of acetophenone, although they had never experienced the unpleasant shocks to their feet. Moreover, the children's children, the grandchildren of the original mice, showed the same startle response. The experimenters found that in DNA

samples taken from the original mice, a region related to the relevant olfactory receptor had become demethylated, and it was also demethylated
in the sperm of the male mice, and even in their sons' sperm. A fear had
been transmitted across the Weismann barrier. (Note, though, that this
example of epigenetic change is actually more like Weismann's misrepre-
sentation of Lamarckian inheritance than like Lamarck's own idea: It
seems to reflect a sudden change inflicted from outside.)

Parents can of course transmit fears to their children in all kinds
of ways. But findings like these, suggesting that traumatic experiences
might leave physical traces in the body that are inherited by subsequent
generations, have given rise to much discussion of the notion of epige-
netically inherited "intergenerational trauma." Researchers have found
corresponding epigenetic and behavioral changes in the human children
of survivors of great traumas such as the concentration camps of World
War II, a six-month famine in the Netherlands, and the terrorist attacks
on September 11, 2001. A more hopeful finding is that it appears that
recovery can also happen at an epigenetic level. When veterans with post-
traumatic stress disorder received therapy, their DNA showed changes in
the methylation patterns associated with trauma. And when the mice
were retrained to lose their fear of the sweet smell, their next litters of
offspring were unafraid too.

Still more examples of epigenetic changes involve the responses of
various kinds of animals to climate change. For instance, marine biologists have recently found that the metabolism of the spiny chromis, a
slender damselfish inhabiting the coral reefs of the Pacific, is responding to the higher water temperatures brought on by climate change by
transforming in myriad ways, and that these changes are inherited from
generation to generation. The spiny chromis isn't alone. Scientists in
Australia have recently found that the reef-building coral that provides

the damselfish's home also responds to changing ocean temperatures and salinity with epigenetic changes that it passes on to its offspring. German evolutionary geneticists have found wild guinea pigs subjected to heat stress show heritable alterations in their livers, which play a crucial role in metabolic regulation, and biologists in the United States and Kenya have shown that the metabolisms of baboons also heritably change in response to differences in their food supply. In fact, it seems increasingly apparent that all kinds of living beings respond to their changing environments with heritable epigenetic transformations. These might provide means of rapid adaptation that work in advance of slower genetic changes.

The field of epigenetics is quite young and much remains unknown, such as how long epigenetic modifications can persist from generation to generation, and—importantly—whether these changes ever show demonstrable connections to animal behaviors or habits, as in Lamarck's theory. But already scientists have gathered evidence of epigenetic inheritance in all forms of life—bacteria, fungi, protozoans, plants, invertebrates, fish, amphibians, reptiles, birds, mammals—suggesting that epigenetic inheritance is ubiquitous. Not only is it ubiquitous; it's also inseparable from genetic inheritance, since it shapes how genes are expressed in organisms, and natural selection can only select for or against genes as they are expressed in organisms.

. . . .

If you've heard the word "epigenetics" before, you might have started noticing that lately it seems to be appearing increasingly often. But it's not actually that new. It was coined in 1942, the same year that Julian Huxley announced the modern synthesis. The person who coined it was a British embryologist named Conrad Waddington, known to his friends affectionately as Wad. He studied art and philosophy as well as developmental biology and spent the early part of his career at Cambridge University before moving to the University of Edinburgh after World War II. During the 1930s and 1940s, Waddington was at the center of a group of people at Cambridge, Oxford, and the University of London who resisted the movement that developed into the modern synthesis. Calling themselves the Theoretical Biology Club, these renegades pressed for a more complex and nuanced picture of the evolutionary process. In 1935, Waddington traveled to Pasadena, California, to spend some time learning about genetics in Thomas Hunt Morgan's *Drosophila* lab at Caltech (Morgan had moved there from Columbia University several years earlier). While he was there, Waddington also got to know Dobzhansky, who was then Morgan's student.

Despite these modern synthesis influences, Waddington came away more determined than ever to find a way of integrating genetics with developmental biology. He had trouble believing that "even the most statistically minded geneticists" could be "entirely satisfied that nothing more is involved than the sorting of random mutations" by natural selection. This seemed to him wildly implausible. Such a process could never account for the extraordinary precision with which animals fit into their ecological niches. "I want to argue that this theory is an extremist one," he wrote, and also that it neglects relevant knowledge from other areas of biology.

The process that Waddington proposed in place of the "extremist" theory was what he called "canalization," a channeling of natural selection by animals' environments, experiences, and behaviors. An organism, he reasoned, was the outcome of interactions among its DNA, the rest of its body, and its environment. "Naturally," he wrote, "no character can ever be completely environmental in origin, nor yet completely hereditary, since both genotype and environment are necessary components of all development." He described a process in which animals acquired traits by their behavioral responses to a particular environment and these traits later became genetically fixed through natural selection. The traits were "acquired characters" that became converted into inherited characters, and in this sense his theory included a kind of inheritance of acquired characteristics. It worked by animals, through their responsive behaviors, helping to shape the environment in which natural selection operated.

Waddington coined the term "epigenotype" to reconnect the two domains that Weismann had divorced, the domain of germ cells and the domain of body cells. These had been going by the names of "genotype" and "phenotype" since 1911, when Wilhelm Johannsen had proposed these terms in the same paper where he had introduced the word "gene." Waddington now emphasized that "between genotype and phenotype, and connecting them to each other, there lies a whole complex of developmental processes. It is convenient to have a name for this complex: 'epigenotype' seems suitable."

In order to convey what he meant, he drew philosophical pictures (he also wrote a book about the relations between science and painting). One of his pictures shows an abstract rendition of a hilly landscape with a ball at the top, just about to roll down toward the viewer. He explains that this drawing depicts the "epigenetic landscape." The path of the ball represents the process of development and shows how it can take place along various different trajectories. "Although the epigenetic landscape only provides a rough and ready picture of the developing embryo and cannot be interpreted rigorously," Waddington wrote, "it has certain merits for those who, like myself, find it comforting to have some mental picture, however vague, for what they are trying to think about."

His fondness for philosophical drawings reflects another way in which Waddington differed from his contemporaries in evolutionary biology. While most of them were insisting on an absolute divide between science and all other forms of understanding, especially artistic and humanistic ones, Waddington wrote that he thought scientists should acknowledge that "the scientific attitude is as full of passion, as much a function of the whole man and not merely of an intellectual part of him, as any other approach to human action." Like other endeavors, therefore, science involved "imagination and insight," not just a mind "swept clean of preconceived notions." Also unlike many of his fellow biologists, and no doubt connected with his complex, interwoven picture of evolution and development, Waddington rejected the idea of innate, genetically caused differences in intelligence or behavior between groups of humans. "The facts make it impossible to deny the important part played by the environment," he wrote, "whereas there seems to be no compelling reason to attribute much weight to the contribution of heredity. . . . The conclusion one must come to is that we cannot shelve the problems of ethics by referring them to the uncontrollable mutations of our genes."

You may not be surprised to hear that during Waddington's lifetime, and throughout most of the rest of the twentieth century, under the domination of the modern synthesis, his ideas were dismissed by the principal figures in evolutionary biology; the taint of "inheritance of acquired characteristics" clung to them. But around the turn of the twenty-first century, Wad's work and the terms he introduced, notably "epigenetics," began slowly to reemerge. We might match this to the "rediscovery" of Mendel a century earlier, the two reemergences bookending the twentieth century. Perhaps the renewal of Waddington's epigenetics will turn out to have marked the first glimmers of an end to a long, absolutist reign of genetics in evolutionary biology and the beginning of a new, more pluralistic regime.

ANIMAL BEHAVIORS

The more general idea behind Waddington's notion of the epigenome, the idea that organisms shape the course of evolution by their behaviors, is now very current even beyond studies focused on molecular epigenetic structures. As we saw in chapter 30, Ernst Mayr argued for the importance of behavior as a factor in evolution around the middle of the twentieth century (though with regular caveats about behavior being governed by a "genetic program").

Even before Mayr made this argument, scientists had been working along the same lines. In 1944, Charles Bogert, the curator of herpetology at the American Museum of Natural History, and Raymond Bridgman Cowles, a herpetologist at the University of California, Los Angeles went to Coachella Valley, California. This was decades before the festival that made Coachella famous, and Bogert and Cowles were interested not in indie music but rather in lizards and snakes. They conducted a study showing that these animals of the Sonoran Desert were able to regulate their body temperatures through their behavior. In a follow-up article in 1949, Bogert generalized the finding by including field sites in Arizona and Florida, confirming that lizards across the sites, through their behaviors, exerted an "astonishing amount of control" over their temperatures and suggesting that this control must be an important force in shaping the course of evolution. The lizards' control over their own temperatures weakened the selective pressure of the ambient temperature and increased the range of environmental niches in which the animals could thrive.

Bogert and Cowles's work inspired a generation of research, especially in comparative physiology. Then, in the early twenty-first century, three biologists—Raymond Huey at the University of Washington, Paul Hertz at Barnard College, and Barry Sinervo at the University of California, Santa Cruz—revisited Bogert and Cowles's findings and developed a conceptual framework on their basis. Huey, Hertz, and Sinervo gave a new name—"the Bogert effect"—to Bogert's generalization about lizards' behavioral control of their body temperature, inspiring renewed interest in the regulatory role of animal behaviors in evolution.

One person whose research has been informed by these earlier studies is Martha Muñoz, an ecologist and evolutionary biologist at Yale, from whom I learned all of the above information about Bogert, Cowles, Huey, Hertz, Sinervo, and their versatile lizards. Muñoz studies anoles, a kind of lizard closely related to iguanas, in the Dominican Republic. She has shown that as sea-level lizards have moved up into the mountains, reaching as high as three thousand meters, they have adapted to the much colder temperatures by being "behaviorally nimble": basking on the sun-drenched boulders. Rather than showing physiological differences that would give them greater cold tolerance, they have rapidly evolved shorter legs and flatter skulls, enabling them to hide from predators in the crevices of the rock. Their resourceful behavior has slowed down the evolution of cold tolerance and accelerated the evolution of anatomical changes useful for hiding. "Far from being passive vessels at the mercy of their circumstances," says Muñoz, "organisms can influence evolution directly."

According to Muñoz, the role of animal behavior in speeding up evolution in some areas ("behavioral drive") and slowing it down in others ("behavioral buffering" or "behavioral inertia") is ripe for a full exploration: "I think the phenomenon is ready to go more mainstream." Her own research confirms that "behavior is one of the major architects of evolution" and that "organisms can influence natural selection," accelerating it, dampening it, and directing its course.

Darwin's finches, once again, provide another kind of example, related not to thermoregulation but to mating, as Peter and Rosemary Grant have shown. These two evolutionary biologists, both originally British, are now retired from Princeton University, where they spent their careers. Together, they have studied the Galápagos birds since 1973. In one of their extraordinary findings, they witnessed "speciation" in real time, demonstrating the emergence of a new species—they nicknamed it Big Bird because of its imposing size—in only forty years. This enabled them to study just how the process takes place. They showed that the reproductive isolation of a population, which is crucial to the formation of a species—preventing its individual members from interbreeding with other populations—was behavioral rather than genetic. "Darwin's finches do interbreed" and produce fertile offspring, the Grants explain, but rarely. "This shows that in the early stages of speciation, before genetic incompatibilities arise, populations are almost entirely isolated from each other by a ... behavioral barrier" consisting in part, for instance, of differing courtship rituals. This barrier is "culturally inherited": Young birds learn which mates to choose from their parents (as in the case of the great tits and blue tits mentioned above).

"We think of speciation," the Grants conclude, "as an eco-behavioral-genetic process of evolutionary divergence. Genetic incompatibilities arise much later through mutation, and eventually the door to gene exchange closes forever. However, in this young radiation of Darwin's finches all species, apparently, still have their doors open." As Mary Jane West-Eberhard, a theoretical biologist at the University of Michigan, has put it, "Genes are followers, not leaders, in evolution."

ANIMAL CULTURE

The cultural inheritance of mating preferences in Darwin's finches brings us to an especially fascinating category of nongenetic inheritance: animal culture. Many animals have something like what we call culture in reference to human life, and this includes cultural forms that are inherited and passed down. Lamarck knew this; he wrote that intelligent animals, birds and mammals, could create new habits and ways of life that would be maintained over generations. Darwin knew it too; his theory of sexual selection described how animals shaped the world and their descendants according to their senses of beauty. There is overwhelming and definitive evidence that all kinds of animals—birds, primates, whales, fish, insects—learn from one another and have cultural traditions that are handed down from generation to generation. Animals communicate vitally important things to one another including information about food, predators, songs, migration routes, and where to find a mate. Communication and social learning make a crucial difference to survival and reproduction.

Humpback whales, for instance, develop different approaches to hunting from one community to another. In the northeast Pacific Ocean, they form small groups to swim in spirals around their prey, blowing air through their blowholes to trap the fish in a net of bubbles. The individual whales play various roles, some blowing bubbles, others making feeding calls. The humpback whales of the Gulf of Maine, meanwhile, have developed a refinement on bubble-net feeding to catch sand lances (long, skinny fish that look like eels). During the last few decades, the whales in that region have perfected their technique by adding a vigorous tail thumping of the surface of the water, which shocks the sand lances into forming a tight school. The humpbacks have learned this technique from one another, younger whales from older ones, so that it has become a regional cultural tradition. Whales of different communities within a given species have other cultural differences too, such as distinct songs. This sort of cultural knowledge shapes the course of natural selection and the evolution of the genome. Another example is killer whales, who have also developed specialized hunting techniques for the various prey in different areas and have evolved distinct shapes and digestive physiologies with corresponding genetic differences.

The Galápagos woodpecker finch offers another example of an animal cultural tradition that is essential to the birds' survival and has shaped the course of their evolution. Instead of poking their beaks into the bark of trees, these birds use a tool, a piece of twig or a spine from a cactus, to extract delicious grubs. The finches learn this behavior from one another and do it a bit differently in different populations. Because they have created their own niche, natural selection has not caused them to evolve the standard woodpecker-like arched, pointy beaks and long, snaking tongues. Instead, Galápagos woodpecker finches have quite small, stout beaks and short tongues. But they're handy with a chopstick.

NICHE CONSTRUCTION

The various kinds of extra-genetic changes just described—the inherited microbiome, parental effects, epigenetic inheritance, animal behavior, and animal culture—also overlap with the theory of "niche construction" that came up in the last chapter: the idea that organisms contribute to shaping their environments, creating the ecological niches that exert selective pressures on them. Charles Darwin offered a beautiful example of niche construction in his last major work, published just the year before he died: a study of earthworms. Here he examined the earthworm's ability, through its humble and minute mode of activity, to play a role of prodigious importance in shaping the landscape. In fact, earthworms shaped not just their own ecological niche, according to Darwin, but the English countryside itself. The agricultural economy of England depended on the actions of earthworms because of the importance of the "vegetable mould" or humus that they produced. Moreover, the earthworms' activity was itself conjoined with slightly larger creatures, such as moles, and even smaller ones, burrowing larvae and insects, especially ants, that brought fine earth up to the surface.

Darwin thought earthworms exhibited a "degree of intelligence," rather than acting by "mere blind instinctive impulse," as was evident in the fact that they varied their behavior to suit the context. For example, they positioned leaves and other objects to plug up the mouths of their burrows according to the shapes of the spaces and the objects in question. The work of earthworms, in addition to being essential to ecology and to agriculture, was a work of art. "When we behold a wide, turf-covered expanse," Darwin concluded, "we should remember that its smoothness, on which so much of its beauty depends, is mainly due to all the inequalities having been slowly levelled by worms. It is a marvellous reflection that the whole of the superficial mould over any such expanse has passed, and will again pass, every few years through the bodies of worms. The plough is one of the most ancient and most valuable of man's inventions; but long before he existed the land was in fact regularly ploughed, and still continues to be thus ploughed by earthworms."

The worms create the soil conditions in which they live. Moreover, in themselves, they are not structurally well adapted to living on dry land; a freshwater habitat would be more suited to their physiology. But they survive and thrive in the earth by their industrious activity of "tunneling, exuding mucus, eliminating calcite, and dragging leaf litter below ground." Natural selection acts upon them within an environment of their own making.

Because organisms take an active part in shaping their environments, John Odling-Smee, the person who coined the term "niche construction," has written that biologists should rethink what the theory of evolution should be about: not just natural selection acting upon organisms in given environments but "the coevolution of organisms with their environments." In recent years, in addition to the "extended evolutionary synthesis" and "niche construction," scientists have used various other names and phrases to describe this kind of approach to evolutionary theory. For instance, "evo-devo" refers to the union of evolutionary and developmental biology, and "eco-evo" to the union of ecology and evolutionary biology. Odling-Smee thinks it's all part of the same story: "Let's call it an 'eco-evo-devo synthesis.'"

Bird nests, beaver dams, termite mounds, mole burrows, spiderwebs, beehives—the evidence of organisms creating their own environments is ubiquitous. And those are just the dramatic instances of animal architecture and engineering. The subtler examples are legion too, such as the eucalyptus trees and the misty microclimate my family and I encountered on our walk in the East Bay Hills, or the California redwoods and the coastal fog. Organisms aren't just in the environment; they make the environment and they are the environment.

. . . .

From the start, Lamarck's science of self-making, world-creating living beings antagonized the powerful. In contrast, the dominant tradition in evolutionary biology from the end of the nineteenth century, largely abiding still today, renders living organisms as the objects of outside forces, posing no threat to gods or monarchs or other worldly powers. Evolutionary science became the reductive account of an essentially passive living world. By this standard, Lamarck's vision of a world continually in the making, shaped by its own mortal inhabitants, wasn't even wrong: It just wasn't science.

But science, like living beings, is always changing. Evolutionary biology today is full of exhilarating unrest. For their part, the Mojave woodrat and its friendly gut bacteria, the fostered baby blue tits, the delicate, resourceful damselfish, the city finches and country finches and selectively mating finches and tool-using finches of the Galápagos, the humpback whales in the Gulf of Maine, the nimble mountain lizards of the Dominican Republic, and legions of other living things—all of them, in fact—exemplify what Lamarck knew and mainstream evolutionary biology has long denied: Living beings are active, creative, self-making, and world-making.

"There . . . exists in nature a peculiar, powerful, and ever active cause, which has the faculty of forming combinations, of increasing and varying them. . . . Now this powerful cause . . . resides in the organic activity of living bodies. . . . [Living] organization, whether animal or vegetable, has, by means of the power of life, gradually composed and complicated itself, from the greatest simplicity to the greatest complexity."

Lamarck was right.

Acknowledgments

This book's evolution has depended upon the collaborative, creative agency of many living, intelligent, and highly generous beings.

Thanks for advice, guidance, expert information, critical responses, and readings of the manuscript to Keith Michael Baker, David A. Bell, Deborah Coen, Jennifer P. Daly, Matthew Eddy, Marcus Feldman, Paula Findlen, Michael Gordin, Deborah M. Gordon, the late Jonathan Hodge, Eva Jablonka, Jill Lepore, George Levine, Kevin Padian, Mark Peterson, Greg Priest, Gregory Radick, Robert J. Richards, Emma Rothschild, Silvia Sebastiani, Irina Sirotkina, Lincoln Taiz, Stéphane Tirard, Olivia Weeks, and Caroline Winterer.

Thanks to Richard W. Burkhardt for his careful reading of the entire text, important challenges, constructive criticism, factual corrections, and for drawing upon his extraordinary store of knowledge to answer my deluge of questions.

Thanks to Pietro Corsi, friend, mentor, and model of intellectual generosity, who tirelessly supplied help and advice, magicked me into archives, offered pivotal suggestions, read and commented on the entire unwieldy draft, and brought the protagonists of this story to life by gossiping about them with me over lunch in the *Jardin des plantes*.

Acknowledgments

Thanks to Ludmilla Jordanova, whose friendship is one of the lucki-est things that writing this book has brought me, after she responded to my late request for help with extraordinary gifts of time, expertise, criti-cal insights, and hospitality.

Thanks to Martha Muñoz, who answered an out-of-the-blue query from an unknown historian by spending a morning explaining her path-breaking research on the evolution of reptiles, amphibians, and fishes. Her work represents the present and the future for which, I hope, this book provides a richer, more nuanced past.

Thanks to Jean-Robert Callon de Lamarck and to Anne Lécrivain, who so kindly provided information about their great-great-great-great-grandfather. Anne Lécrivain invited me to her house near Paris to see the beautiful portrait of Lamarck that appears in the frontispiece of this book and that has remained in the family ever since Charles Thévenin painted it around 1802, as well as the other books and papers in her col-lection. She also showed me Lamarck's ceremonial sword, which he re-ceived as a founding member of the Legion of Honor, and which I was enchanted to see her suddenly draw forth from where it was hidden in plain view among the umbrellas by her front door.

Thanks to Toby Mundy of Aevitas Creative Management for per-suading me to try something new, and for helping me find my way to doing it using his signature combination of wisdom, kindness, irrever-ence, and wit. At Aevitas, thanks also to Augustus Brown, and to Elena Steiert for excellent editorial suggestions including the title.

Thanks to Courtney Young at Riverhead Books for her keen edito-rial vision and unerring ability to lift the storyline out of the verbiage and help me see how to give it light and air. At Riverhead, thanks also to Jason Booher for the beautiful Lamarckian cover; to Corinne Leong for shepherding the manuscript through many stages; and to Denise Boyd,

Ryan Boyle, Michael Brown, Viviann Do, Tom Dussell, Ashley Garland, Mary Kate Garmire, Grace Han, Kitanna Hiromasa, Steven Hoffman, Geoff Kloske, Maggie Kutz, Hannah Lopez, Jynne Dilling Martin, Christina Nguyen, Gabby Nugent, Ingrid Sterner, Lisa Thornbloom, Anna Scheithauer, Melissa Solis, Shailyn Tavela, Claire Vaccaro, and Helen Yentus.

Thanks to the Stanford School of Humanities and Sciences and the Stanford History Department for some extra funding that allowed me to travel to libraries, archives, museums, and gardens, and that gave me precious time to think and to write.

Thanks to the librarians and archivists who helped me find sources and images, especially Mme. Maud Haon Maatouk of the Muséum national d'histoire naturelle in Paris, Mme. Alicia Leclercq of the Musée et parc Buffon in Montbard, and Dr. Anna Klyukina and Scientific Secretary Irina Khmelnitskaia of the State Darwin Museum in Moscow.

Thanks to my students and colleagues at Stanford whose ongoing conversation about history and science informs every page; and to the audiences at many lectures and colloquia on Lamarck who have asked me hard questions and made revelatory suggestions.

Thanks to my comrades in the American Association of University Professors who have been courageously fighting to defend higher education, science, and scholarship.

Thanks to my family and friends. Myra Jehlen has read every word of the book-in-progress, and her questions, suggestions, challenges, and sharp pencil have been crucial. Carl Riskin has been an essential sounding board, prose stylist, humor consultant, and editorial voice. Madeleine Riskin-Kutz, in addition to giving me a starting point, has enriched the story throughout with pivotal literary and linguistic connections. Oliver Riskin-Kutz's rigorously critical historical sense has been a polestar, often illuminating for me my own unthinking assumptions. Alain Jehlen has been a constant source of deep-thinking responses and counterpoints. Jac Boutaud and Annie Imbaud's magnificent life-made world at the Petite Loiterie was both inspiration and refuge. Harry Bernstein and Caren Meghreblian came to my rescue at a critical juncture with a perfect writing sanctuary. Matthias Beekmann and Sophie Godin-Beekmann have shown me what science with a conscience looks like and how it can persist in the face of every countervailing force.

Thanks finally to Christopher Kutz: you patiently read everything a hundred times over in minutely revised versions, you offer crucial advice gently, you weather the explosive response philosophically, and then . . . you do it all again. This one, at last, is for you.

LIST OF ILLUSTRATIONS

Page vi: Heliogravure of *Jean-Baptiste de Monet Chevalier de Lamarck* by Charles Thévenin, ca. 1802. Source: Museum national d'histoire naturelle, Paris, PO2034. Original: oil on canvas, 20 x 25 cm, collection of Anne Lécrivain.

Page 7: *Pistacia vera* by Pierre-Joseph Redouté, ca. 1801–1819, color lithograph. For Henri-Louis Duhamel de Monceau, *Traité des arbres et arbustes que l'on cultive en France en pleine terre*, vol. 4, plate 17, p. 70, folio. Source: Muséum national d'histoire naturelle, Paris, IC BOT/Anaca/Pistacia/7.

Page 12: *En 1734 le célebre Bernard de Jussieu planta le cedre du Liban au Jardin des plantes*, engraving by Charles-Étienne Gaucher, 1798, 100 x 56 mm, after a drawing by Charles Monnet, in Jauffret, *Voyage au Jardin*, p. 156. Source: Bibliothèque nationale de France, département Philosophie, histoire, sciences de l'homme, 8-Z LE SENNE-10875.

Page 16: *Le Lion du Muséum d'histoire naturelle avec son chien* by Nicolas Maréchal, an 2 (1794/5), for the frontispiece to Georges Toscan, *Histoire du lion*, an 3 (1795), reproduced in *La Décade philosophique, littéraire et politique*, an 3, no. 116 (10 vendemiaire [October 1, 1794]), plate between pp. 126 and 127. Source: Bibliothèque nationale de France, département Littérature et art, Z-23188-23229.

Page 19: *Les Éléphants représentés dans l'instant de premières caresses qu'ils se sont faites après qu'on leur a fait entendre de la musique*, drawing and engraving by J.-P. Houel, from Houel, *Histoire naturelle des éléphans*, an 12 (1803), plate 15, between pp. 94 and 95. Source: Bibliothèque nationale de France, département Réserve des livres rares, S-1440.

Page 26: *Study of the Giraffe Given to Charles X by the Viceroy of Egypt* by Nicolas Huet, 1827, watercolor over black chalk, on paper, 19.4 x 25.4 cm. Source: the Morgan Library & Museum, New York. 1994.1. Purchased on the Sunny Crawford von Bülow Fund 1978. Photo credit: The Morgan Library & Museum, New York.

Page 247: *Kanguroo*, an 8 (1799–1800), no information given about artist or engraver, from Buffon's *Histoire naturelle, générale et particulière: Histoire naturelle des quadru-pèdes*, edited by C.S. Sonnini, vol. 32 (an 8), plate 19. Source: Bibliotheque nationale de France, département Sciences et techniques, S-10176.

Page 315: *Hadenoecus subterraneus*, cave cricket, 1879, no information given about the artist(s), from Packard and Putnam, *The Mammoth Cave*, 13. Source: HathiTrust.

A NOTE ON THE NOTES

All the sources that informed this book are listed in the bibliography with full citations, while the endnotes contain abbreviated citations. Both the bibliography and the endnotes are divided into primary sources (the raw material of history, including books and other writings by Lamarck and the other protagonists of the story, journals, letters, archival documents, images, and so on) and secondary sources (works by historians and other scholars). In the endnotes for each chapter, most of the sources appear in a paragraph above the numbered notes in roughly the order of their relevance to the chapter but also grouped by author. Each chapter also has a few numbered notes containing commentary and further sources. All translations of quoted passages, except where otherwise indicated, are my own.

BIBLIOGRAPHY

LIST OF ABBREVIATIONS

AN: Archives nationales de France

    PVAP: Procès-verbaux des assemblées des professeurs du Muséum

AP: *Archives parlementaires de 1787–1799 (première série)*

AS: Archives de l'Académie des sciences

    PVAS: Procès-verbaux des séances

BMHN: Bibliothèque centrale du Muséum national d'histoire naturelle

BVMFB: Bibliothèque de la ville de Mayenne, fonds Bissy

DCP: Darwin Correspondence Project

DP: Darwin Papers

IF: Institut de France

    GC: Papiers et correspondance du Baron Georges Cuvier

    PVIF: Registre des procès-verbaux de l'assemblée générale

BIOGRAPHIES AND EULOGIES OF LAMARCK

PRIMARY

Bourdon, Isidore. "Lamarck." In *Illustres médecins et naturalistes des temps modernes*, 335–54. Paris: Comptoir des Imprimeurs-Unis, 1844.

Bourguin, Louis Auguste. "Lamarck." In "Les grands naturalistes français au commencement du XIXe siècle: Lamarck." *Annales de la Société Linnéenne de Maine-et-Loire* 6 (1863): 194–221.

Cuvier, Georges. "Éloge de M. de Lamarck, lu à l'Académie des sciences, le 26 novembre 1832." *Mémoires de l'Académie royale des sciences de l'Institut de France* 8 (1935): i–xxxi.

Duval, Mathias. "Le transformiste français Lamarck." *Bulletins et mémoires de la Société d'anthropologie de Paris*, no. 12 (1889): 336–74.

Geoffroy Saint-Hilaire, Étienne. "Discours prononcé aux funérailles de Lamarck," 1829. In *Fragments biographiques*, 209–19.

———. *Fragments biographiques, précédés d'études sur la vie, les ouvrages et les doctrines de Buffon*, 80–82. Paris: F. D. Pillot, 1838.

Hamy, Ernest-Théodore. *Les débuts de Lamarck, suivis de recherches sur Adanson, Jussieu, Pallas, Geoffroy Saint-Hilaire, Georges Cuvier, etc.* Paris: E. Guilmoto, 1909.

Hermanville, F. J. F. "Notice biographique sur Lamarck." In *Mémoires de la Société académique d'archéologie, sciences, et arts du département de l'Oise*, vol. 17, part 2 (1899): 443–80.

Lamarck, Auguste, to Eugène de Lamarck, June 11, 1865, reproduced in *Bulletin du Muséum national d'histoire naturelle* 13 (1907): 375–79.

Lamarck, Jean-Baptiste. *Considérations en faveur du chevalier de la Marck*. Paris: Gueffier, 1789. Reproduced in Landrieu, *Lamarck*, 34–36.

Latreille, Pierre-André. "Discours prononcé aux obsèques de M. le chevalier de Lamarck." *Moniteur universel*, Dec. 26, 1829, 1941.

Martins, Charles. "Introduction biographique." In Jean-Baptiste Lamarck, *Philosophie zoologique*, 1:v–lxxxiv. Paris: F. Savy, 1873.

Panckoucke, Charles-Louis-Fleury, ed. "Lamarck." In *Dictionnaire des sciences médicales. Biographie médicale*, 5: 483–88. Paris: CLF Panckoucke, 1822.

SECONDARY

Bange, Raphaël, and Pietro Corsi. "Les auditeurs de Lamarck." Accessed July 17, 2025. redi.imss.fi.it/lamarck/auditeurs/presentation.php?lang=fr.

———. "Chronologie de la vie de Lamarck." Accessed July 17, 2025. https://redi.imss.fi.it/lamarck/chronologie.

———. "The Students of Lamarck in Numbers." Accessed July 17, 2025. redi.imss.fi.it/lamarck/ice/ice_book_detail.php?lang=fr&type=text&bdd=lamarck&table=bio_lamarck&bookId=21&typeofbookId=4&num=.

Bourdier, Franck. "Esquisse d'une chronologie de la vie de Lamarck." Paris: École pratique des hautes études, 1971. Accessed Aug. 26, 2025. https://archive.org/details/LamarckBiographie/page/n23/mode/2up.

Burkhardt, Richard W. *The Spirit of System: Lamarck and Evolutionary Biology*. Harvard University Press, 1977. 2nd ed. Cambridge, Mass.: Harvard University Press, 1995.

Corsi, Pietro. *Lamarck: Genèse et enjeu du transformisme, 1770–1830*. Paris: CNRS, 2001.

Corsi, Pietro, Jean Gayon, Gabriel Gohau, and Stéphane Tirard, eds. *Lamarck, philosophe de la nature*. Paris: Presses universitaires de France, 2006.

Delange, Yves. *Jean-Baptiste Lamarck: Sa vie et son oeuvre*. Paris: Actes Sud, 2002.

Elliot, Hugh. Introduction to *The Zoological Philosophy: An Exposition with Regard to the Natural History of Animals*, by Jean-Baptiste Lamarck, translated by Hugh Elliot, xviii–xcii. London: Macmillan, 1914.

Grimoult, Cédric. *Lamarck*. Paris: Ellipses, 2020.

Jordanova, Ludmilla. *Lamarck*. Oxford: Oxford University Press, 1984.

Landrieu, Marcel. *Lamarck, le fondateur du transformisme.* Paris: Société zoologique de France, 1909.

Laurent, Goulven, ed. *Jean-Baptiste Lamarck, 1744–1829.* Paris: Éditions du CTHS, 1997.

———. *La naissance du transformisme. Lamarck entre Linné et Darwin.* Paris: Vuibert et ADAPT, 2001.

Packard, Alpheus Spring. *Lamarck, the Founder of Evolution: His Life and Work.* London: Longmans, 1901.

GENERAL PRIMARY SOURCES

ARCHIVAL

Archives de l'Académie des Sciences [**AS**]

> Pochettes des séances.
>
> *Procès-verbaux des séances de l'Académie tenues depuis la fondation de l'Institut jusqu'au mois d'août 1835: Publiés conformément à une décision de l'Académie par MM. les secrétaires perpétuels.* 10 volumes. Institut de France, Académie des sciences. Hendaye: Impr. de l'Observatoire d'Abbadia, 1910–1922 (Hendaye). [**PVAS**]

Archives Nationales de France [**AN**]

> *Inventaire dissolutif de la communauté d'entre le Cen Lamarck et le Cnne Reverdy son épouse, 19 prairial an IX,* MC/ET/CXII/840. Transcribed at Pietro Corsi, ed., *Oeuvres et rayonnement de Jean-Baptiste Lamarck,* https://redi.imss.fi.it /lamarck/index.php.
>
> Lamarck to J.H. Bernardin de Saint-Pierre, April 24, 1793. In *Plan du travail à faire pour mettre en ordre les herbiers du Cabinet national d'histoire naturelle,* AJ/15/512 folio 503, also at the *Electronic Enlightenment,* www.e -enlightenment.com.
>
> Procès-verbaux des assemblées des professeurs du Muséum, AJ/15/96–AJ/15/128 (1790–1830). [**PVAP**]

Archives parlementaires de 1787 à 1860: recueil complet des débats législatifs et politiques des Chambres françaises, première série (1787–1799). Edited by Jérôme Mavidal and Émile Laurent. 82 volumes. Paris: Paul Dupont, 1867–1913. [**AP**]

Bibliothèque Centrale du Muséum National d'Histoire Naturelle [**BMHN**]

> Manuscrits de Jean-Baptiste Pierre-Antoine de Monet de Lamarck (1744–1829).
>
> Monuments Lamarck au Muséum et à Bazentin (Somme), MS 2228 / 5.

Bibliothèque de la Ville de Mayenne, Fonds Bissy [**BVMFB**]

> Bissy, Jacques François. *Animaux sans vertèbres*. Notes prises aux leçons de Lamarck au Muséum d'histoire naturelle du 29 floréal au 25 thermidor an X. Transcribed at Pietro Corsi, ed., *Oeuvres et rayonnement de Jean-Baptiste Lamarck*, https://redi.imss.fi.it/lamarck/index.php.

Darwin Correspondence Project, Cambridge University Library, www.darwinproject.ac .uk [**DCP**] The project has also published the letters in Frederick Burkhardt, James A. Secord et al., eds., *The Correspondence of Charles Darwin*. 30 vols. Cambridge, U.K.: Cambridge University Press, 1985–2023. [**DCP**]

Darwin Papers, darwin-online.org.uk [**DP**]

> Darwin, C. R. 1838–1868. "Books to Be Read/Books Read" notebook. CUL-DAR119. Edited by John van Wyhe.

> [Gautry, P. J.]. 1861. Darwin library. List of books received in the University Library Cambridge, March–May 1961.

Institut de France [**IF**]

> Papiers et correspondance du Baron Georges Cuvier [**GC**]

>> Auguste Lamarck [Guillaume Emmanuel Auguste] to Georges Cuvier, Feb. 8, 1830, MS 3252 / f. 358–59.

>> Auguste Lamarck [Guillaume Emmanuel Auguste] to Georges Cuvier, Feb. 20, 1830, MS 3156 / no. 5; also published in Hamy, *Les débuts de Lamarck*, 5–19.

>> Rosalie Lamarck, "Note de Mlle Lamarck au sujet de son père," MS 3156 / no. 9. (Also published in Hamy, *Les débuts de Lamarck*, 26–28.

>> Registre des procès-verbaux de l'assemblée générale, an X (mars 1803–mai 1813). [**PVIF**]

PRINT

Agassiz, Louis. "Sketch of the Natural Provinces of the Animal World and Their Relation to the Different Types of Man." In Samuel George Morton, *Types of Mankind*, lviii–lxxvi. Philadelphia: Lippincott, Grambo, 1854.

Antommarchi, François Carlo. *Mémoires du docteur F. Antommarchi, ou Les derniers momens de Napoléon*. 2 vols. Paris: Barrois l'aîné, 1825.

Apollodorus. *The Library*. Edited and translated by Sir James George Frazer. 2 vols. Cambridge, Mass.: Harvard University Press, 1921.

Arago, François. *Histoire de ma jeunesse*. In *Oeuvres complètes de François Arago*. 13 vols. Edited by J.-A. Barral. Vol. 1, 1–102. Paris: Gide, 1854–62.

Aristotle. *Historia Animalium* (3rd century BC). Translated by D'Arcy Wentworth Thompson. Oxford: Clarendon, 1910.

Audouin, Jean-Victor. *Histoire des insectes nuisibles à la vigne et particulièrement de la Pyrale qui dévaste les vignobles des départements de la Côte-d'Or...* Paris: Fortin, Masson, 1842.

Bachaumont, Louis Petit de. *Mémoires secrets pour servir à l'histoire de la République des Lettres en France depuis 1762 jusqu'à nos jours.* Vol. 21. London: John Adamson, 1783.

Balch, W. M., et al. "Factors Regulating the Great Calcite Belt in the Southern Ocean and Its Biogeochemical Significance." *Global Biogeochemical Cycles* 30 (2016): 1124–44. doi:10.1002/2016GB005414.

Baldwin, James Mark. "Autobiography of James Mark Baldwin." In *History of Psychology in Autobiography*, edited by Carl Murchison, 1:1–30. Worcester, Mass.: Clark University Press, 1930.

———. *Darwin and the Humanities.* Baltimore: Review Publishing, 1909.

———. *Mental Development in the Child and the Race.* 3rd ed. New York: Macmillan, 1906.

———. "A New Factor in Evolution." *American Naturalist* 30, no. 354 (1896): 441–51, and "A New Factor in Evolution, Continued," 30, no. 355 (1896): 536–53.

Balzac, Honoré de. "Avant-propos de la *Comédie humaine*." In *Œuvres complètes de H. de Balzac*, 1:17–32. Paris: A. Houssieux, 1855.

———. *Louis Lambert.* Gosselin, 1832. Reprinted by Flammarion, 1905.

———. *Le père Goriot; Histoire Parisienne.* Librairie de Werdet, 1835. Reprinted by Calmann Lévy, 1891.

Barbeu-Dubourg, Jacques. *Le botaniste français, ou Manuel d'herborisation.* 2 vols. Paris: Lacombe, 1767.

Bastian, Henry Charlton. *The Beginnings of Life, Being Some Account of the Nature, Modes of Origin and Transformations of Lower Organisms.* London: Macmillan, 1872.

Bateson, Beatrice, ed. *William Bateson, F.R.S., Naturalist: His Essays and Addresses, Together with a Short Account of His Life.* Cambridge, U.K.: Cambridge University Press. 1928.

Bateson, William. "Common-Sense in Racial Problems: The Galton Lecture." *Eugenics Review* (1919). In B. Bateson, *William Bateson*, 371–88.

———. "Dr. Kammerer's Testimony to the Inheritance of Acquired Characters." *Nature*, July 3, 1919, 344–45.

———. "Heredity and Variation in Modern Lights." In B. Bateson, *William Bateson*, 215–31.

———. *Mendel's Principles of Heredity: A Defence.* Cambridge, U.K.: Cambridge University Press, 1902.

———. "Presidential Address to the Zoological Section." In B. Bateson, *William Bateson*, 233–59.

———. *Problems of Genetics.* New Haven, Conn.: Yale University Press, 1913.

———. "Problems of Heredity as a Subject for Horticultural Investigation." *Journal of the Royal Horticultural Society* 25 (1900–1901): 54–61.

———. "The Progress of Genetic Research." In *Royal Horticultural Society, Report of the Third International Conference 1906 on Genetics*, 90–97. London: Royal Horticultural Society, 1907.

Baumé, Antoine. *Chymie expérimentale et raisonnée.* Vol. 4. Paris: Didot, 1774.

Bernardin de Saint Pierre, Jacques Henri. *Mémoire sur la nécessité de joindre une ménagerie au Jardin des plantes* (1792). In *Oeuvres completes de Jacques-Henri-Bernardin de Saint-Pierre*, edited by L. Aimé-Martin, 12: 631–69. Paris: Méquignon-Marvis, 1818.

Blainville, Henri de. "Palaéontologie." *Journal de physique, de chimie, d'histoire naturelle et des arts* 94 (Jan. 1822): liv–lv.

———. "Sur une femme de la race Hottentote." In *Bulletin du Societé Philomatique de Paris*, 183–90. Paris: Imprimerie de Plassan, 1816.

Blainville, Henri de, and F. L. M. Maupied. *Histoire des sciences de l'organisation et de leurs progrès*. Vol. 3. Paris: Perisse Frères, 1845.

Boerhaave, Hermann. "Préface, où l'on traite de l'ouvrage et de la vie de l'autheur." In Vaillant, *Botanicon Parisiense*, 1727.

Bogert, C. M. "Thermoregulation, a Factor in Reptile Evolution." *Evolution* 3 (1949): 195–211.

Bonnet, Charles. *Œuvres d'histoire naturelle et de philosophie*. Vol. 4. Paris: Fauche, 1781.

Bonneville de Marsangy, Louis. *Mme. Campan à Écouen: Étude historique et biographique*. Paris: H. Champion, 1879.

Bory de Saint-Vincent, Jean-Baptiste. "Création." In *Dictionnaire classique d'histoire naturelle* 5 (1824): 40–47.

Bougainville, Louis-Antoine de. *Voyage autour du monde par la frégate du roi La Boudeuse et la flute l'Étoile*. Paris: Sasillant & Nyon, 1771.

Bourdon, Isidore. *Mémoire sur le vomissement*. Paris: Méquignon-Marvis, 1819.

———. *Notions d'hygiène pratique*. 2nd ed. Paris: Hachette, 1860.

———. *Principes de physiologie médicale*. Paris: Baillère, 1828.

Brown-Séquard, Charles-Édouard. "Faits nouveaux établissant l'extrême fréquence de la transmission, par hérédité, d'états organiques morbides, produits accidentellement chez des ascendants." In *Archives de physiologie normale et pathologique*. Paris: Gauthier-Villars, 1882.

Bruguière, Jean Guillaume, et al. *Tableau encyclopédique et méthodique des trois règnes de la nature: Vers, coquilles, mollusques et polypiers*. 3 volumes. Paris: Agasse, 1827 (first edition in 4 volumes 1791–1827; Lamarck supervised volumes 3 and 4).

Buchez, Philippe. *Traité de politique et de science sociale*. 2 vols. Paris: Amyot, 1866.

Buffon, Georges. *Correspondance inédite*. 2 vols. Edited by Henri Nadault de Buffon. Paris: Hachette, 1860.

———. *Des époques de la nature. Histoire naturelle, générale et particulière*, supplement. Vol. 5 (1778).

———. "Discours préliminaire" (1749). In *Histoire naturelle, générale et particulière*, 1:1–62.

———. "Discours sur la nature des oiseaux." In *Histoire naturelle des oiseaux*, 1:1–60. Paris: Imprimerie royale, 1770.

———. "Discours sur le style, prononcé à l'Académie Française, par M. de Buffon, le jour de sa reception." Aug. 25, 1753. In *Suppléments à l'Histoire naturelle, générale et particulière*, 4:1–13 (1777).

——. *Histoire naturelle, générale et particulière.* 43 vols. Paris: Imprimerie royale, 1749–1803.

——. *Histoire naturelle des minéraux.* 5 vols. Paris: Imprimerie royale, 1783–88.

Burton, Robert, Jane Burton, and Kim Taylor. *Bird Behavior.* New York: Knopf, 1985.

Cabanis, Pierre-Jean-Georges. *Rapports du physique et du moral de l'homme.* Vol. 1. Paris: Béchet, 1824.

Cambry, Jacques. *Description du département de l'Oise.* Paris: P. Didot l'aîné, 1803.

Camerarius, Rudolf Jacob. *De sexu plantarum epistola.* Tübingen: Martin Rommey, 1694.

Camper, Petrus. *Dissertation physique de M. Pierre Camper, sur les différences réelles que presentent les traits du visage chez les hommes de différents pays et de différents âges.* Translated by Denis-Bernard Quatremere d'Isjonval. Utrecht: Chez B. Wild & J. Altheer, 1791.

Carter, D. E., and Eckerman, D. A. "Symbolic Matching by Pigeons: Rate of Learning Complex Discriminations Predicted from Simple Discriminations." *Science* 187, no. 4177 (1975): 662–64.

Chagniot, Jean, and Hervé Drévillon. "La vénalité des charges militaires sous l'Ancien Régime." *Revue historique de droit français et étranger* 86, no. 4 (2008): 483–522.

[Chambers, Robert]. *Vestiges of the Natural History of Creation.* London: John Churchill, 1844 [1st ed.]; 1853 [10th ed.]; 1860 [11th ed., final text].

Chan, M. A., H. J. Cleaves II, and P. J. Boston. "What Would Earth Be Like Without Life?" *Eos*, March 12, 2018.

Chaptal, Jean Antoine. *Mes souvenirs sur Napoléon par le Cte Chaptal publié par son arrière-petit-fils le Vte A. Chaptal.* Paris: Plon, 1893.

Claveau, Anatole. Review of *Correspondance inédite*, by Buffon. *Revue contemporaine*, 11e année, 2e série, T. 27, Vol. 62 de la collection, 571–80. Paris: Bureau de la Revue Contemporaine, 1862.

Condillac, Étienne Bonnot de. *Essai sur l'origine des connaissances humaines.* 1746. Paris: Imprimerie Auguste Delalain, 1822.

Cope, Edward Drinker. *The Origin of the Fittest: Essays on Evolution.* New York: Appleton, 1887.

——. "The Progress of Discovery of the Laws of Evolution." *American Naturalist* 10, no. 4 (1876): 218–21.

——. "Two Perils of the Indo-European." *Open Court*, no. 126 (1890): 2052–54, and no. 127 (1890): 2070–71.

Correns, Carl. "G. Mendel's Law Concerning the Behavior of Progeny of Varietal Hybrids." Translated by Leonie Kellen Piernick. *Genetics* 35, no. 5, part 2 (1950): 33–41. Translation of "G. Mendel's Regel über das Verhalten der Nachkommenschaft der Rassenbastarde." *Berichte der Deutschen Botanischen Gesellschaft* 18 (1900): 158–68.

Coulomb, Charles. "Second mémoire sur l'électricité et le magnétisme." In *Mémoires de l'Académie des sciences pour l'année 1785*, 578–611. Paris: Imprimerie royale, 1788.

Cowles, Raymond Bridgman, and C. M. Bogert. "Preliminary Study of the Thermal Requirements of Desert Reptiles." *Bulletin of the American Museum of Natural History* 83

(1944): 261–96. Reprinted in 1974 by the Society for the Study of Amphibians and Reptiles, Miscellaneous Publications, Facsimile Reprints in Herpetology. Reprinted in *Iguana* 13, no. 1 (2006).

Cuvier, Frédéric, ed. *Dictionnaire des sciences naturelles.* 60 vols. of text, 11 vols. of plates. Paris: Le Normant, 1816–29.

———. "Essais sur les facultés des brutes" (Jan. 1812). Report of a presentation in the *Nouveau bulletin des sciences par la Société philomatique,* no. 65 (Feb. 1813): 217–18.

Cuvier, Georges. "Considérations sur les mollusques, et en particulier sur les céphalopodes. Lues à l'Académie royale des sciences, 22 février 1830." Extrait des *Annales des sciences naturelles* (March 1830): 1–19.

———. *Discours sur la théorie de la terre servant d'introduction aux recherches sur les ossemens fossiles.* Paris: Dufour et D'Ocagne, 1821.

———. "Extrait d'observations faites sur le cadaver d'une femme connue à Paris et à Londres sous le nom de Vénus Hottentote." In *Mémoires du Muséum d'histoire naturelle* 3 (1817): 259–74.

———. "Femme de race Boschismanne." In Geoffroy Saint-Hilaire and F. Cuvier, *Histoire naturelle des mammifères,* 1:1–7.

———. *Histoire des sciences naturelles depuis leur origine jusqu'à nos jours. 2e partie, Comprenant les 16e et 17e siècles.* Paris: Fortin, Masson, 1841.

———. "Mémoire sur la structure interne et externe, et sur les affinités des animaux auxquels on a donné le nom de vers, lu à la Société d'histoire naturelle, le 21 floréal de l'an III, par G. Cuvier." *La décade philosophique* 5 (1795): 385–96.

———. *Mémoires pour servir à l'histoire et à l'anatomie des mollusques.* Paris: Déterville, 1817.

———. *Recherches sur les ossemens fossiles de quadrupèdes, où l'on rétablit les caractères de plusieurs espèces d'animaux que les révolutions du globe paroissent avoir détruites.* 4 vols. Paris: Déterville, 1812.

———. *Recherches sur les ossemens fossiles, où l'on rétablit les caractères de plusieurs animaux dont les révolutions du globe ont détruit les espèces.* 5 vols. Paris: Chez G. Dufour et E. d'Ocagne, 1821–24.

———. *Le règne animal distribué d'après son organisation: Pour servir de base a l'histoire naturelle des animaux et d'introduction a l'anatomie comparée.* 4 vols. [vol. 3 with Pierre André Latreille]. Paris: Déterville, 1817.

———. *Tableau élémentaire de l'histoire naturelle des animaux.* Paris: Baudoin, 1798.

Darwin, Charles. *The Autobiography of Charles Darwin, 1809–1882. With the Original Omissions Restored.* Edited by Nora Barlow. London: Collins, 1958.

———. *De l'origine des espèces; ou, Des lois de progrès chez les êtres organisés.* Translated by Clémence-Auguste Royer. Paris: Guillaumin, Masson, 1862.

———. *The Descent of Man, and Selection in Relation to Sex.* 2nd ed. London: John Murray, 1877.

———. *The Formation of Vegetable Mould, Through the Action of Worms, with Observations on Their Habits.* London: John Murray, 1881.

———. *Journal of Researches into the Natural History and Geology of the Countries Visited During the Voyage of H.M.S. Beagle Round the World.* 2nd ed. London: John Murray, 1845.

———. *The Movements and Habits of Climbing Plants.* 2nd ed. London: John Murray, 1875.

———. Notebook B: Transmutation (1837–38). CUL- DAR121. In The Complete Work of Charles Darwin Online, edited by John van Wyhe.

———. *On the Origin of Species by Means of Natural Selection; or, The Preservation of Favoured Races in the Struggle for Life.* London: John Murray, 1859 [1st ed.], 1861 [3rd ed.], 1869 [5th ed.], 1876 [6th ed., final text].

———. *The Power of Movement in Plants.* London: John Murray, 1880.

———. "Preliminary Notice." In Ernst Krause, *Erasmus Darwin*, translated by W. S. Dallas, 1–129. London: John Murray, 1879.

———. *The Variation of Animals and Plants Under Domestication.* 2nd ed. 2 vols. London: John Murray, 1875.

Darwin, Erasmus. *The Botanic Garden. A Poem in Two Parts. Part I. Containing the Economy of Vegetation. Part II. The Loves of the Plants. With Philosophical Notes.* London: J. Johnson, 1791.

———. *The Temple of Nature; or, The Origin of Society.* London: For J. Johnson by T. Bensley, 1803.

———. *Zoonomia; or, The Laws of Organic Life.* 2 vols. London: J. Johnson, 1794–96.

Darwin, Leonard. *The Need for Eugenic Reform.* 1921. New York: D. Appleton, 1926.

Daubenton, Louis Jean-Marie. "Memoire sur les differences de la situation du grand trou occipital dans l'homme et dans les animaux." In *Mémoires de l'Académie des sciences pour l'année 1764*, 568–75. Paris: Imprimerie royale, 1767.

Décembre-Alonnier [Joseph Décembre]. "France." In *Dictionnaire populaire illustré d'histoire, de géographie, de biographie, de technologie, de mythologie, d'antiquités, des beaux-arts et de littérature.* 2nd ed. 3 vols. Vol. 2, 1078–91. Paris: s.n., n.d.

Décret de la Convention nationale n. 1020, 10 June 1793 relatif à l'organisation du Jardin national des Plantes du Cabinet d'Histoire naturelle, sous le nom du Muséum d'histoire naturelle. gallica.bnf.fr/ark:/12148/bpt6k6472815x/f3.item.texteImage.

Delamétherie, Jean Claude. *Théorie de la terre.* Vol. 5. 2nd ed. Paris: Maradan, 1797.

———. *Vues physiologiques sur l'organisation animale et végétale.* Paris: Didot, 1780.

De Maillet, Benoît. *Telliamed; or, Conversations Between an Indian Philosopher and a French Missionary on the Diminution of the Sea.* Translated and edited by Albert V. Carozzi. Urbana: University of Illinois Press, 1968.

Description de l'Égypte; ou, Recueil des observations et des recherches qui ont été faites en Égypte pendant l'expedition de l'Armée française. Paris: L'Imprimérie impériale, 1809.

Desmarest, Anselme Gaëtan. "Kanguroo ou Kangurou." In Déterville, *Nouveau dictionnaire* (1803), 12:354–58.

Déterville, Jean-François-Pierre, ed. *Nouveau dictionnaire d'histoire naturelle appliquée aux arts: Principalement à l'agriculture et à l'économie rurale et domestique.* 36 vols. Paris: Déterville, 1816–19.

De Vries, Hugo. *Intracellular Pangenesis: Including a Paper on Fertilization and Hybridization.* 1889. Translated by C. Stuart Gager. Chicago: Open Court, 1910.

———. "Sur la loi de disjonction des hybrides." *Comptes rendus hebdomadaires des séances de l'Académie des sciences,* Vol. 30 (1900), 845–47.

Dias, Brian G., and Kerry J. Ressler, "Parental Olfactory Experience Influences Behavior and Neural Structure in Subsequent Generations." *Nature Neuroscience* 17, no. 1 (2014): 89–96.

Diderot, Denis. "Lettre sur les sourds et muets, à l'usage de ceux qui entendent et qui parlent." s.n., 1751.

———. *Rêve de d'Alembert* (1769). Reprinted in Paris: Bossard, 1921.

———. *Supplément au Voyage de Bougainville; ou, Dialogue entre A et B sur l'inconvénient d'attacher des idées morales à certaines actions physiques qui n'en comportent pas* (1796). Paris: Belin, 2011.

Dobzhansky, Theodosius. *Mankind Evolving: The Evolution of the Human Species.* New Haven, Conn.: Yale University Press, 1962.

Dolomieu, Déodat de. "Lettre [à Picot de Lapeyrouse] . . . sur un genre de pierres calcaires très peu effervescentes avec les acides et phosphorescentes par la collision." *Journal de physique,* Jan. 30, 1791.

———. "Mémoire sur les pierres composées et sur les roches," extrait du *Journal de physique,* Nov. 1791.

Duclos-Guyot, Pierre, and Philibert Commerson. *Journaux de Philibert Commerson, médecin botanist embarqué sur l'Étoile, et de Pierre Duclos-Guyot, volontaire sur la Boudeuse, puis sur l'Étoile.* In Taillemite, *Bougainville et ses compagnons,* 2:419–522.

Eckermann, John Peter. *Conversations of Goethe with Eckermann and Soret.* Translated by John Oxenford. London: George Bell & Sons, 1875.

Einstein, Albert. *The Collected Papers of Albert Einstein.* 16 vols. Princeton, N.J.: Princeton University Press, 1987–2021.

Faye, Hervé. *Sur l'origine du monde: Théories cosmogoniques des anciens et des modernes.* Paris: Gauthier-Villars, 1884.

Fisher, R. A. *The Genetical Theory of Natural Selection.* Oxford: Clarendon, 1930.

Flourens, Pierre. *Éloge historique de Marie-Henri* [sic] *Ducrotay de Blainville.* Paris: Firmin-Didot, 1854.

———. *Examen du livre de M. Darwin sur la phrénologie.* Paris: Garnier, 1864.

———. "Notice sur la Vénus hottentotte." *Journal complémentaire du Dictionnaire des sciences médicales* 4 (1819): 146.

———. *Ontologie naturelle; ou, Étude philosophique des êtres.* Paris: Garnier, 1861.

———. *Recueil des éloges historiques lus dans les séances publiques de l'Academie des Sciences.* 2nd series. Vol. 2. Paris: Garnier Frères, 1857.

Fourcroy, Antoine-François. *Philosophie chimique: Vérités fondamentales de la chimie moderne, disposées dans un nouvel ordre.* 2nd ed. Paris: Du Pont, 1795.

Franklin, Benjamin. *Experiments and Observations on Electricity.* London: E. Cave, 1751.

———. *The Papers of Benjamin Franklin.* Edited by Ellen Cohn. New Haven, Conn.: Yale University Press, 1959–.

Gall, Franz Joseph, and Johann Kaspar Spurzheim. *Anatomie et physiologie du système nerveux en général, et du cerveau en particulier, avec des observations sur la possibilité de reconnaître plusieurs dispositions intellectuelles et morales de l'homme et des animaux, par la configuration de leurs têtes.* 2 vols. Paris: F. Schoell, 1810.

Galton, Francis. "The Average Contribution of Each Several Ancestor to the Total Heritage of the Offspring." *Proceedings of the Royal Society of London* 61 (1897): 401–13.

———. "A Diagram of Heredity." *Nature* 57 (1898): 293.

———. *Hereditary Genius: An Inquiry into Its Laws and Consequences.* London: Macmillan, 1869.

———. *Inquiries into Human Faculty and Its Development.* London: Macmillan, 1883.

Geoffroy Saint-Hilaire, Étienne. "Description des crocodiles de l'Égypte." In *Description de l'Égypte,* 185–265.

———. *Fragments biographiques précédés d'études sur la vie, les ouvrages et les doctrines de Buffon.* Paris: F. D. Pillot, 1838.

———. "Histoire naturelle des poissons du Nil." In *Description de l'Égypte,* 1–52.

———. "Mémoire où l'on se propose de rechercher dans quels rapports de structure organique et de parenté sont entre eux les animaux des âges historiques, et vivant actuellement, et les espèces antédiluviennes et perdues." *Mémoires du Muséum national d'histoire naturelle* 17 (1828): 209–29.

———. "Mémoire sur le dégré d'influence du monde ambiant pour modifier les formes animals; question intéressant l'origine des espèces téléosauriennes et successivement celle des animaux de l'époque actuelle." *Mémoires de l'Académie des sciences de l'Institut de France* 12 (1833): 63–92.

———. "Observations sur l'affection mutuelle de quelques animaux et particulièrement sur les services rendus au Requin par le Pilote." *Annales du Muséum d'histoire naturelle* 9 (1807): 469–76.

———. *Philosophie anatomiques des monstruosités humaines.* Paris: Chez l'auteur, 1822.

———. *Principes de philosophie zoologique.* Paris: Pichon et Didier, 1830.

———. "Quelques considérations sur la giraffe." *Annales des sciences naturelles* 11 (1827): 210–23.

———. "Recherches sur l'organisation des gavials." *Mémoires du Muséum d'histoire naturelle* 12 (1825): 97–155.

———. *Traité sommaire des coquilles, tant fluviatiles que terrestres, qui se trouvent aux environs de Paris.* Paris: Musier, 1767.

Geoffroy Saint-Hilaire, Étienne, and Frédéric Cuvier. *Histoire naturelle des mammifères.* 7 vols. Paris: Belin, 1824–42.

Geoffroy Saint-Hilaire, Étienne, and Georges Cuvier. "Histoire naturelle des Orangs-Outans." *Magasin encyclopédique, ou, Journal des sciences, des lettres et des arts* 3 (1795): 451–63.

Bibliography

Geoffroy Saint-Hilaire, Isidore. "Histoire naturelle des poisons de la Mer Rouge et de la Méditerranée." In *Description de l'Égypte*, 311–43.

———. "Suite de l'histoire naturelle des poissons du Nil." In *Description de l'Égypte*, 265–310.

———. *Vie, travaux et doctrine scientifique d'Étienne Geoffroy Saint-Hilaire*. Paris: P. Bertrand, 1847.

Goethe, Johann Wolfgang von. "Dernières pages de Goethe expliquant à l'Allemagne les sujets de philosophie naturelle controversies au sein de l'Académie des sciences de Paris." *Revue encyclopédique* 53, part 1 (1832): 563–73, and *Revue encyclopédique* 54, part 2 (1832): 54–68.

Gouges, Olympe de. "Déclaration des droits de la femme et de la citoyenne." Paris, 1791.

———. "Réflexions sur les hommes nègres." In *Oeuvres de Madame de Gouges*. Vol. 3. Paris: Chez l'auteur, 1788.

Grant, Peter R., and B. Rosemary Grant. *40 Years of Evolution: Darwin's Finches on Daphne Major Island*. Princeton, N.J.: Princeton University Press, 2014.

Gray, Asa. *Darwiniana: Essays and Reviews Pertaining to Darwinism*. New York: Appleton, 1876.

Grew, Nehemiah. *The Anatomy of Plants: Idea of a Philosophical History of Plants*. London: W. Rawlins, 1682.

Guillaume, James, ed. *Procès-verbaux du Comité d'instruction publique de la Convention Nationale*. 6 vols. Paris: Imprimerie nationale, 1891–1907.

Gumbleton, W.-E. "Une introduction difficile." *Revue horticole: Journal d'horticulture pratique*, no. 74 (1902): 103–4.

Haeghens, J., Charles Martins, and A. Bérigny. *Annuaire météorologique de la France pour 1849*. Paris: Gaume Frères, 1848.

Haldane, J. B. S. *Everything Has a History*. London: Allen and Unwin, 1951.

———. *Heredity and Politics*. London: George Allen, 1938.

———. *The Inequality of Man*. London: Chatto and Windus, 1932. Harmondsworth: Penguin, 1938.

———. "Mathematical Darwinism: A Discussion of the Genetical Theory of Natural Selection." *Eugenics Review* 23, no. 2 (1931): 115–17.

———. "The Origin of Life." *Rationalist Annual* 3 (1929): 3–10.

Houel, Jean-Pierre-Louis-Laurent. *Histoire naturelle des deux éléphans, male et femelle, du Museum de Paris, venus de Hollande en France en l'an VI*. Paris: Chez l'auteur, 1803.

Howard, Luke. *Essay on the Modifications of Clouds*. 1803. 2nd ed. London: Harvey & Darton, 1832.

Huey, Raymond B., Paul E. Herz, and B. Sinervo. "Behavioral Drive Versus Behavioral Inertia in Evolution: A Null Model Approach." *American Naturalist* 161, no. 3 (2003): 357–66.

Humbert-Bazile, Nicolas-Edme-Marie, and Henri Nadault de Buffon. *Buffon, sa famille, ses collaborateurs et ses familiers*. Paris: Jules Renouard, 1863.

Hutton, James. *An Investigation into the Principles of Knowledge*. 3 vols. Edinburgh: Strahan and Cadell, 1794.

———. "Theory of the Earth." *Transactions of the Royal Society of Edinburgh* 1 (1788): 209–306.

———. *Theory of the Earth*. 2 vols. Edinburgh: Creech, 1795.

Huxley, Julian. "The Case for Eugenics." *Sociological Review* 18 (1926): 279–90.

———. *Essays in Popular Science*. New York: Knopf, 1927.

———. *Evolution: The Modern Synthesis*. London: George Allen and Unwin, 1942.

———. "The Humanist Frame." In *The Humanist Frame*, edited by Julian Huxley. New York: Harper and Brothers, 1961.

———. *If I Were Dictator*. New York: Harper, 1934.

———. *New Bottles for New Wine*. London: Chatto and Windus, 1957.

———. *Soviet Genetics and World Science: Lysenko and the Meaning of Heredity*. London: Chatto & Windus, 1949.

———. *UNESCO: Its Purpose and Philosophy*. Preparatory Commission of the United Nations Educational, Scientific, and Cultural Organisation. New York: UNESCO, 1946.

Huxley, Julian, and A. C. Haddon. *We Europeans: A Survey of Racial Problems*. 1935. Harmondsworth: Penguin, 1939.

Huxley, Thomas Henry. "On the Hypothesis That Animals Are Automata, and Its History." 1874. In *Collected Essays*, 1:199–250. London: Macmillan, 1894.

———. "On the Reception of the Origin of Species." In *The Life and Letters of Charles Darwin*, edited by Francis Darwin, 2:179–205. London: John Murray, 1887.

Jardine, William. "Memoir of Latreille." In *The Naturalist's Library (Entomology 5)*, 32:17–60. Edinburgh: W. H. Lizars, 1841.

Jauffret, Louis-François. *Voyage au Jardin des plantes . . . avec l'histoire des deux éléphans, et celle des autres animaux de la ménagerie nationale*, Sixième Journée, 205–40. Paris: Chez Houel, 1798. archive.org/details/voyageaujardinde00jauf/mode/2up.

Johannsen, Wilhelm. "The Genotype Conception of Heredity." *American Naturalist* 45, no. 531 (1911): 129–59.

Jones, T. B., and A. C. Kamil. "Tool-Making and Tool-Using in the Northern Blue Jay." *Science* 180, no. 4090 (1973): 1076–78.

Jussieu, Antoine de. *Discours sur le progrès de la botanique au Jardin royal de Paris, suivi d'une introduction à la connaissance des plantes, prononcez à l'ouverture des demonstrations publiques, le 31 may 1718*. Paris: Étienne Ganeau, 1718.

Jussieu, Antoine Laurent de. *Genera Plantarum*. Paris: Herissant, 1789.

———. "Observations on the Natural Order of Nyctaginaea." In *Annals of Botany*, edited by Charles Konig and John Sims, 2:278–87. London: R. Taylor, 1806. Original: "Observations sur la famille des plantes nyctaginées." *Annales du Muséum national d'histoire naturelle* 2 (1803): 269–79.

Kamil, A., and R. Balda. "Cache Recovery and Spatial Memory in Clark's Nutcrackers (*Nucifraga columbiana*)." *Journal of Experimental Psychology: Animal Behavior Processes* 11 (1985): 95–111.

Kammerer, Paul. *The Inheritance of Acquired Characteristics.* Translated by A. Paul Maerker-Branden. New York: Boni and Liveright, 1924.

——. "Paul Kammerer's Letter to the Moscow Academy of Sciences." *Science* 64, no. 1664 (1926): 493–94.

Karénine, Wladimir. *George Sand: Sa vie et ses œuvres.* 4 vols. Paris: Plon, 1899–1926.

Kellogg, Vernon L. *Darwinism Today: A Discussion of Present-Day Scientific Criticism of the Darwinian Selection Theories.* New York: Henry Holt, 1907.

Kennedy, M. J., M. Droser, L. M. Mayer, D. Pevear, and D. Mrofka. "Late Precambrian Oxygenation: Inception of the Clay Mineral Factory." *Science* 311 (2006): 1446–49.

Klein, Jacob Theodor. *Naturalis dispositio echinodermatum.* . . . Danzig: Schreiber, 1734.

Kroodsma, Donald. *Acoustic Communication in Birds: Production, Perception, and Design Features of Sounds.* Amsterdam: Elsevier Science, 1983.

Lacépède, Bernard Germain, comte de. *Histoire naturelle de l'homme, par M. le comte de Lacépède, précédée de son éloge historique par M. le baron G. Cuvier.* Paris: Levrault, 1827.

——. Introduction to *La Ménagerie du Muséum national d'histoire naturelle; ou, Description et histoire des animaux qui y vivent ou qui y ont vécu; par les citoyens Lacépède et Cuvier, avec des figures peintes d'après nature, par le citoyen Maréchal, peintre du Muséum,* by Bernard Germain Lacépède and Georges Cuvier, 1:1–9. Paris, 1801.

——. "Rapport des professeurs du Muséum, sur les collections d'histoire naturelle rapportées d'Égypte, par É. Geoffroy." *Annales du Muséum national d'histoire naturelle* 1 (1802): 234–41.

Lalande, Jérôme. "Éloge de M. Commerson." *Journal de physique, de chimie, d'histoire naturelle et des arts* 5 (1775): 89–120.

——. "Histoire de l'astronomie pour 1801." Extrait du *Magasin encyclopédique.* Paris: Chez Fuchs, n.d.

——. "Notices sur la vie et les ouvrages de Lavoisier." *Magasin encyclopédique, ou, Journal des sciences, des lettres et des arts* 5 (1795): 174–84.

Lamarck, Eugène de Monet de. *Lettres d'un marin.* Evreux: Auguste Hérissey, 1871.

Lamarck, Jean Baptiste. *Annuaires météorologiques pour l'an VIII [X, XI, XI, XIII, XIV, 1807, 1808, 1809, 1810] de la République.* Paris: Chez l'auteur au Muséum d'histoire naturelle, 1800–1809.

——. "Classes." In *Encyclopédie méthodique: Botanique,* 2: 29–36. Paris: Panckoucke, 1783.

——. "Conchyliologie." In *Nouveau dictionnaire d'histoire naturelle, appliquée aux arts, à l'agriculture, à l'économie rurale et domestique, à la médecine,* 7: 412–28. Paris: Déterville, 1817.

——. "Coquille." In *Nouveau dictionnaire d'histoire naturelle, appliquée aux arts, à l'agriculture, à l'économie rurale et domestique, à la médecine,* 7: 556–83. Paris: Déterville, 1817.

——. "De l'influence de la lune sur l'atmosphère terrestre." *Journal de physique* (1798): 428–35.

———. "Discours d'ouverture du cours des animaux sans vertèbres, prononcé dans le Muséum d'Histoire naturelle, en mai 1806." In *Discours d'ouverture (An VIII, An X, An XI et 1806)*, edited by A. Giard, 107–57. Paris, 1907.

———. *Discours d'ouverture prononcé le 21 floréal an 8.* Paris: Déterville, 1801.

———. *Extrait du cours de zoologie sur les animaux sans vertèbres.* Paris: D'Hautel & Gabon, 1812.

———. "Extrait d'un mémoire sur l'influence de la lune sur l'atmosphère terrestre, lu à l'Institut national, par le citoyen Lamarck." *Magasin encyclopédique* 2 (1798): 145–48.

———. *Flore française; ou, Description succincte de toutes les plantes qui croissant naturellement en France.* 3 vols. Paris: Imprimerie royale, 1778.

———. *Histoire naturelle des animaux sans vertèbres.* 7 vols. Paris: Déterville, 1815–22.

———. *Histoire naturelle des végétaux.* Vol. 1. Paris: Déterville, 1803.

———. "Homme." In *Nouveau dictionnaire d'histoire naturelle, appliquée aux arts, à l'agriculture, à l'économie rurale et domestique, à la médecine*, 15: 270–76. Paris: Déterville, 1817.

———. *Hydrogéologie; ou, Recherches sur l'influence qu'ont les eaux sur la surface du globe terrestre.* Paris: Chez l'auteur, 1802.

———. *Inédits de Lamarck.* Edited by Max Vachon, Georges Rousseau, and Yves Laissus. Paris: Masson, 1972.

———. *Mémoires de physique et d'histoire naturelle.* Paris: Chez l'auteur, 1797.

———. "Mémoire sur le mode de rédiger et de noter les observations météorologiques, afin d'en obtenir des résultats utiles, et sur les considérations que l'on doit avoir en vue pour cet objet." *Journal de physique, de chimie, d'histoire naturelle et des arts* 51 (1800): 419–26.

———. "Mémoire sur le projet du comité des Finances" (1789). In Landrieu, *Lamarck*, 36–39.

———. "Mémoire sur les cabinets d'histoire naturelle et particulièrement sur celui du Jardin des plantes" (1790). Reproduced in Landrieu, *Lamarck*, 42–52.

——— [Louis Cotte]. "Mémoire sur les principaux phénomènes de l'atmosphère. Sentiment de M. Le Chevalier de la Marck." [Cotte's summary based upon Lamarck's manuscript of his presentation to the Académie des sciences in 1777.] In *Mémoires sur la météorologie, pour server de suite et de supplement au Traité de météorologie, publié en 1774*, 1: 205–15. Paris: Imprimerie royale, 1788.

———. "Météorologie." In *Nouveau dictionnaire d'histoire naturelle, appliquée aux arts, à l'agriculture, à l'économie rurale et domestique, à la médecine, etc. Par une société de naturalistes et d'agriculteurs*, 20: 451–77. Paris: Déterville, 1818.

———. "Nouvelle définition des termes que j'emploie pour exprimer certaines formes de nuages." *Annuaire météorologique pour l'an XIII de l'ère de la république française* (1805): 112–25.

———. *Philosophie zoologique; ou, Exposition des considérations relatives à l'histoire naturelle des animaux.* 2 vols. Paris: Dentu, 1809.

———. *Recherches sur les causes des principaux faits physiques.* 2 vols. Paris: Maradan, 1794.

———. *Recherches sur l'organisation des corps vivants.* Paris: Maillard, 1802.

———. *Réfutation de la théorie pneumatique.* Paris: Agasse, 1796.

———. "Sur la forme des nuages." *Annuaire météorologique pour l'an X* (1802): 149–64.

———. *Système analytique des connaissances positives de l'homme.* Paris: A. Belin, 1820.

———. *Système des animaux sans vertèbres.* Paris: Déterville, 1801.

———. *Tableau encyclopédique et méthodique. Botanique.* Paris: Panckoucke/Agasse, 1791–1823.

Lamothe-Langon, Étienne-Léon de. *Mémoires de Napoléon.* 6 vols. Paris: Charles Gosselin, 1834–35.

Landolphe, Jean-François. *Memoires, contenant l'histoire de ses voyages pendant 36 ans, aux cotes d'Afrique et aux Ameriques; red. sur son manuscrit, par Jacques-Salbigoton Quesne.* 2 vols. Paris: Bertrand & Pillet, 1823.

Lankester, E. Ray. "Functionless Organs." Letter in *Nature*, Aug. 16, 1888, 364.

Laplace, Pierre Simon. *Exposition du système du monde.* 6th ed. Paris: Bachelier, 1835.

———. *Œuvres complètes de Laplace.* 14 vols. Paris: Gauthier-Villars, 1878–1912.

———. *A Philosophical Essay Concerning Probabilities.* Translated from the 6th French edition by Frederick Wilson Truscott. New York: John Wiley, 1902. First published in 1814.

Latreille, Pierre-André. *Histoire naturelle, générale et particulière, des crustacés et des insectes.* 14 vols. Paris: F. Dufart, 1802–5.

———. *Précis des caractères génériques des insectes, disposés dans un ordre naturel.* Brive: F. Bourdeaux, 1796 (?).

Lavoisier, Antoine Laurent. *Traité élémentaire de chimie.* 2 vols. Paris: Cuchet, 1789.

Lavoisier, Antoine Laurent, and Pierre Simon Laplace. "Mémoire sur la chaleur." Paris: Gauthier-Villars, 1920. First published in *Mémoires de l'Académie des sciences pour l'année 1780* (1783): 355–408.

Lea, A. J., J. Altmann, S. C. Alberts, and J. Tung. "Resource Base Influences Genome-Wide DNA Methylation Levels in Wild Baboons (*Papio cynocephalus*)." *Molecular Ecology* 25, no. 8 (2016): 1681–96.

Leeuwenhoek, Antony van. "Part of a Letter from Mr. Antony van Leeuwenhoek, F.R.S., Concerning Green Weeds Growing in Water, and Some Animalcula Found About Them." *Philosophical Transactions of the Royal Society of London* 23, no. 283 (1703): 1304–11.

Legallois, César Julien Jean. *Expériences sur le principe de vie, notamment sur celui des mouvements du cœur, et sur le siège de ce principe, suivies du rapport fait à ta première classe de l'Institut sur celles relatives aux mouvements du cœur.* Paris: d'Hautel, 1812.

Leibniz, G. W., and Samuel Clarke. *Correspondence.* Edited by Roger Ariew. Indianapolis: Hackett, 2000.

Le Preux, Paul-Gabriel. *Éloge de M. de Jussieu, docteur régent de la Faculté de médecine de Paris . . . lu dans la séance publique de ladite Faculté, le 5 octobre 1778.* S.l., s.n., 1779.

Levins, Richard, and Richard Lewontin. "The Problem of Lysenkoism." In *The Dialectical Biologist*, 163–97. Cambridge, Mass.: Harvard University Press, 1985.

Lewontin, Richard. "Organism and Environment." In *Learning, Development and Culture*, edited by H. C. Plotkin, 151–70. Chichester, U.K.: Wiley, 1982.

———. *The Triple Helix: Gene, Organism, and Environment*. Cambridge, Mass.: Harvard University Press, 2000.

Liew, Y. J., et al. "Intergenerational Epigenetic Inheritance in Reef-Building Corals." In *Nature Climate Change* 10 (2020): 254–59.

Linnaeus, Carl. *Systema naturae*. 10th ed. (1758–59). 7 vols. Translated by William Turton. London: Lackington, Allen, 1802–6.

Louchart, Antoine, and Laurent Viriot, "From Snout to Beak: The Loss of Teeth in Birds." *Trends in Ecology & Evolution* 26, no. 12 (2011): 663–73.

Lubbock, Constance, ed. *The Herschel Chronicle: The Life-Story of William Herschel and His Sister Caroline Herschel*. Cambridge, U.K.: Cambridge University Press, 1933.

Lucretius. *De Rerum Natura*. Loeb Classical Library edition. Edited by E. Capps, T. E. Page, and W. H. D. Rouse. Translated by W. H. D. Rouse. New York: G. P. Putnam's Sons, 1924.

Lyell, Charles. *The Geological Evidences of the Antiquity of Man, with Remarks on Theories of the Origin of Species by Variation*. London: John Murray, 1863.

———. *Principles of Geology, Being an Attempt to Explain the Former Changes of the Earth's Surface, by Reference to Causes Now in Operation*. Vol. 2, 1st ed. London: John Murray, 1832. Vol. 2, 12th ed. London: John Murray, 1875.

Lysenko, Trofim Denisovich. *The Science of Biology Today*. New York: International Publishers, 1948.

MacBride, E. W. *Introduction to the Study of Heredity*. London: Williams and Norgate, 1924.

Mayr, Ernst. *Animal Species and Evolution*. Cambridge, Mass.: Harvard University Press, 1963.

———. "The Biology of Race and the Concept of Equality." In "On Equality." Special issue, *Daedalus* 131, no. 1 (2002): 89–94.

———. "The Emergence of Evolutionary Novelties." In *Evolution After Darwin*. Vol. 1, *The Evolution of Life, Its Origin, History, and Future*, edited by Sol Tax, 349–80. Chicago: University of Chicago Press, 1960.

———. *The Growth of Biological Thought: Diversity, Evolution, and Inheritance*. Cambridge, Mass.: Belknap Press of Harvard University Press, 1982.

———. *Toward a New Philosophy of Biology: Observations of an Evolutionist*. Cambridge, Mass.: Belknap Press of Harvard University Press, 1988.

———. *What Makes Biology Unique?* Cambridge, Mass.: Harvard University Press, 2004.

———. "Where Are We?" *Cold Spring Harbor Symposium on Quantitative Biology* 24 (1959): 1–14.

Mendel, Gregor. *Experiments in Plant Hybridisation*. 1865. Translated by the Royal Horticultural Society of London. Cambridge, Mass.: Harvard University Press, 1925.

Mérat de Vaumartoise, François Victor. "Anecdote sur le Cèdre du Liban." *Annales de la Société royale d'horticulture de Paris* (1836): 112–14.

Michaud, Louis Gabriel. "Tournefort." In *Biographie universelle, ancienne et moderne*, 2nd ed. (Paris: Delagrave, 1843–1865), vol. 42: 41–47.

Millikan, G. C., and R. I. Bowman. "Observations on Galápagos Tool-Using Finches in Captivity." *Living Bird* 6 (1967): 23–41.

Moreau de la Sarthe, Jacques Louis. *Histoire naturelle de la femme, suivie d'un traité d'hygiène.* Vols. 1–2. Paris: Duprat, 1803.

Morgan, Thomas Hunt. *A Critique of the Theory of Evolution.* Princeton, N.J.: Princeton University Press, 1916.

Morveau, Guyton de, and Antoine-Laurent Lavoisier, Claude Berthollet, and Antoine-François Fourcroy. *Méthode de nomenclature chimique.* Paris: Chez Cuchet, 1787.

Muñoz, Martha. "The Bogert Effect, a Factor in Evolution." *Evolution: International Journal of Organic Evolution* 76, no. S1 (2022): 49–66.

Neill, Patrick, ed. *Journal of a Horticultural Tour Through Some Parts of Flanders, Holland, and the North of France, in the Autumn of 1817, by a Deputation of the Caledonian Horticultural Society.* Edinburgh: Bell & Bradfute, 1823.

Newton, Isaac. *The Chronology of Ancient Kingdoms Amended.* London: Tonson, 1728.

———. "General Scholium." In *The Mathematical Principles of Natural Philosophy*, translated by Andrew Motte, 2:387–93. London: Benjamin Motte, 1729.

———. *Opticks; or, A Treatise of the Reflections, Refractions, Inflections, and Colours of Light.* 4th ed. London: William Innys, 1730.

Nicard, Pol. *Étude sur la vie et les travaux de M. Ducrotay de Blainville.* Paris: Baillière, 1890.

Noble, G. Kingsley. "Kammerer's Alytes." *Nature*, Aug. 7, 1926, 209–11.

Nollet, Jean-Antoine. *Essai sur l'électricité des corps.* 2nd ed. Paris: Guerin and Mathurin, 1775.

———. "Observations sur quelques nouveaux phénomènes d'electricité." In *Mémoires de l'Académie royale des sciences pour l'année 1746*, 1–23. Paris: Imprimerie royale, 1751.

Odling-Smee, F. J. "Niche Constructing Phenotypes." In *The Role of Behavior in Evolution*, edited by H. C. Plotkin, 73–132. Cambridge, Mass.: MIT Press, 1988.

———. *Niche Construction: How Life Contributes to Its Own Evolution.* Cambridge, Mass.: MIT Press, 2024.

Osborn, Henry Fairfield. *Impressions of Great Naturalists: Reminiscences of Darwin, Huxley, Balfour, Cope, and Others.* New York: C. Scribner's Sons, 1924.

Othon, Charles-Nicholas, Prince of Nassau-Siegen. *Journal de Charles-Othon de Nassau-Siegen, passager sur la Boudeuse.* In Taillemite, *Bougainville et ses compagnons*, 2: 371–418.

Owen, Richard. *On the Anatomy of Vertebrates.* Vol. 3. London: Longmans, Green, 1868.

Packard, Alpheus Spring. "A Century's Progress in American Zoology." *American Naturalist* 10 (1876): 591–98.

[Packard, Alpheus Spring]. Introduction to *The Standard Natural History.* Vol. 1, *Lower Invertebrates*, edited by John Sterling Kingsley. Boston: S. E. Cassino, 1885.

Packard, Alpheus Spring, and F. W. Putnam. *The Mammoth Cave and Its Inhabitants.* Salem: Salem Press, 1879.

Pasquier, Étienne. *Éloge de M. le baron Georges Cuvier. Chambre des Pairs. Séance du 17 décembre 1832.* Paris: s.n., 1832.

Pasteur, Louis. "II. Générations dites spontanées (1860–1866)"; "Discussion avec le Dr. Bastian sur les générations dites spontanées"; "Rapport sur les expériences relatives à la génération spontanée." In *Œuvres de Pasteur.* Vol. 2, *Fermentations et générations dites spontanées,* 185–358, 459–77, 637–47. Paris: Masson, 1922.

———. *Oeuvres de Pasteur.* 7 vols. Paris: Masson, 1922.

Pennisi, Elizabeth. "Natural Inspiration," *Science* 7, no. 6613 (2022).

Pfaff, Cathrin, Julia A. Schultz, and Rico Schellhorn. "The Vertebrate Middle and Inner Ear: A Short Overview." *Journal of Morphology* 280, no. 8 (2019): 1098–105.

Pouchet, Félix-Archimède. *Hétérogénie; ou, Traité de la generation spontanée.* Paris: Baillière, 1859.

———. "Note sur des proto-organismes végétaux et animaux, nés spontanément dans l'air artificiel et dans le gaz oxygène." *Comptes rendus de l'Académie des sciences* 47 (1858): 979–84.

———. *Théorie positive de l'ovulation spontanée et de la fécondation des mammifères et de l'espèce humaine: Basée sur l'observation de toute la série animale.* Paris: Baillière, 1847.

Poulton, Edward B. "Dr. Romanes' Article in the Contemporary Review for June." *Nature,* July 26, 1888, 295–96.

———. "Lamarckism Versus Darwinism." Letter in *Nature* 38, no. 982 (1888): 388.

Prévost, Constant. "Funérailles de M. Blainville: Discours de M. Constant Prevost . . . le mardi 7 mai 1850." Institut de France, Académie des sciences.

Priestley, Joseph. *The History and Present State of Electricity.* London: Dodsley, 1767.

Quatrefages de Bréau, Jean Louis Armand de. *Charles Darwin et ses précurseurs français: Étude sur le transformisme.* Paris: Germer Baillière, 1870.

Quinet, Edgar. "Discours sur Geoffroy Saint-Hilaire." In *Oeuvres complètes d'Edgar Quinet,* 339–42. Paris: Pagnerre, 1857.

Rabbe, Alphonse, Claude Augustin Vieilh de Boisjolin, and Charles-Augustin Sainte-Beuve, eds. "Buffon." In *Biographie universelle et portative des contemporains; ou, Dictionnaire historique des hommes vivants et des hommes morts depuis 1788 jusqu'à nos jours.* Vol. 1. Paris: Chez l'éditeur, 1836.

Réaumur, René-Antoine Ferchault de. *Lettres élémentaires sur la botanique et receuil des plantes colorées [Œuvres complètes de Rousseau].* Paris: Poinçot, 1789.

———. *Memoires pour servir a l'histoire des insectes.* 6 vols. Paris: De l'imprimerie royale, 1734–42.

Rivarol, Antoine de. *Discours sur l'universalité de la langue française.* Paris: Bailly, Dessenne, 1784.

Romanes, George J. "Lamarckism *Versus* Darwinism." Letter in *Nature* 38, no. 983 (1888): 413.

Sainte-Beuve, Charles Augustin. *Volupté.* Bruxelles: Charpentier, 1835.

Saint-Germain, Louis-Antoine de. "Le routier inédit d'un compagnon de Bougainville." *La géographie* 35, no. 3 (1921): 15–250.

Sand, George. *Histoire de ma vie.* 10 vols. Paris: M. Lévy, 1856.

Saussure, Horace-Bénédict de. *Voyage dans les Alpes.* 4 vols. Neuchâtel: Samuel Fauche, 1779–96.

Savigny, Jules-César. *Mémoires sur les animaux sans vertèbres.* Paris: Déterville, 1816.

Schrödinger, Erwin. "Mind and Matter." 1956. In *What Is Life?,* 91–164.

———. *What Is Life?* 1944. Cambridge, U.K.: Cambridge University Press, 1967.

Scott, John P. *Animal Behavior.* Chicago: University of Chicago Press, 1972.

Shaw, George Bernard. *Back to Methuselah: A Metabiological Pentateuch.* London: Constable, 1921.

Slagsvold, Tore, and Karen L. Wiebe. "Social Learning in Birds and Its Role in Shaping a Foraging Niche." *Philosophical Transactions of the Royal Society B* 366, no. 1567 (2011): 969–77.

Smith, Thomas. *The Wonders of Nature and Art; or, A Concise Account of Whatever Is Most Curious and Remarkable in the World.* Vol. 11. London: J. Walker, 1804.

Spencer, Herbert. *Essays: Scientific, Political, and Speculative.* 3 vols. London: Williams and Norgate, 1891.

———. *Principles of Biology.* Vol. 1. London: Williams and Norgate, 1864.

Spurzheim, Johann Gaspar. *Observations sur la phraenologie; ou, La connaissance de l'homme moral et intellectuel, fondée sur les fonctions du système nerveux.* Paris: Treuttel & Würtz, 1818.

Stahl, Georg Ernst. *Zymotechnia Fundamentalis, Seu Fermentationis Theoria Generalis.* Halle: Christoph Salfeld, 1697.

Stresemann, Erwin. *Ornithology from Aristotle to the Present.* 1951. Edited by G. William Cottrell. Translated by Hans J. Epstein and Cathleen Epstein. With a foreword and epilogue by Ernst Mayr. Cambridge, Mass.: Harvard University Press, 1975.

Taillemite, Étienne, ed. *Bougainville et ses compagnons autour du monde.* 2 vols. Paris: Imprimerie nationale, 1977.

Thewissen, J. G. M., et al., "Evolutionary Aspects of the Development of Teeth and Baleen in the Bowhead Whale." *Journal of Anatomy* 230, no. 4 (2017): 549–66.

Thibaudeau, Antoine Clair. *Mémoires sur le Consulat, 1799 à 1804. Par un ancien Conseiller d'état.* Paris: Chez Ponthieu, 1827.

Toaldo, Giuseppe. *Essai météorologique sur la véritable influence des astres, des saisons et changemens de tems.* Translated by Joseph D'Aquin. Chambery: Gorrin, 1784.

Toscan, Georges. "Du pouvoir de la musique sur les animaux, et du concert donné aux éléphans." *Décade philosophique, littéraire et politique,* no. 32 (1798): 257–64, and no. 33 (1798): 321–29.

———. *Histoire du lion de la Ménagerie du Muséum national d'histoire naturelle et de son chien.* Paris: Cuchet, 1795.

Toulmin, G. H. *The Antiquity and Duration of the World.* London: Cadell, 1780.

Tournefort, Joseph Pitton de. *Relation d'un voyage du Levant fait par ordre du Roy.* 3 vols. Paris: Imprimerie royale, 1717.

Trembley, Abraham. *Mémoires pour servir à l'histoire d'un genre de polypes d'eau douce, à bras en forme de cornes.* Leyden: Chez Jean & Herman Verbeek, 1744.

Trembley, Maurice. *Correspondance inédite entre Réaumur et Abraham Trembley comprenant 113 lettres recuillies et annotées.* Genève: Georg, 1943.

Vaillant, Sébastien. *Discours sur la structure des fleurs . . . prononcé à l'ouverture du Jardin royal de Paris le Xe jour du mois de juin 1717.* English translation: "Sebastian Vaillant's 1717 Lecture on the Structure and Function of Flowers," edited and translated by Paul Bernasconi and Lincoln Taiz, *Huntia: A Journal of Botanical History* 11, no. 2 (2002): 105–18. www.huntbotanical.org/admin/uploads/02hibd-huntia-11-2-pp97-128.pdf.

Veilleux, H. D., T. Ryu, J. M. Donelson, L. van Herwerden, L. Seridi, Y. Ghosheh, M. L. Berumen, W. Leggat, T. Ravasi, and P. L. Munday. "Molecular Processes of Transgenerational Acclimation to a Warming Ocean." *Nature Climate Change* 5 (2015): 1074–78.

Villermé, Louis-René. "Mémoire sur la taille de l'homme en France." *Annales d'hygiène publique,* no. 1 (1829): 351–95.

Virey, Julien-Joseph. "Homme." In *Nouveau dictionnaire d'histoire naturelle, appliquée aux arts, à l'agriculture, à l'économie rurale et domestique, à la médecine,* 15:1–270. Paris: Déterville, 1817.

Vivès, François. *Journal de François Vivez, chirurgien sur l'Étoile.* In Taillemite, *Bougainville et ses compagnons,* 2:179–291.

Waddington, C. H. *Behind Appearance: A Study of the Relations Between Painting and the Natural Sciences in This Century.* Cambridge, Mass.: MIT Press, 1969.

———. "Canalization of Development and the Inheritance of Acquired Characteristics." *Nature,* Nov. 14, 1942, 563–65.

———. "The Epigenotype." *Endeavor* 1 (1942): 18–20. Reprinted in the *International Journal of Epidemiology,* no. 41 (2012): 10–13.

———. *Science and Ethics.* London: George Allen and Unwin, 1942.

———. *The Scientific Attitude.* West Drayton, Middlesex: Penguin, 1941.

———. *The Strategy of the Genes: A Discussion of Some Aspects of Theoretical Biology.* 1957. London: Routledge, 2014.

Wallace, Alfred Russel. "On the Tendency of Varieties to Depart Indefinitely from the Original Type." *Zoological Journal of the Linnean Society,* Aug. 20, 1858, 46–50.

Ward, Lester F. *Lester Frank Ward and the Welfare State.* Edited by Henry Steele Commager. Indianapolis: Bobbs-Merrill, 1967.

———. *Neo-Darwinism and Neo-Lamarckism.* Washington, D.C.: Gedney and Roberts, 1891.

———. *Pure Sociology: A Treatise on the Origin and Spontaneous Development of Society.* New York: Macmillan, 1903.

Weismann, August. *Essays upon Heredity and Kindred Biological Problems.* Translated by Edward B. Poulton, Selmar Schönland, and Arthur E. Shipley. Oxford: Clarendon, 1889.

———. *The Germ-Plasm: A Theory of Heredity.* 1892. Translated by W. Newton Parker and Harriet Rönnfeldt. New York: Scribner's, 1893.

———. "On the Mechanical Conception of Nature." In *Studies in the Theory of Descent*, edited and translated by Raphael Meldola, 2:634–718. London: Sampson, Low, Marston, Searle, and Rivington, 1882.

Weldon, W. F. R. "On the Ambiguity of Mendel's Categories." *Biometrika* 2 (1902): 44–55.

Weyrich, A., D. Lenz, and J. Fickel. "Environmental Change-Dependent Inherited Epigenetic Response." *Genes* 10, no. 1 (2018): 4.

Whitehead, Hal, and Luke Rendell. *The Cultural Lives of Whales and Dolphins*. Chicago: University of Chicago Press, 2015.

Wilson, E. O. *On Human Nature*. 1978. Cambridge, Mass.: Harvard University Press, 2004.

———. *Sociobiology: The New Synthesis*. Cambridge, Mass.: Harvard University Press, 1975.

Wright, Charles W. *A Guide Manual to the Mammoth Cave of Kentucky*. Louisville, Ky.: Bradley and Gilbert, 1860.

Wrisberg, Heinrich August. *Observationum de animalculis infusoriis*. Göttingen: Vandenhoek, 1765.

Yehuda, Rachel. "How Parents' Trauma Leaves Biological Traces in Children." *Scientific American*, July 1, 2022.

Zhou, Xiumei, and Yongsheng Liu. "Hybridization by Grafting: A New Perspective?" *HortScience* 50, no. 4 (2015): 520–21.

Zirkle, Conway. *Evolution, Marxian Biology, and Social Science*. Philadelphia: University of Pennsylvania Press, 1959.

SECONDARY SOURCES

Adams, Mark B., ed. *The Evolution of Theodosius Dobzhansky: Essays on His Life and Thought in Russia and America*. Princeton, N.J.: Princeton University Press, 1994.

Alder, Ken. *The Measure of All Things: The Seven-Year Odyssey and Hidden Error That Transformed the World*. New York: Free Press, 2002.

Allen, Garland. *Thomas Hunt Morgan, the Man and His Science*. Princeton, N.J.: Princeton University Press, 1978.

American Philosophical Society Papers of Robert M. Patterson. "Biographical/Historical Note." as.amphilsoc.org/repositories/2/resources/1736 accessed 7/1/2025.

Anderson, Mary Perle. "The Cedar of Lebanon." *Torreya* 8, no. 12 (1908): 287–92.

Andress, David. *The Terror: The Merciless War for Freedom in Revolutionary France*. New York: Farrar, Straus, Giroux, 2006.

Appel, Toby A. *The Cuvier-Geoffroy Debate: French Biology in the Decades Before Darwin*. New York: Oxford University Press, 1987. Available online through Stanford.

Ashby, Jack. "The Earliest Strange Creatures: Europe's First Meetings with Marsupials." *UCL Culture Blog*, May 5, 2015. blogs.ucl.ac.uk/museums/2015/05/05/the-earliest-strange-creatures-europes-first-meetings-with-marsupials.

Ashworth, William B. "Jean Baptiste d'Omalius d'Halloy." Scientist of the Day, Linda

Hall Library, Feb. 16, 2021. www.lindahall.org/about/news/scientist-of-the-day/jean
-baptiste-domalius-dhalloy.

Asma, Stephen. "Darwin's Causal Pluralism." *Biology and Philosophy* 11, no. 1 (1996): 1–20.

Audelin, Louise. "Les Jussieu, une dynastie de botanistes au XVIIIe siècle, 1680–1789."
2 vols. Thesis, École national des chartes, 1987.

Baczko, Bronisław. *Ending the Terror: The French Revolution After Robespierre*. Translated
by Michael Petheram. Cambridge, U.K.: Cambridge University Press, 1994. First published as *Comment sortir de la Terreur*. Gallimard, 1989.

Badou, Gérard. *L'énigme de la Vénus hottentote*. Vanves: J.-C. Lattès, 2000.

Baker, John R. *Abraham Trembley of Geneva: Scientist and Philosopher, 1710–1784*. London: Edward Arnold, 1952.

Barlier, Jean-Pierre. *La Société des amis des noirs, 1788–1791: Aux origines de la première abolition de l'esclavage, 4 février 1794*. Paris: Éditions de l'Amandier, 2010.

Barnett, Lydia. *After the Flood: Imagining the Global Environment in Early Modern Europe*.
Baltimore: Johns Hopkins University Press, 2019.

Bashford, Alison. *The Huxleys: An Intimate History of Evolution*. Chicago: University of
Chicago Press, 2022.

Bell, David A. *Napoleon: A Concise Biography*. New York: Oxford University Press, 2015.

Bénichou, Claude, and Claude Blanckaert. *Julien-Joseph Virey: Naturaliste et anthropologue*.
Paris: Vrin, 1988.

Benson, Keith R. "Why American Marine Stations? The Teaching Argument." *American
Zoologist* 28, no. 1 (1988): 7–14.

Bernasconi, Paul, and Lincoln Taiz. "Sebastian Vaillant's 1717 Lecture on the Structure and
Function of Flowers." *Huntia: A Journal of Botanical History* 11, no. 2 (2002): 97–104.
www.huntbotanical.org/admin/uploads/02hibd-huntia-11-2-pp97–128.pdf.

———, eds. and trans. "Claude-Joseph Geoffroy's 1711 Lecture on the Structure and Uses
of Flowers." *Huntia: A Journal of Botanical History* 13, no. 1 (2006): 5–86. www.hunt
botanical.org/admin/uploads/02hibd-huntia-13-1-pp5-86.pdf.

Blanc, Olivier. *Marie-Olympe de Gouges, 1748–1793: Des droits de la femme à la guillotine*.
Paris: Tallandier, 2014.

Blanckaert, Claude. "Les conditions d'émergence de la science des races au début du XIXe
siècle." In *L'idée de "race" dans les sciences humaines et la littérature (XVIIIe–XIXe siècles). Actes du colloque international de Lyon (16–18 novembre 2000)*, edited by Sarga
Moussa, 133–49. Paris: L'Harmattan, 2003.

———. "La mesure de l'intelligence. Jeu des forces vitales et réductionnisme cérébral selon
les anthropologues français (1860–1880)." *Ludus Vitalis: Revista de filosofía de las cien-
cis de la vida* 2, no. 3 (1994): 35–68.

———. "'Les vicissitudes de l'angle facial' et les débuts de la craniométrie (1765–1875)." *Revue de synthèse* 108, no. 3–4 (1987): 417–53.

———, ed. *La Vénus hottentote*. Paris: Publications scientifiques du Muséum, 2013.

Borrello, Mark, and David Sepkoski. "Ideology as Biology." *New York Review of Books*,
Feb. 5, 2022.

Bourdier, Franck. "Lamarck et Geoffroy Saint-Hilaire face au problème de l'évolution biologique." *Revue d'histoire des sciences* 25, no. 4 (1972): 311–25.

Bowler, Peter. *The Eclipse of Darwinism: Anti-Darwinian Evolution Theories in the Decades Around 1900.* Baltimore: Johns Hopkins University Press, 1983.

———. "E. W. MacBride's Lamarckian Eugenics and Its Implications for the Social Construction of Scientific Knowledge." *Annals of Science* 41, no. 3 (1984): 245–60.

Brauer, Fae. "'L'Art eugénique': Biopower and the Biocultures of Neo-Lamarckian Eugenics." *L'esprit créateur* 52, no. 2 (2012): 42–58.

Brin, Gershon. *The Concept of Time in the Bible and the Dead Sea Scrolls.* Leiden: Brill, 2001.

Buchwald, Jed, and Mordechai Feingold. *Newton and the Origin of Civilization.* Princeton, N.J.: Princeton University Press, 2013.

Buffetaut, Éric. "Henri-Marie Ducrotay de Blainville inventa la paléontologie." *Pour la science*, no. 541 (Nov. 2022): 72–78.

Burkhardt, Richard W. "Constructing the Zoo: Science, Society, and Animal Nature at the Paris Menagerie, 1794–1831." In *Animals in Human Histories: The Mirror of Nature and Culture*, edited by Mary J. Henninger-Voss, 231–57. Rochester, N.Y.: University of Rochester Press, 2002.

———. "Lamarck, Cuvier, and Darwin on Animal Behavior and Acquired Characters." In Gissis and Jablonka, *Transformations of Lamarckism*, 33–44.

———. "Lamarck, Evolution, and the Inheritance of Acquired Characters." *Genetics* 194, no. 4 (2013): 793–805.

———. "Lamarck, Evolution, and the Politics of Science." *Journal of the History of Biology* 3, no. 2 (1970): 275–98.

———. *The Leopard in the Garden: Animal and Human Lives in Paris at the First Public Zoo of the Modern Era.* Chicago: University of Chicago Press, 2025.

———. "The Leopard in the Garden: Life in Close Quarters at the Muséum d'Histoire Naturelle." *Isis* 98, no. 4 (2007): 675–94.

Caffin-Carcy, Odile. "Que devint Versailles après le depart de la Cour?" *Revue historique* 286, no. 1(579) (1991): 53–80.

Cahan, Emily D. "The Genetic Psychologies of James Mark Baldwin and Jean Piaget." *Developmental Psychology* 20, no. 1 (1984): 128–35.

Cap, Paul Antoine. *Philibert Commerson, naturaliste, voyageur: Étude biographique, suivie d'un appendice (comprenant le testament, la correspondance et des fragments de Commerson).* Paris: V. Masson & fils, 1861.

Carey, Nessa. *The Epigenetics Revolution: How Modern Biology Is Rewriting Our Understanding of Genetics, Disease, and Inheritance.* New York: Columbia University Press, 2012.

Catalogue of Life. https://www.catalogueoflife.org. Accessed Aug. 18, 2025.

Celestin, Louis-Cyril. *Charles-Edouard Brown-Séquard: The Biography of a Tormented Genius.* Heidelberg: Springer, 2014.

Charlton, Noel G. *Understanding Gregory Bateson: Mind, Beauty, and the Sacred Earth.* Albany: State University of New York Press, 2008.

Churchill, Frederick B. *August Weismann: Development, Heredity, and Evolution.* Cambridge, Mass.: Harvard University Press, 2015.

Clark, William. *Academic Charisma and the Origins of the Research University.* Chicago: University of Chicago Press, 2006.

Cockerell, T. D. A. "Biographical Memoir of Alpheus Spring Packard, 1839–1905." In *Biographical Memoirs.* Vol. 9. Washington, D.C.: National Academy of Sciences, 1920.

Corsi, Pietro. *The Age of Lamarck: Evolutionary Theories in France, 1790–1830.* Berkeley: University of California Press, 1988.

———. "Biologie." In Corsi et al., *Lamarck, philosophe de la nature,* 37–64.

———. "Idola Tribus: Lamarck, Politics, and Religion in the Early Nineteenth Century." In *The Theory of Evolution and Its Impact,* edited by Aldo Fasolo, 11–39. Milan: Springer, 2012.

———. "Julien-Joseph Virey, the First Critic of Lamarck." In *Histoire du concept d'espèce dans les sciences de la vie: Colloque international, mai 1985,* 176–87. Paris: Fondation Singer Polignac, 1987.

———. "The Revolutions of Evolution: Geoffroy and Lamarck, 1825–1840." *Bulletin du Musée d'Anthropologie préhistorique de Monaco* 51 (2011): 97–122.

Crais, Clifton, and Pamela Scully. *Sarah Baartman and the Hottentot Venus: A Ghost Story and a Biography.* Princeton, N.J.: Princeton University Press, 2008.

Dal Prete, Ivano. *On the Edge of Eternity: The Antiquity of the Earth in Medieval and Early Modern Europe.* Oxford: Oxford University Press, 2022.

Damkaer, David M. *A Copepodologist's Cabinet: A Biographical and Bibliographical Dictionary,* 124–32. Philadelphia: American Philosophical Society, 2002.

Damrosch, Leo. *Jean-Jacques Rousseau: Restless Genius.* New York: Houghton Mifflin, 2005.

Daston, Lorraine. "Cloud Physiognomy." In "Description Across Disciplines." Special issue, *Representations,* no. 135 (Summer 2016): 45–71.

Dawkins, Richard. *The Extended Phenotype: The Long Reach of the Gene.* 1982. Oxford: Oxford University Press, 1999.

———. *The God Delusion.* London: Bantam, 2006.

———. *The Selfish Gene.* 1976. Oxford: Oxford University Press, 2006.

Dawson, Virginia. *Nature's Enigma: The Problem of the Polyp in the Letters of Bonnet, Trembley, and Réaumur.* Philadelphia: American Philosophical Society, 1987.

Day, David. *Claiming a Continent: A New History of Australia.* New York: Harper Collins, 2001.

deJong-Lambert, William. *The Cold War Politics of Genetic Research: An Introduction to the Lysenko Affair.* Dordrecht: Springer, 2012.

Delange, Yves. "Les phénomènes de l'atmosphère et la météorologie de Lamarck." In *Jean-Baptiste Lamarck, 1744–1829,* edited by G. Laurent. Paris: Éditions du Comité des travaux historiques et scientifiques, 1997.

Deligeorges, Stéphane, Alexandre Gady, and Frédérique Labalette. *Le Jardin des plantes et le Muséum national d'histoire naturelle.* Paris: Éditions du Patrimoine, Centre des Monuments Nationaux, 2004.

Dennett, Daniel. *Darwin's Dangerous Idea: Evolution and the Meanings of Life.* New York: Simon & Schuster, 1995.

d'Estienne, Jeanne Mesmin. "La Maison de Charenton du XVIIe au XXe siècle: Construction du discours sur l'asile." *Revue d'histoire de la protection sociale,* no. 1 (Dec. 2008): 19–35.

Dobell, Clifford. *Antony van Leeuwenhoek and His "Little Animals," Being Some Account of the Father of Protozoology and Bacteriology and His Multifarious Discoveries in These Disciplines.* New York: Harcourt Brace, 1932.

Ducoulombier, Henri. "L'aigle et le pou: Le typhus dans la Grande Armée." *Histoire des sciences médicales* 48, no. 3 (2014): 351–60.

Dupuis, Claude. "Pierre-André Latreille (1762–1833): The Foremost Entomologist of His Time." *Annual Review of Entomology* 19 (1974): 1–14.

Duvernay-Bolens, Jacqueline. "L'homme zoologique. Races et racisme chez les naturalistes de la première moitié du XIXe siècle." *L'homme* 35, no. 133 (1995): 9–32.

Englund, Steven. *Napoleon: A Political Life.* New York: Scribner, 2004.

Faucheux, Michel. *Olympe de Gouges.* Paris: Gallimard, 2018.

Gayon, Jean. "From Mendel to Epigenetics: History of Genetics." In "Trajectories of Genetics, 150 Years After Mendel / Trajectoire de la génétique, 150 après Mendel," edited by Bernard Dujon and Georges Pelletier. Special issue, *Comptes rendus biologies* 339 (2016): 225–30.

Geison, Gerald. *The Private Science of Louis Pasteur.* Princeton, N.J.: Princeton University Press, 1995.

Gelbart, Nina Rattner. *Minerva's French Sisters: Women of Science in Enlightenment France.* New Haven, Conn.: Yale University Press, 2021.

Gillispie, Charles Coulston. "Scientific Aspects of the French Egyptian Expedition, 1798–1801." *Proceedings of the American Philosophical Society* 133, no. 4 (1989): 447–74.

Gissis, Snait B. *Lamarckism and the Emergence of "Scientific" Social Sciences in Nineteenth-Century Britain and France.* New York: Springer, 2024.

Gissis, Snait B., and Eva Jablonka, eds. *Transformations of Lamarckism: From Subtle Fluids to Molecular Biology.* Cambridge, Mass.: MIT Press, 2011.

Gliboff, Sander. "The Case of Paul Kammerer: Evolution and Experimentation in the Early 20th Century." *Journal of the History of Biology* 39, no. 3 (2006): 525–63.

Gordin, Michael D. "Lysenkoism." In *Encyclopedia of the History of Science,* Feb. 2022. doi:10.34758/d5bq-k368.

Gould, Stephen Jay. *The Hedgehog, the Fox, and the Magister's Pox: Mending the Gap Between Science and the Humanities.* Cambridge, Mass.: Harvard University Press, 2011.

———. *The Mismeasure of Man.* New York: Norton, 1981.

———. *Ontogeny and Phylogeny.* Cambridge, Mass.: Harvard University Press, 1977.

———. *The Panda's Thumb: More Reflections in Natural History.* 1980. New York: Norton, 1992.

———. "The Pleasures of Pluralism." *New York Review of Books,* June 26, 1997.

———. *Time's Arrow, Time's Cycle: Myth and Metaphor in the Discovery of Geological Time.* Cambridge, Mass.: Harvard University Press, 1987.

———. "A Tree Grows in Paris: Lamarck's Division of Worms and Revision of Nature." In *The Lying Stones of Marrakesh: Penultimate Reflections in Natural History.* New York: Harmony, 2000.

———. "Zealous Advocates." *Science* 176 (1972): 623–25.

Graham, Loren. *Lysenko's Ghost: Epigenetics and Russia.* Cambridge, Mass.: Harvard University Press, 2016.

Guan Quanzhong, Dong Dazhong, Zhang Hualing, Sun Shasha, Zhang Surong, and Guo Wen. "Types of Biogenic Quartz and Its Coupling Storage Mechanism in Organic-Rich Shales: A Case Study of the Upper Ordovician Wufeng Formation to Lower Silurian Longmaxi Formation in the Sichuan Basin, SW China." *Petroleum Exploration and Development* 48, no. 4 (2021): 813–23.

Guilhaumou, Jacques. "'La terreur à l'ordre du jour': Un parcours en révolution (1793–1794)." In *Dictionnaire des usages socio-politiques (1770–1815),* fasc. 2, "Notions-concepts: Héroïsme, libéral, liberté, loi agraire république, terreur, tyrannie, vertu," 127–60. Paris: INALF / Klincksieck, 1987.

Haffer, Jürgen. *Ornithology, Evolution, and Philosophy: The Life and Science of Ernst Mayr, 1904–2005.* Berlin: Springer, 2008.

Hagan, M., J. Forbes, and A. Richmond. "Atmospheric Tides." In *Encyclopedia of Atmospheric Sciences,* edited by James R. Holton, Judith A. Curry, and John A. Pyle, 159–65. Amsterdam: Academic Press, 2002.

Hahn, Roger. *The Anatomy of a Scientific Institution: The Paris Academy of Sciences, 1666–1803.* Berkeley: University of California Press, 1971.

———. *Pierre Simon Laplace, 1749–1827: A Determined Scientist.* Cambridge, Mass.: Harvard University Press, 2005.

Hamy, Ernest-Théodore. "Les derniers jours du Jardin du Roi et la fondation du Muséum d'histoire naturelle." In *Centenaire de la fondation du Muséum d'histoire naturelle,* 1–162. Paris: Imprimerie nationale, 1893.

Hartman, Saidiya V. *Scenes of Subjection: Terror, Slavery, and Self-Making in Nineteenth-Century America.* New York: Oxford University Press, 1997.

Hazen, Robert M., Dominic Papineau, Wouter Bleeker, Robert T. Downs, John M. Ferry, Timothy J. McCoy, Dimitri A. Sverjensky, and Hexiong Yang. "Mineral Evolution." *American Mineralogist* 93 (2008): 1693–720.

Heilbron, J. L. *Electricity in the Seventeenth and Eighteenth Centuries.* Berkeley: University of California Press, 1979.

Hodge, M. J. S. "Against 'Revolution' and 'Evolution.'" *Journal of the History of Biology* 38, no. 1 (2005): 101–21.

———. "Lamarck's Great Change of Mind." In *Before and After Darwin: Origins, Species, Cosmogonies, and Ontologies* (London: Routledge, 2008), 101–11.

———. "Lamarck's Science of Living Bodies." *British Journal for the History of Science* 5, no. 4 (1971): 323–52.

Holmes, Matthew. *The Graft Hybrid: Challenging Twentieth-Century Genetics.* Pittsburgh: University of Pittsburgh Press, 2024.

Holmes, Rachel. *The Hottentot Venus: The Life and Death of Saartjie Baartman: Born 1789–Buried 2002.* London: Bloomsbury, 2007.

Honour, Hugh. *The Image of the Black in Western Art.* Vol. 4, *From the American Revolution to World War I.* 1989. Cambridge, Mass.: Harvard University Press, 2012.

Howard, Thomas A. *Protestant Theology and the Making of the Modern German University.* Oxford: Oxford University Press, 2006.

Jack, Belinda Elizabeth. *George Sand: A Woman's Life Writ Large.* New York: Knopf, 2000.

Joravsky, David. *The Lysenko Affair.* Cambridge, Mass.: Harvard University Press, 1970.

Jouannet, François. "Notice historique de Cyprien-Prosper Brard, ingénieur civil des mines." Périgueux: Dupont, 1839.

Kampourakis, Kostas. *Making Sense of Genes.* Cambridge, U.K.: Cambridge University Press, 2017.

Kauffman, Stuart. *At Home in the Universe: The Search for the Laws of Self-Organization and Complexity.* New York: Oxford University Press, 1996.

Keller, Evelyn Fox. *The Century of the Gene.* Cambridge, Mass.: Harvard University Press, 2002.

Kevles, Daniel J. *In the Name of Eugenics: Genetics and the Uses of Human Heredity.* Berkeley: University of California Press, 1985.

Kingsley, J. S. "Edward Sylvester Morse (1838–1925)." *Proceedings of the American Academy of Arts and Sciences* 61, no. 12 (1926): 549–55.

Koestler, Arthur. *The Case of the Midwife Toad.* New York: Random House, 1971.

Kotar, S. L., and J. E. Gessler. *Ballooning: A History, 1782–1900.* London: McFarland, 2011.

Kwok, Roberta. "How Life and Luck Changed Earth's Minerals." *Quanta Magazine*, Aug. 11, 2015. www.quantamagazine.org/is-mineral-evolution-driven-by-chance-20150811/#.

Laband, John. *The Land Wars: The Dispossession of the Khoisan and AmaXhosa in the Cape Colony.* Cape Town: Penguin Books, 2020.

Lack, H. Walter. "The Discovery, Naming, and Typification of *Bougainvillea spectabilis* (Nyctaginaceae)." *Willdenowia* 42, no. 1 (2012): 117–26.

Lagueux, Olivier. "Geoffroy's Giraffe: The Hagiography of a Charismatic Mammal." *Journal of the History of Biology* 36, no. 2 (2003): 225–47.

Laissus, Yves, and Jacques Petter. *Les animaux du Muséum, 1793–1993.* Paris: Muséum national d'histoire naturelle/Imprimerie nationale, 1993.

Lala [formerly Laland], Kevin, Tobias Uller, Nathalie Feiner, Marcus W. Feldman, and Scott F. Gilbert. *Evolution Evolving: The Developmental Origins of Adaptation and Biodiversity.* Princeton, N.J.: Princeton University Press, 2025.

Laland [subsequently Lala], Kevin, Tobias Uller, Marcus W. Feldman, Kim Sterelny, Gerd B. Müller, Armin Moczek, Eva Jablonka, and John Odling-Smee. "The Extended Evolutionary Synthesis: Its Structure, Assumptions, and Predictions." *Proceedings of the Royal Society B* 282 (2015): 20151019.

Lanham, Url. *The Bone Hunters.* New York: Columbia University Press, 1973.

Latour, Bruno. "Pasteur et Pouchet: Hétérogenèse de l'histoire des sciences." In *Eléments d'histoire des sciences,* edited by Michel Serres, 423–45. Paris: Bordas, 1989.

Laurent, Goulven. *Jean-Baptiste Lamarck, 1744–1829.* Paris: Éditions du CTHS, 1997.

Lauzun, Philippe. *Les Lamouroux: Une famille agenaise.* Agen: Veuve Lamy, 1893.

Lebleu, Olivier. *In the Footsteps of Zarafa, First Giraffe in France: A Chronicle of Giraffomania, 1826–1845.* Lanham, Md.: Rowman and Littlefield, 2020.

Lecourt, Dominique. *Proletarian Science? The Case of Lysenko.* Introduction by Louis Althusser. Translated by Ben Brewster. New York: Schocken Books, 1977.

Lenoir, Timothy. *Instituting Science: The Cultural Production of Scientific Disciplines.* Stanford, Calif.: Stanford University Press, 1997.

Lever, Maurice. *Sade: A Biography.* Translated by Arthur Goldhammer. New York: Farrar, 1993.

Levine, George. *Darwin the Writer.* Oxford: Oxford University Press, 2011.

———. *Darwin Loves You: Natural Selection and the Re-enchantment of the World.* Princeton: Princeton University Press, 2008.

Lewalski, Barbara K. *The Life of John Milton: A Critical Biography.* Oxford: Blackwell, 2003.

Logan, Cheryl A. *Hormones, Heredity, and Race: Spectacular Failure in Interwar Vienna.* New Brunswick, N.J.: Rutgers University Press, 2013.

Lovejoy, Arthur. *The Great Chain of Being: A Study of the History of an Idea.* 1936. Cambridge, Mass.: Harvard University Press, 1971.

Lyons, Joy Medley. *Making Their Mark: The Signature of Slavery at Mammoth Cave.* Fort Washington, Pa.: Eastern National, 2006.

Madriñán, Santiago. *Nikolaus Joseph Jacquin's American Plants.* Boston: Brill, 2013.

Mah, Harold. "The Epistemology of the Sentence: Language, Civility, and Identity in France and Germany, Diderot to Nietzsche." *Representations,* no. 47 (Summer 1994): 64–84.

Maienschein, Jane, "Epigenesis and Preformationism." In *The Stanford Encyclopedia of Philosophy* (Spring 2017 ed.), edited by Edward N. Zalta. plato.stanford.edu/archives /spr2017/entries/epigenesis.

Marck, Bernard. *Elles ont conquis le ciel: 100 femmes qui ont fait l'histoire de l'aviation et de l'espace.* Paris: Arthaud, 2009.

McNew, Sabrina M., Daniel Beck, Ingrid Sadler-Riggleman, Sarah A. Knutie, Jennifer A. H. Koop, Dale H. Clayton, and Michael K. Skinner. "Epigenetic Variation Between Urban and Rural Populations of Darwin's Finches." *BMC Evolutionary Biology* 17 (2017).

McWilliam, Neil. *Dreams of Happiness: Social Art and the French Left, 1830–1850.* Princeton, N.J.: Princeton University Press, 2017.

Mead, A. D. "Alpheus Spring Packard." *Popular Science Monthly,* May 1905, 43–48.

Morrissey, John F., and James L. Sumich. *An Introduction to the Biology of Marine Life.* 9th ed. Sudbury, Mass.: Jones and Bartlett, 2009.

Morton, Alan G. *History of Botanical Science: An Account of the Development of Botany from Ancient Times to the Present Day*. London: Academic Press, 1981.

Mousset, Sophie. *Olympe de Gouges et les droits de la femme*. Paris: Le Félin, 2003.

Mukherjee, Siddhartha. *The Gene: An Intimate History*. New York: Scribner, 2016.

Neukirchen, Florian. *The Formation of Mountains*. Berlin: Springer, 2022.

Noyes, Deborah. *Lady Icarus: Balloonomania and the Brief, Bold Life of Sophie Blanchard*. New York: Random House, 2022.

Numbers, Ronald L. *Darwinism Comes to America*. Cambridge, Mass.: Harvard University Press, 1998.

Nussac, Louis de. "Les débuts d'un savant naturaliste, le Prince de l'Entomologie: Pierre André Latreille à Brive de 1762–1798." *Bulletin de la Société scientifique, historique, et archéologique de la Corrèze* 28 (Jan.–March 1906): 97–224.

Outram, Dorinda. *Georges Cuvier: Vocation, Science, and Authority in Post-Revolutionary France* (1984). Abingdon, U.K.: Routledge, 2022.

Pääbo, Svante. *Neanderthal Man: In Search of Lost Genomes*. New York: Basic Books, 2014.

Pallary, Paul. "Marie Jules-César Savigny: Sa vie et son oeuvre." *Mémoires présentés à l'Institut d'Égypte* 17 and 23 (1931–34), in three parts.

Palmer, R. R. *Twelve Who Ruled: The Year of the Terror in the French Revolution* (1941. Princeton, N.J.: Princeton University Press, 2005.

Panese, Francesco. "3. La fabrique du 'Nègre' au cap du XIXe siècle: Petrus Camper, Johann Friedrich Blumenbach et Julien-Joseph Virey." In *L'invention de la race: Des représentations scientifiques aux exhibitions populaires*, edited by Nicolas Bancel, Thomas David, and Dominic Thomas, 59–73. Paris: La Découverte/Recherches, 2014.

Pelseneer, Paul. "Les premiers temps de l'idée évolutionniste: Lamarck, Geoffroy Saint-Hilaire et Cuvier." In *Annales de la Société royale zoologique et malacologique de Belgique* 50 (1914–19): 53–89.

Péquignot, A. "The Rhinoceros (fl. 1770–1793) of King Louis XV and Its Horns." *Archives of Natural History* 40, no. 2 (2013): 213–27.

Perrin, Claude. *La vie rocambolesque d'André Garnerin pionnier du parachute*. Paris: Messène, 2000.

Petit-Perrin, François. "La météorologie de Lamarck." Master's thesis, Université Paris I Pantheon-Sorbonne, 2004–5.

Pfeifer, Edward J. "The Genesis of American Neo-Lamarckism." *Isis* 56, no. 2 (1965): 156–67.

Pieters, Florence F. J. M. "Notes on the Menagerie and Zoological Cabinet of Stadholder William V of Holland, Directed by Aernout Vosmaer." *Journal of the Society for the Bibliography of Natural History* 9, no. 4 (1980): 539–63.

Pigliucci, Massimo, and Gerd B. Müller, eds. *Evolution, the Extended Synthesis*. Cambridge, Mass.: MIT Press, 2010.

Pilbeam, Pamela. *The Constitutional Monarchy in France, 1814–48*. Abingdon, U.K.: Routledge, 1999.

Pizzetta, Jules. "Ducrotay de Blainville." *Galérie des naturalistes, Histoire des sciences naturelles*. Paris: Hennuyer, 1894.

Porter, Theodore M. *The Rise of Statistical Thinking, 1820–1900*. Princeton, N.J.: Princeton University Press, 1988.

Pouillard, Violette. "Visions of Concord: Wild Animals and the Garden of the Revolution." *Journal for the History of Environment and Society* 4 (2009): 11–40.

Price, Roger. *The French Second Empire: An Anatomy of Political Power*. 2001. Cambridge, U.K.: Cambridge University Press, 2004.

———. *The French Second Republic: A Social History*. Ithaca, N.Y.: Cornell University Press, 1972.

Pronteau, Jeanne. *Edme Verniquet (1727–1804): Architecte et auteur du grand plan de Paris (1785–1791)*. Paris: Commission des travaux historiques de la ville de Paris, 1986.

Provine, William B. *The Origins of Theoretical Population Genetics* (1971). Chicago: University of Chicago Press, 2001.

Putnam, Walter. "Captive Audiences: A Concert for the Elephants in the Jardin des Plantes." *TDR* 51, no. 1 (2007): 154–60.

Qureshi, Sadiah. "Displaying Sara Baartman, the 'Hottentot Venus.'" *History of Science* 42, no. 2 (2004): 233–57.

Radick, Gregory. "Animal Agency in the Age of the Modern Synthesis: W. H. Thorpe's Example." *British Journal for the History of Science: Themes* 2 (2017): 35–56.

———. "The Baldwin Effect and the Potentialities for Thoughtful Darwinism Around 1900." In *The Riddle of Organismal Agency*, edited by Alejandro Fábregas-Tejeda, Jan Baedke, Guido I. Prieto, and Gregory Radick. London: Routledge, 2024.

———. *Disputed Inheritance: The Battle over Mendel and the Future of Biology*. Chicago: University of Chicago Press, 2023.

Rafferty, Edward C. *Apostle of Human Progress: Lester Frank Ward and American Political Thought, 1841–1913*. New York: Rowman & Littlefield, 2003.

Randolph, Helen F. *Mammoth Cave and the Cave Region of Kentucky*. Louisville, Ky.: Standard Printing Company, 1924.

Reynaud-Paligot, Carole. "Anthropologie raciale et savoirs biologiques." *Arts et savoirs*, Dec. 3, 2020, 1–11.

Richards, Richard A. *Biological Classification: A Philosophical Introduction*. Cambridge, U.K.: Cambridge University Press, 2016.

Richards, Robert J. "The Foundations of Archetype Theory in Evolutionary Biology: Kant, Goethe, and Carus." *Republics of Letters* 6, no. 1 (2018).

———. *The Tragic Sense of Life: Ernst Haeckel and the Struggle over Evolutionary Thought*. Chicago: University of Chicago Press, 2008.

Ridley, Glynis. *The Discovery of Jeanne Baret: A Story of Science, the High Seas, and the First Woman to Circumnavigate the Globe*. New York: Crown, 2010.

Ringer, Fritz. *The Decline of the German Mandarins: The German Academic Community, 1890–1933*. Cambridge, Mass.: Harvard University Press, 1969.

Riskin, Jessica. *The Restless Clock: A History of the Centuries-Long Argument over What Makes Living Things Tick*. Chicago: University of Chicago Press, 2016.

———. *Science in the Age of Sensibility: The Sentimental Empiricists of the French Enlightenment.* Chicago: University of Chicago Press, 2002.

Ritvo, Harriet. *The Platypus and the Mermaid, and Other Figments of the Classifying Imagination.* Cambridge, Mass.: Harvard University Press, 1997.

Robert, Adolphe, and Gaston Cougny. *Dictionnaire des parlementaires français de 1789 à 1889.* 5 vols. Paris: Edgar Bourloton, 1889–91.

Roberts, Andrew. *Napoleon: A Life.* New York: Viking, 2014.

Robinson, Agnes Mary Frances. "Buffon in His Garden." In *The French Ideal*, 233–74. New York: E. P. Dutton, 1911.

Roger, Jacques. *Buffon: A Life in Natural History.* Translated by Sarah Lucille Bonnefoi. Ithaca, N.Y.: Cornell University Press, 1997.

Role, André. "Vie aventureuse d'un savant: Philibert Commerson, martyr de la botanique (1727–1773)." *Histoire des sciences médicales* 8 (1974): 151–72.

Rookmaaker, Kees. "Mauled by a Rhinoceros: The Final Years of Alfred Duvaucel (1793–1824) in India." *Zoosystema* 41, no. 1 (2019): 259–67.

Rudwick, M. J. S. *The Meaning of Fossils: Episodes in the History of Palaeontology.* New York: Elsevier, 1972.

Sauvigny, Guillaume de Bertier de. *The Bourbon Restoration.* Translated by Lynn M. Case. Philadelphia: University of Pennsylvania Press, 1966.

Schiavon, Martina, and Laurent Rollet. *Pour une histoire du Bureau des longitudes, 1795–1932.* Nancy: PUN–Éditions Universitaires de Lorraine, 2017.

Schierwater, Bernd, and Rob DeSalle. *Invertebrate Zoology: A Tree of Life Approach.* Boca Raton, Fla.: CRC Press, 2021.

Schneider, William. "Toward the Improvement of the Human Race: The History of Eugenics in France." In "Sex, Science, and Society in Modern France." Special issue, *Journal of Modern History* 54, no. 2 (1982): 268–91.

Schopp, Claude. *Alexandre Dumas: Genius of Life.* Translated by A. J. Koch. New York: Franklin Watts, 1988.

Scurr, Ruth. *Napoleon: A Life Told in Gardens and Shadows.* London: Chatto & Windus, 2021.

Sebastiani, Silvia. "Enlightenment Humanization and Dehumanization, and the Orangutan." In *The Routledge Handbook of Dehumanization*, edited by M. L. Frick, 187–200. Abingdon, U.K.: Routledge, 2021.

———. "A 'Monster with Human Visage': The Orangutan, Savagery, and the Borders of Humanity in the Global Enlightenment." In "Knowing Savagery: Humanity in the Circuits of Colonial Knowledge." Special issue, *History of the Human Sciences* 32, no. 4 (2019): 80–99.

Secord, James A. *Victorian Sensation: The Extraordinary Publication, Reception, and Secret Authorship of "Vestiges of the Natural History of Creation."* Chicago: University of Chicago Press, 2000.

Séguin, Philippe. *Louis Napoléon le Grand.* Paris: Grasset, 1990.

Sillerud, Laurel O. *Abiogenesis: The Physical Basis for Living Systems.* Cham, Switzerland: Springer, 2024.

Simmons, Dana. *Vital Minimum: Need, Science, and Politics in Modern France.* Chicago: University of Chicago Press, 2015.

Smith, Justin E. H. "The Ibis and the Crocodile: Napoleon's Egyptian Campaign and Evolutionary Theory in France, 1801–1835." *Republics of Letters* 6, no. 1 (2016).

———. *Nature, Human Nature, and Human Difference.* Princeton, N.J.: Princeton University Press, 2015.

Smocovitis, Betty [Vassiliki]. *Unifying Biology: The Evolutionary Synthesis and Evolutionary Biology.* Princeton, N.J.: Princeton University Press, 1996.

Spang, Rebecca L. "'And They Ate the Zoo': Relating Gastronomic Exoticism in the Siege of Paris." *MLN* 107, no. 4 (1992): 752–73.

Spary, E. C. *Utopia's Garden: French Natural History from Old Regime to Revolution.* Chicago: University of Chicago Press, 2000.

Staum, Martin S. *Labeling People: French Scholars on Society, Race, and Empire, 1815–1848.* Montreal: McGill-Queen's University Press, 2003.

Steinberg, Ronen. *The Afterlives of the Terror: Facing the Legacies of Mass Violence in Post-Revolutionary France.* Ithaca, N.Y.: Cornell University Press, 2019.

Stocking, George W. *Race, Culture, and Evolution: Essays in the History of Anthropology.* Chicago: University of Chicago Press, 1982.

Subramanian, Samanth. *A Dominant Character: The Radical Science and Restless Politics of J. B. S. Haldane.* New York: Norton, 2020.

Szabo, Franz A. J. *The Seven Years War in Europe, 1756–1763.* Abingdon, U.K.: Routledge, 2013.

Taschwer, Klaus. *The Case of Paul Kammerer: The Most Controversial Biologist of His Time.* Translated by Michal Schwartz. Montreal: Bunim & Bannigan, 2019.

Thomas, Hugh. *The Slave Trade: The Story of the Atlantic Slave Trade, 1440–1870.* New York: Simon & Schuster, 2013.

Tirard, Stéphane. "Générations spontanées." In Corsi et al., *Lamarck, philosophe de la nature,* 65–104.

———. "Lamarck's Conception of Origins of Life." In *Encyclopedia of Astrobiology,* 904–5. Berlin: Springer, 2011.

Tozzer, Alfred M. "Frederic Ward Putnam, 1839–1915." In *National Academy of Sciences Biographical Memoirs.* Vol. 16, 4th memoir (1933).

Turner, Roy Steven. "Humboldt in North America." In *Humboldt International: Der Export des deutschen Universitätsmodells im 19. und 20. Jahrhundert,* edited by Christoph Schwinges, 289–312. Basel: Schwabe, 2001.

Vargas, Alexander O. "Did Paul Kammerer Discover Epigenetic Inheritance? A Modern Look at the Controversial Midwife Toad Experiments." *Journal of Experimental Zoology, Part B, Molecular and Developmental Evolution* 312, no. 7 (2009): 667–78.

Vartanian, Aram. "Trembley's Polyp, La Mettrie, and Eighteenth Century French Materialism." *Journal of the History of Ideas* 11, no. 3 (1950): 259.

Ward, Peter. *Lamarck's Revenge: How Epigenetics Is Revolutionizing Our Understanding of Evolution's Past and Present.* New York: Bloomsbury, 2018.

Weber, Bruce H., and David J. Depew, eds. *Evolution and Learning: The Baldwin Effect Reconsidered.* Cambridge, Mass.: MIT Press, 2003.

Weindling, Paul. "Julian Huxley and the Continuity of Eugenics in Twentieth-Century Britain." *Journal of Modern European History* 10, no. 4 (2012): 480–99.

West-Eberhard, Mary Jane. *Developmental Plasticity and Evolution.* New York: Oxford University Press, 2003.

Wheeler, William Morton. "On Instincts." *Journal of Abnormal Psychology* 15 (1920–21): 295–318.

Williams, Roger L. *Botanophilia in Eighteenth-Century France: The Spirit of the Enlightenment.* Dordrecht: Kluwer, 2001.

Winterer, Caroline. *How the New World Became Old: The Deep Time Revolution in America.* Princeton, N.J.: Princeton University Press, 2024.

Wiscart, Jean-Marie. *La noblesse de la Somme au dix-neuvième siècle.* Amiens: Encrage, 1994.

Woloch, Isser. *Napoleon and His Collaborators: The Making of a Dictatorship.* New York: Norton, 2001.

Yates, F., and K. Mather. "Ronald Aylmer Fisher, 1890–1962." *Biographical Memoirs of Fellows of the Royal Society* 9 (1963): 91–129.

Zimmer, Carl. *She Has Her Mother's Laugh: The Powers, Perversions, and Potentials of Heredity.* New York: Dutton, 2018.

Zirkle, Conway. *Evolution, Marxian Biology, and the Social Scene.* Philadelphia: University of Pennsylvania Press, 1959.

NOTES

PROLOGUE: THE GARDEN WHERE IT ALL BEGAN
....

MAIN PRIMARY SOURCES:

Darwin, *Origin*, 1st ed., 134–39 ("use and disuse"); Huxley, "Reception," 188 ("unscientific"); Weismann, *Essays*, 447 ("scientific men"); Spencer, *Essays*, 1:6 ("scientific men"), 244 ("scientific method," "scientific truth"), 2:58 ("scientific spirit"), 143 ("scientific minds," quoting J. S. Mill); Tournefort, *Voyage*, 1:376–77; Cuvier, *Histoire des sciences*, 2:474, 493–94 (Tournefort); Cuvier, "Éloge," xx (poetry); Vaillant, "Sébastien Vaillant's 1717 Lecture," 106 (plant sex scene); Jussieu, *"Discours sur le progrès,"* 18; Linnaeus, *Praeludia Sponsaliorum Plantarum*, 1729, quoted in Bernasconi and Taiz, "Vaillant's Lecture," 99; Mérat, "Anecdote" and Flourens, *Recueil*, 52 (Jussieu's hat); Lamarck, "Mémoire sur les cabinets" and Décret de la Convention, 1793 (Jardin name change); Jauffret, *Voyage*, 208, 222, and Pieters, "Notes," 541 (elephants); Lacépède, "introduction to *La Ménagerie*," 17–18 ("on the ruins"); E. Geoffroy Saint-Hilaire, "Quelques considérations," 219 (giraffe's good nature); E. Geoffroy Saint-Hilaire, *Principes*, 66–67n (function before structure); Sainte-Beuve, *Volupté*, 1:227–30; Lamarck, *Histoire naturelle des végétaux*, 1:88–89 (Vaillant).

MENAGERIE:

I. Geoffroy Saint-Hilaire, *Vie*, 45–52, 45 ("beautiful dromedary"), 48 (unexpected news); AN: AJ/15/844, "Rapport des citoyens Geoffroy et Lamarck professeurs de Zoologie aux citoyens professeurs composant le Directoire du Muséum d'histoire naturelle," 1 germinal an 2 [March 21, 1794] (list of animals); Bernardin de Saint Pierre, *Mémoire*, 635 ("feels, loves, thinks"), 638–40 (lion and dog), 645–47 (cockatoo and spaniel), 646 ("dog-lions"), 647 (rhinoceros), 658 ("fleeing and trembling"); Toscan, *Histoire du lion*, 28 (joys of friendship), 31 ("natural gaiety"), 31–32 ("pressed his friend"); Guillaume, *Procès-verbaux*, 2:794, 814–21, and 3:319–20 (Committee of Public Instruction establishment of menagerie).

ELEPHANT CONCERTS:

Toscan, "Du pouvoir de la musique," 257 (armed with music), 259 ("curiosity, astonishment," "tendre musette" scene), 261–62 ("Ça ira"), 321–22 (Haydn, *Nina*); Houel, *Histoire naturelle*, 95–104, 97 (Gluck), 98 ("soft harmony"), 101–2 (cacophony), 103 (Hans too immature). The concert program with field notes is in AN AJ/15/581 [Séance extraordinaire du 27 prairial an 6].

SECONDARY SOURCES:

Gould, *Panda's Thumb*, 54 185, 193 (causal pluralism); **Asma**, "Darwin's Causal Pluralism"; **Deligeorges et al.**, *Le Jardin des plantes*; **Spary**, *Utopia's Garden*; **Williams**, *Botanophilia*, chap. 1 (Vaillant); **Bernasconi and Taiz**, "Vaillant's Lecture"; **Michaud**, "Tournefort," 45; **Burkhardt**, "Constructing the Zoo"; Burkhardt, *Leopard*; **Caffin-Carcy**, "Que devint Versailles" (transfer of animals); **Laissus and Petter**, *Les animaux du Muséum*; **Péquignot**, "Rhinoceros"; **Pouillard**, "Visions of Concord" (menagerie); **Putnam**, "Captive Audiences" (elephant concerts); **Lebleu**, *Footsteps*, chaps. 4, 7 (giraffe); **Lagueux**, "Geoffroy's Giraffe"; **Landrieu**, *Lamarck*, 226; **Bourdier**, "Esquisse," 32; **Bange and Corsi**, *Chronologie*; Bange and Corsi, "Les auditeurs de Lamarck."

1. Spencer endorsed Lamarck's and Darwin's idea that animals transformed themselves hereditarily by their behaviors (*Principles*, 1:406) but emphasized that these arose ultimately from the "universal laws of the re-distribution of matter and motion" (*Principles*, 1:464).

2. Grew, *Anatomy of Plants*, 171–72; Camerarius, *De sexu plantarum epistola*; Boerhaave, "Préface," 8th page of each text, Latin and French. Plant sexuality was also advanced by John Ray and others, including Claude-Joseph Geoffroy, whom Vaillant accused of "harvesting" the idea from earlier authors. See Bernasconi and Taiz, "Vaillant's Lecture" and "Geoffroy's Lecture." The story of Vaillant and the pistachios has been much repeated, but the dearth of original sources leads Bernasconi and Taiz to suggest it "may be the stuff of folklore" ("Geoffroy's Lecture," 40).

3. Christian version: Charles Alexander Johns, *The Forest Trees of Great Britain* (1847), cited in Anderson, "Cedar," 291; prosaic version: Gumbleton, "Une introduction difficile," 104n1. Some versions of the story have Bernard de Jussieu receiving the cedar from Peter Collinson; see Neill, *Journal*, 442.

4. Buffon, *Histoire naturelle*, 4:382–83. Buffon prudently added that Revelation precluded the idea of all animals originating from one since all participated equally in the grace of creation. The only form of transformation he explicitly endorsed was degeneration: *Histoire naturelle*, 14:311–74, 5:226–27. On Buffon's attempts at a zoo: Bachaumont, *Mémoires secrets*, 21:31.

5. "The darkest hours": History of the Ménagerie, the Zoo of the Jardin des Plantes, accessed July 9, 2025, www.jardindesplantesdeparis.fr/en/history-of-the-menagerie-the-zoo-of-the-jardin-des-plantes; Spang, "'And They Ate the Zoo.'" The animals that got eaten weren't permanent residents of the Garden of Plants menagerie but refugees transferred at the beginning of the siege from the Jardin d'acclimatation in the Bois de Boulogne.

6. "La ménagerie du Jardin des plantes de Paris peut-elle se vanter de posséder Nénette, une femelle orang-outan enfermée depuis 1972?" *Le monde des idées*, July 4, 2021; see also "Jardin des Plantes Menagerie Unfit for Animals' Wellbeing?" sortiraparis.com, Jan. 5, 2021.

CHAPTER 1: THE LITTLE PRIEST

MAIN PRIMARY SOURCES:

A. Lamarck to Cuvier, Feb. 20, 1830 (9: "not the least vocation," "le petit abbé"); A. Lamarck to Eugène de Lamarck, June 11, 1865, 377 ("won't be a priest"); **Geoffroy Saint-Hilaire**, "Discours prononcé," 209; **Cuvier**, "Éloge," iii.

SECONDARY SOURCES:

Packard, *Lamarck*, chap. 1, 5 ("almost a ruin," "other occupations"); **Landrieu**, *Lamarck*, 13–20.

1. Price of military commissions: Chagniot and Drévillon, "La vénalité des charges." Franklin's interventions are in Franklin, *Papers*: Ctesse de Lameth to Franklin, May 10, 1778, July 12, 1778; De Bout to Franklin, May 10, 1778, June 2, 1778, July 4, 1778, July 12, 1778; Franklin to the Ctesse de Lameth, July 12, 1778.

CHAPTER 2: THE ERSATZ CHEVALIER

MAIN PRIMARY SOURCES:

A. Lamarck to Cuvier, Feb. 20, 1830, 10–14 (Battle of Vellinghausen); A. Lamarck to Eugène de Lamarck, June 11, 1865, 377; **R. Lamarck**, "Note de Mlle. Lamarck," 27 (music notebooks trade); **Cuvier**, "Éloge," iii–v; **Geoffroy Saint-Hilaire**, "Discours prononcé," 209–10; **Panckoucke**, "Lamarck," 484; **Bourdon**, "Lamarck."

SECONDARY SOURCES:

Landrieu, *Lamarck*, 21–23; **Szabo**, *Seven Years War*, 9, 74 (Battle of Vellinghausen).

1. The comtesse de Lameth, née Marie-Thérèse de Broglie, lived in the Château d'Hénencourt: Wiscart, *La noblesse de la Somme*, 22. Comte François IV de Lastic, aged thirty-one, was just assuming command of an infantry regiment under de Broglie that would become the regiment de Beaujolais.
2. Chevalier de Saint Martin: Lamarck's paternal grandfather brought the title north to Picardy from Bearn, in the opposite corner of France. Landrieu, *Lamarck*, 15. N.B.: We have only Auguste Lamarck's word for his father's dramatic entrance into military life, with no corroborating evidence in the French military archives.
3. De Broglie acted against the orders of Louis XV's minister of state, Choiseul, who had placed a restriction on the promotion of soldiers. When, the month following Lamarck's appointment as ensign, he was further promoted to lieutenant, Choiseul apparently cracked down: Lamarck was un-promoted, but only to the rank of sublieutenant. Bourdon, "Lamarck," 337; Landrieu, *Lamarck*, 23nn1, 2.

CHAPTER 3: THE VIOLIST, NO, BANKER, NO . . .

MAIN PRIMARY SOURCES:

A. Lamarck to Cuvier, Feb. 20, 1830, 15 (neck cure); A. Lamarck to Eugène de Lamarck, June 11, 1865, 377 ("a total change in his life"); **R. Lamarck**, "Note de Mlle de Lamarck," 27 (not banking!); **Geoffroy Saint-Hilaire**, "Discours prononcé"; **Cuvier**, "Éloge," v (neck derangement); **Panckoucke**, "Lamarck," 485; **Bourdon**, "Lamarck," 338–39, 344 (ignoring Rousseau); **Bourguin**, "Lamarck"; **Rousseau**, *Lettres élémentaires*.

SECONDARY SOURCES:

Landrieu, *Lamarck*, 23–28; **Packard**, *Lamarck*, 13–17; **Bourdier**, "Esquisse," 26; **Damrosch**, *Jean-Jacques Rousseau*, 472.

1. Bourguin, "Lamarck," 195, 220 (Bourguin doesn't cite his sources, so we have to take his word for it).
2. Bourdier, and Bange and Corsi, place Lamarck in the countryside retreat with his eldest brother, Louis Philippe, not Bazentin, but they also say that Louis Philippe was "four years older," which was the case for Bazentin, whereas Louis Philippe was fifteen years older. Bourguin, the original source of the story ("Lamarck," 195), says the brother in question was Bazentin. Timing: Bourdier says 1771; Bange and Corsi say 1770; Bourguin doesn't say but implies it was after several years of medical and botanical study in Paris, which would be the mid-1770s. Panckoucke ("Lamarck," 485) puts the Rousseau episode several years later, after Lamarck's return from travels with Buffon's son, but by then Rousseau had been dead for more than three years.

CHAPTER 4: THE ROMANTIC BOTANIST

MAIN PRIMARY SOURCES:

Lamarck, *Recherches sur l'organisation*, 86 (Vicq d'Azyr's course); [Louis Cotte], "Mémoire sur les principaux" (Lamarck's memoir on clouds as recorded by Cotte); Lamarck, *Flore française*, 1:xxvii (chaos), lxi–lxv and lxxx (dichotomous key), 7 ("multiple resources"), 105–6 ("greenery enameled"), 2:iv (acknowledging Haüy); **A. Lamarck** to Cuvier, Feb. 20, 1830, v (could see only clouds); A. Lamarck to Eugène de Lamarck, June 11, 1865, 377–78 (nomenclature game in the Garden); **Buffon**, "Discours préliminaire," 15–24 (rejecting Linnaeus); Buffon, *Correspondance inédite*, 1:274 ("King of Sentences"); Buffon, "Discours sur le style," 11 ("style"); **R. Lamarck**, "Note de Mlle de Lamarck," 25; **Geoffroy Saint-Hilaire**, "Discours prononcé," 212–13; **Cuvier**, "Éloge," v–vii; **Panckoucke**, "Lamarck," 484–85; **Bourdon**, "Lamarck," 338–41; **Bourguin**, "Lamarck," 196 (Lamarck's inelegant pen).

SECONDARY SOURCES:

Packard, *Lamarck*, 79; **Landrieu**, *Lamarck*, 21–30, 67, 133; **Bourdier**, "Esquisse," 26–28 (shell collection), 32–33; **Spary**, *Utopia's Garden*, 13, 16, 25–33 (Buffon and Daubenton), 41.

1. Bourdier ("Esquisse," 32) says the couple married in September 1792, just five days before Rosalie de la Porte's death. More recent authors have picked this up, but no marriage license survives, and Bange and Corsi, *Chronologie*, point out that the information in the children's baptismal records indicates the parents were already married. There is frustratingly little on which to base an image of Rosalie de la Porte and her relationship with Lamarck. What we know is that they were together for at least a decade and a half, through many dramatic changes of fortune.
2. The technique is akin to earlier methods in mathematics and philosophy: Gould, *Hedgehog*, 123–25.
3. Daubenton came from Montbard, the same Burgundy town as Buffon, and their families were close. When Buffon needed help with his monumental natural history of everything, the ultimately thirty-six-volume *Natural History, General and Particular*, he sum-

moned Daubenton, who was a decade younger, to come to the King's Garden and write anatomical descriptions based upon exhaustive dissections and observations. Daubenton contributed almost two hundred descriptions of quadrupeds, and Buffon designated him demonstrator attached to the royal natural history cabinet. Buffon and Daubenton later had a falling-out, and Philippe Guéneau de Montbeillard took Daubenton's place as co-author.

CHAPTER 5: THE TRAVELING NATURALIST

MAIN PRIMARY SOURCES:

Lamarck, "Considérations" (trip with two Andrés); Lamarck, *Recherches sur les causes,* 2:284–373, 296 (bulb in carafe), 298–99 (Houdini turnip, apartment peas), 371 (descending into the mine), 372 (clayey, glassy), 373 ("half-transformed"); for cited passages, see also Lamarck, "Classes," 35; Lamarck, *Mémoires,* 298; **Buffon** to Gilles Germain Richard de Ruffey, April 5, 1769, in *Correspondance inédite,* 1:121–22 ("cruel wound"); Buffon to comte de Buffon fils, Aug. 26, 1783, in *Correspondance inédite,* 2:172 ("my dear son"); Buffon to Guéneau de Montbeillard, May 11, 1781, in *Correspondance inédite,* 2:97 ("leaves tomorrow"), and June 6, 1781, in *Correspondance inédite,* 2:99 ("in Amsterdam"); **Panckoucke,** "Lamarck," 485; **A. Lamarck** to Cuvier, Feb. 20, 1830, 15; **Cuvier,** "Éloge," viii; **Bourdon,** "Lamarck," 343–44 (Buffonet splattered ink); **Bourguin,** "Lamarck," 199.

SECONDARY SOURCES:

Packard, *Lamarck,* 20–21; **Landrieu,** *Lamarck,* 31; **Robinson,** "Buffon in His Garden," 260–62; **Jordanova,** *Lamarck,* 5; **Bourdier,** "Esquisse," 28; **Spary,** *Utopia's Garden,* 42–43, 56; **Hodge,** "Lamarck's Great Change of Mind"; **Morton,** *History of Botanical Science,* 241 (Gleditsch); **Madriñán,** introduction to *Jacquin's North American Plants.*

1. The work in which Lamarck describes the apartment peas, *Recherches sur les causes,* was "paraphé" (initialed) by the secretary of the academy in 1781, the year Rosalie Joséphine turned three and André was born; they had moved to the rue Copeau in 1778 soon after Rosalie Joséphine's birth.
2. Buffonet anecdotes and Buffon's relations with him: Claveau, review of Buffon's *Correspondance inédite,* 576; Rabbe et al., "Buffon," 1:679.
3. In Vienna, Lamarck and Buffonet visited Nikolaus von Jacquin, professor of botany and chemistry and correspondent of Linnaeus. Jacquin's daughter Franziska took piano lessons with Mozart.
4. Lamarck's account of the visit to the silver mine in Kremnica has not survived; it is mentioned in Humbert-Bazile, *Buffon,* 194–95.

CHAPTER 6: THE STEALTH GEOLOGIST

MAIN PRIMARY SOURCES:

Lamarck, "Considerations," 2–4; Lamarck, "Classes," 34–35 (topsoil chart).

MOUNTAINS MADE *BY* LIVING BEINGS:

Lamarck, *Recherches sur les causes*, 2:306, 315, 376; Lamarck, *Hydrogéologie*, chap. 4, 91–186; **Cuvier**, "Éloge," 8–11; **Bourdon**, "Lamarck," 344; **Bourguin**, "Lamarck," 199–201; **Guan et al.**, "Types of Biogenic Quartz"; **Balch et al.**, "Factors" (white cliffs of Dover); **Neukirchen**, *Formation of Mountains*, 3–8, 73, 82, 282, 387 (limestone mountain ranges); **Dolomieu**, "Mémoire," 11 and "Lettre," 9 (religious view of the natural history of mountains).

MOUNTAINS MADE *OF* LIVING BEINGS:

Linnaeus, *Systema naturae*, 16:19; **Buffon**, *Histoire naturelle des minéraux*, 1:422. See also **Baumé**, *Chymie expérimentale*, 4:105; **Bonnet**, *Oeuvres*, 4:45; **E. Darwin**, *Botanic Garden*, note 16, 165–67; **Delamétherie**, *Théorie de la terre*, 5:26–28, 248, 452; **Saussure**, *Voyage dans les Alpes*, 1:248, 283.

SECONDARY SOURCES:

Hahn, *Anatomy*, 130–31 ("classes of the academy); **Toulmin**, *Antiquity*, 100–101, 105 (mountains made of living things); **Barnett**, *After the Floods* (biblical interpretations of marine fossils on mountaintops).

1. Hazen et al., "Mineral Evolution," 1693, 1704–9, 1712; Kwok, "How Life and Luck Changed Earth's Minerals" (free oxygen interacted with existing minerals, "origin of minerals depends on life"); Kennedy et al., "Late Precambrian Oxygenation" (clay mineral factory). See also Chan et al., "What Would Earth Be Like Without Life?" Thanks to Carol Cleland and Henderson Cleaves for help understanding the current state of geology with regard to the role of living organisms.

CHAPTER 7: THE CROSS-DRESSING, WORLD-TRAVELING BOTANIST AND HER LOVER

MAIN PRIMARY SOURCES:

Commerson to Louis Gérard, April 11, 1758, in Cap, *Philibert Commerson* (hereafter cited as Cap, *PC*), 74 ("botanomania"), Oct. 25, 1758, 78 (courtship); Commerson to M. Bernard, June 8, 1762, and Jan. 1767, in Cap, *PC*, 80 (bereavement), 81 (generous captain's quarters); Commerson to M. Beau, Oct. 20, 1766, in Cap, *PC*, 82–85 ("in the slipperiest circumstance"); Commerson to M. Beau, May 28, 1767, in Cap, *PC*, 97–100; Commerson to M. Beau, Sept. 7, 1767, in Cap, *PC*, 100–104; Commerson to Georges Marie Commerson (n.d.), in Cap, *PC*, 105–7; **Duclos-Guyot and Commerson**, *Journaux*, 485 (discovery of Barret); **Lalande**, "Éloge," 118 (descriptions of Commerson, Tahiti discovery of Barret's secret, "indefatigable courage"), 100 ("fury to see"), 105 ("apathy"); **Bougainville**, *Voyage*, 253–54 ("piqued her curiosity," "frozen slopes," "experienced botanist"); **Lamarck**, "Mémoire sur le projet," 39 (plants from all over the world); Lamarck, *Mémoires*, 350–51 ("very convincing proof"); **Cuvier**, "Éloge," 10 (Sonnerat's gift).

SECONDARY SOURCES:

I rely throughout especially on **Gelbart**, *Minerva's French Sisters*, see 108–13, 125–53, and 291n20 (Jussieu's botanical expeditions). Otherwise: **Cap**, "Notice biographique," in Cap, *PC*, 7–30; **Role**, "Vie aventureuse"; **Saint-Germain**, "Routier inédit," 238; **Ridley**, *Discovery*, 4–5, 11, 141–42 (prince's aristocratic attire); **Lack**, "Discovery," 121 (*Bougainvillea* discovery).

    1. For Jussieu's botanical expeditions, see Le Preux, *Éloge de M. de Jussieu*, 12, 15–16; and Audelin, "Les Jussieu," 2:337–42 (Audelin cites André Thouin, "Bibl. Mus. Fonds Thouin"). Jussieu's snakebite cure: Barbeu-Dubourg, *Le botaniste*, 215–16n1.

    2. Vivès, *Journal*, 267–68, 237–41, quoted in translation in Gelbart, *Minerva's French Sisters*, 127; Othon, *Journal*, 408; Diderot, *Supplément au Voyage de Bougainville*, 32.

    3. Commerson's description of Barret comes from his Latin dedication of a plant to her, "*Baretia*," in a big folio describing many plants: BMHN, MS 198 (YL51). English translation adapted from Gelbart, *Minerva's French Sisters*, 136. See also Cap, *PC*, 24, 37. The genus *Baretia* is now obsolete, its members having been included in the genus *Turraea*.

    4. A. L. Jussieu, "Observations," 284; A. L. Jussieu, *Genera Plantarum*, 90–91. Antoine-Laurent was heir to his uncle's system of botanical classification, upon which he based his own, more glorious career. The family of plants is now called Nyctaginaceae.

    5. Commerson to Cossigny, April 19, 1770, in Cap, *PC*, 129; English translation of Commerson's letter to Cossigny: Gelbart, *Minerva's French Sisters*, 135.

CHAPTER 8: REVOLUTION IN THE GARDEN

MAIN PRIMARY SOURCES:

Lamarck to J. H. Bernardin de Saint-Pierre, April 24, 1793 [*Plan du travail à faire pour mettre en ordre les herbiers du Cabinet national d'histoire naturelle*], AN AJ/15/512 folio 503, also at the *Electronic Enlightenment* (Lamarck's proposal for the herbarium); Lamarck, *Considérations* and *Mémoire sur le projet du comité* (two desperate pamphlets pleading for the herbarium); **A. Lamarck** to Cuvier, Feb. 20, 1830, 16–18 (Auguste's stories about André); **Landolphe**, *Mémoires*.

SECONDARY SOURCES:

Roger, *Buffon*, 425–26 (Buffon's death); **Landrieu**, *Lamarck*, 30 (Lamarck's professional advancements), 53 (resentful botanists), 52–53 and 52n2 (Bernardin's assessment of the situation), 51 (officers' new plan for the garden), 68 (fifth child, Auguste), 67n1 (move to 4, rue du chemin de Gentilly); **Bange and Corsi**, "Chronologie" (Cornélie's birth in 1792); **Spary**, *Utopia's Garden*, 164–78; **Bourdier and Orliac**, "Lamarck," 30–34, 40 (third and fourth children); **Jordanova**, *Lamarck*, 6–9; **Thomas**, *Slave Trade*, 361 (Landolphe).

    1. For conservator of insects, Lamarck had a friend in mind, the entomologist Guillaume-Antoine Olivier; for worms, he thought of another friend, the conchologist and snail specialist Jean-Guillaume Bruguière; for quadrupeds and birds, he proposed Daubenton; for reptiles and fish, Bernard-Germain de Lacépède; and for mineralogy, the geologist and vulcanologist Barthélemy Faujas de Saint-Fond, who, like Lamarck, had held a tenuous

position, as adjunct keeper of the natural history cabinet, which the Assembly had proposed to eliminate. Lamarck, "Mémoire sur les cabinets."

2. BMHN MS 2228 / 5: Monuments Lamarck au Muséum et à Bazentin (Somme).
3. French Dominicans were known as "Jacobins" because their first monastery in Paris was in the rue Saint Jacques.
4. Guillaume, *Procès-verbaux*, 1:480–86; Landrieu (*Lamarck*, 52–53) traces the appointment of Lamarck to the professorship of insects and worms to a pamphlet written by André Thouin in September 1790; this is also recorded in Hamy, "Les derniers jours," 128.
5. See Lamarck, *Mémoires*, second tableau (after 314) for his first use of the term. N.B.: Latreille had described an insect as an "animal sans vertèbres" in *Précis*, x–xi.

CHAPTER 9: THE RELUCTANT ZOOLOGIST

MAIN PRIMARY SOURCES:

Guillaume, *Procès-verbaux*, 1:480–86 (turned into zoologist); **AN**, *Inventaire dissolutif* (apartment inventory—5: dining room, Rosalie's room; 5–7: master bedroom; 7, 9: children's rooms, Lamarck's wardrobe; 8: study; 10: watch, cutlery; 11–20: library); **A. Lamarck** to Eugène de Lamarck, June 11, 1865, 376 (hospitality to visitors); **Lamarck**, *Flore française*, 1:105 ("innocent welcome").

SECONDARY SOURCES:

Packard, *Lamarck*, 43; **Landrieu**, *Lamarck*, 63, 67–73, 94 (hospitality); **Bourdier**, "Esquisse," 32–33, 35 (hospitality); **Bange and Corsi**, "Chronologie"; **Andress**, *Terror*; **Guilhaumou**, "'La terreur à l'ordre du jour'"; **Palmer**, *Twelve Who Ruled*; **Baczko**, *Ending the Terror*; **Steinberg**, *Afterlives*, chap. 1 (Thermidor). Thanks to Pietro Corsi and Dorinda Outram for help thinking through Lamarck's likely household arrangements.

1. PVAP 24 messidor an 3 (July 12, 1795): AN AJ/15/97, p. 56, no. 67/208; Burkhardt, "Leopard," 683n16. These estimates of the children's ages are based on their years of birth and the year of the move and are therefore approximate.
2. AP tome 73, 419–20; Bertrand Barère de Vieuzac pronounced the phrase "let terror be the order of the day," but Jacques Alexis Thuriot de la Rozière, who was presiding over the session, made it instead "courage and justice" that were "the order of the day."
3. I'm surmising this is Rosalie's room: It's the most formal after the master bedroom and the only one with a single occupant.
4. I'm guessing the clock, vases, and seals were on the mantelpiece based upon the order of description.
5. Pronteau, *Edme Verniquet*, 256–64. Lamarck must not have taught in the amphitheater during his first year, since it was undergoing renovations: 263–64.

CHAPTER 10: NEMESIS

MAIN PRIMARY SOURCES:

Cuvier, *Recherches* (1812), 1:58 ("ensemble"), vol. 2, §4: 20 ("destined"), §6: 20 ("infallibility"), §9: 11 ("15 feet"), 30 ("general cause," "the flood"), 138 ("aqueous"), 139 ("sudden"); §10: 3 ("nippled teeth"); §12: 3, 4 (new creations); **Lamarck**, "Conchyliologie," 416–17 ("so-called . . . catastrophes"); Lamarck, "Discours d'ouverture pour le cours de 1816," BMHN MS 742, elements 78–88, on 1–2 ("What interest . . ."); **Geoffroy Saint-Hilaire**, *Fragments*, 213 (Bruguière's conversation).

SECONDARY SOURCES:

Burkhardt, "Leopard," 687 (Cuvier non-hiring); **Nicard**, *Étude*, 107 (Cuvier's bribe); **Pelseneer**, "Les premiers temps," 70–71 (Cuvier's maneuvering and career); **Robert and Cougny**, "Creuzé-Latouche," in *Dictionnaire*, 2:221; **Packard**, *Lamarck*, 37–40 (shell collection); **Bange and Corsi**, "Les auditeurs" (Lamarck's students); **Landrieu**, *Lamarck*, 73–79 (Beauregard house); **Outram**, *Georges Cuvier*, 28–29, 222, 415; **Rudwick**, *Meaning of Fossils*, 104–15 (Cuvier's interpretations).

1. Lamarck, "Présentation à l'assemblée d'un candidat," Dec. 24. 1794, AN, AJ/15/578. Desmoulins made sense for the job, and Cuvier's ambitions were higher than aide-naturalist.
2. Cambry, *Description*, 67; Hermanville, "Notice biographique," 470–80; see Lamarck to Lucien Bonaparte, 5 brumaire an 9 [Oct. 27, 1800], BMHN Mc93, in which Lamarck was still negotiating the sale of his cabinet to the French state for the Muséum d'histoire naturelle.

CHAPTER 11: THE INVENTION OF INVERTEBRATES

MAIN PRIMARY SOURCES:

Lamarck, *Philosophie zoologique*, 1:121 ("chaos," "monstrous class"); Lamarck, *Système des animaux*, 2–3 (overthrows tripartite taxonomy), 5 (plants/animals), 7 (vertebrates/invertebrates), 8 (vertebrates classes), 10 (fertile in marvels, invertebrates outnumber vertebrates), 11 (scorn and indifference, lower limit of animality, see also 16), 26–28 (powers of generation, polyps), 33–35 (definition, classes of invertebrates); Lamarck, *Recherches sur les causes*, 1:11 (first origins unknowable), 2:273 ("solely due"); Lamarck, *Hydrogéologie*, 78–79 (bosom of sea); Lamarck, *Mémoires*, 345–46 (oldest mountains); Lamarck, *Histoire naturelle des animaux*, 1:149 (elegant insects), 156–57 (polyp generation); Lamarck, *Discours d'ouverture*, 10 (fertile in marvels), 28 (customs, habits), 31 (amusing and curious, "beautiful considerations"); **Leeuwenhoek**, "Part of a Letter," 1304–5 ("long Tayls"), 1306 (roundish animalculum), 1307–8 (animalculum with horns); **Trembley**, *Mémoires*, 149 (plant or animal), 234 (whole new creature), 246–48 (hydra), 302 (animals, not plants), 306 (horses versus oaks); **M. Trembley**, *Correspondance inédite*, 52–62 (Trembley to Réaumur, March 16, 1741, plant/animal hesitations), 63–66 (Réaumur to Trembley, March 25, 1741, "aquatic insects," "polyps"), 78 (Trembley to Réaumur, June 8, 1741, voracious polyps); **Aristotle**, *History of Animals* 588b4–22 (animal, vegetable, mineral); **Linnaeus**,

Systema naturae: "A general system of nature, through the three grand kingdoms of animals, vegetables, and minerals," 4:691 (polyps among worms); **Apollodorus**, *Library*, bk. 2, §5, ¶2 (Lernean Hydra); **Réaumur**, *Mémoires*, 6:liv (polyps).

SECONDARY SOURCES:

Lovejoy, *Great Chain* (Aristotle's *scala naturae*); **Dobell**, *Antony van Leeuwenhoek*, 277–82; **Baker**, *Abraham Trembley*; **Dawson**, *Nature's Enigma*.

1. Aram Vartanian ("Trembley's Polyp," 1307) writes that Leeuwenhoek classified polyps as plants, but clearly not in the cited passages from Leeuwenhoek's letter.

CHAPTER 12: A CAN OF WORMS

MAIN PRIMARY SOURCES:

Lamarck, *Philosophie zoologique*, 1:51 (cirripedes), 122–24 (mollusks, echinoderms/radiolarians), 124–25 (arachnids, division of worms, annelids), 127 (final taxonomy of invertebrates), 174–75 (arachnids); Lamarck, "Coquille," 564, 583 (beauty of shells), 558 (intimate mixture, living to fossil), 560 (viscous matter), 562 (glands), 561 (muscle attaching to shell; growth by juxtaposition), 569 (on outer surface), 562 (constricted and uncomfortable), 574 (spiral-shelled organisms), 568–69 (cowries), 578–79 (murexes); Lamarck, *Système des animaux*, 9–12 (introduction, description of invertebrates); 20 ("immense" place), 22 ("frightening fecundity"), 24 (astonishment at mountains), 26–27 (invertebrate generation and regeneration), 33–35 (definition and classes of invertebrates), 34 ("infinite smallness"), 320–22 ("intestinal" versus "external" worms); Lamarck, *Extrait du cours*, 9 (fourteen kinds of beings, tripartite division), 42 (external worms); Lamarck, *Mémoires*, 255 (life as movement, irrelevance of religious faith), second tableau (after 314) (first published use of "invertebrates"), 408–10 ("contempt and odious rejection"); Lamarck, *Recherches sur les causes*, 1:1 ("astonishing activity"), 2 ("alternating life and death"), 2:185 (mysterious essence of life), 289 ("two powerful forces"), 290 ("drags toward ruin"), 314 ("without organic beings"), 391; Lamarck, "Conchyliologie," 417–18 (animal creation in action).

SPONTANEOUS GENERATION:

Lamarck, *Philosophie zoologique*, 1:15, 65, 211, 2:63–64, 67 (algae and monads), 80 (heat, light, electricity, humidity); Lamarck, *Recherches sur l'organisation*, 37, 100; Lamarck, *Extrait du cours*, 39, 82; Lamarck, *Histoire naturelle des animaux*, 1:152, 174, 178–80, 455–56, 461.

"MONAD":

Lamarck, "Discours d'ouverture du cours," 138 (globular); Lamarck, *Discours d'ouverture*, 16; Lamarck, *Système des animaux*, 16 ("animated point"); Lamarck, *Philosophie zoologique*, 1:285, 2:67 (algae, monads), 212, 456 (water); Lamarck, *Système analytique*, 148.

POWER OF LIFE:

See chapter 18.

SECONDARY SOURCES:

Gould, "Tree Grows in Paris"; **Burkhardt**, *Spirit of System*, 59, 100–101, 120; **Hodge**, "Lamarck's Science of Living Bodies," 326; Hodge, "Against 'Revolution' and 'Evolution,'" 106; **Tirard**, "Générations spontanées"; Tirard, "Lamarck's Conception of the Origins of Life"; **Bourdier**, "Esquisse," 33; **Bange and Corsi**, "Chronologie."

1. Linnaeus derived "mollusk" from the Latin *molluscus*, "thin-shelled," arising in turn from *mollis*, or "soft": *Systema naturae*, 4:3. Jacob Theodor Klein derived "echinoderm" from the Greek root *echino*, "spiny," and the Latin *derm*, "skin": *Naturalis dispositio Echinodermatum* (1734). Cuvier named "crustaceous insects" from the Latin *crusta*, for "crust" or "hard shell": Cuvier, *Tableau*, 451. For Cuvier's removal of mollusks and echinoderms from worms, see "Mémoire sur la structure," 391–94. Lamarck's acknowledgment: *Mémoires*, "premier tableau" after 314. In the mid-nineteenth century, naturalists reclassified jellies as "coelenterata," from a Greek root meaning "hollow," placing them with coral animals and sea anemones. In the current approach to taxonomy, cladistics, "clades," are defined by most recent common ancestry. See Richards, *Biological Classification*.
2. Lamarck appears to have coined "cirrhipèdes"; see F. Cuvier, *Dictionnaire*, 9:265.
3. "Infusoria": Wrisberg, *Observationum*.
4. Schierwater and DeSalle, *Invertebrate Zoology*, 2, 5.
5. BMHN MS 756, 1er cahier, elements 20–21, excerpted in Burkhardt, "Lamarck, Evolution, and the Politics of Science," 283–84.

CHAPTER 13: THE COMPOSITION OF MATTER

MAIN PRIMARY SOURCES:

Lavoisier, *Traité*, 1:xv–xvii (redefinition of element), 4–9 (states of matter), 5–6 (caloric), 87–102 (water decomposition/recomposition), 2:507, 650–52 (pneumatic chemistry), 327–41 (analysis/synthesis method); **Lalande**, "Notices," 182 ("new chemistry"); **Morveau et al.**, *Méthode* (1787) (Greek names); **Lavoisier and Laplace**, "Mémoire," 12 (agnosticism about heat's cause), 75 (respiration is combustion); **Stahl**, introduction to *Zymotechnia*, 8; **Fourcroy**, *Philosophie chimique*, 125–28 (analysis of organic matter), 172 (nature lends materials); **Lamarck**, *Réfutation*, 357 (not susceptible to analysis), 363 (products of organic function), 409 (instead of reducing, "organic action"), 455 (entire surface of globe), 475 (nature doesn't compose), 482–83 (combination versus aggregation); Lamarck, *Mémoires*, 325–26, 394 (products versus constituents), 340–41 (aggregation).

SECONDARY SOURCES:

Riskin, *Science in the Age of Sensibility*, chap. 7; **Burkhardt**, *Spirit of System*, 38–39 (rueful mystification); **Bange and Corsi**, "Chronologie"; **Bourdier**, "Esquisse," 32–33; **Jordanova**, *Lamarck*, 44 (Lamarck's rejection of reductionism).

1. Lamarck, *Recherches sur les causes* (1794); Lamarck, *Réfutation* (1796); Lamarck, *Mémoires* (1797). Lamarck wrote *Recherches* in 1776 and presented it to the Academy of Sciences in 1780, but the academy postponed its publication.

Notes

2. Lamarck to Agasse, 27 vendémiaire an VI [Oct. 18, 1797]: "Si le projet de décret passe, je suis du nombre de ceux qui seront exportés ou bannis. Je suis cependant républicain très prononcé." katabase.huma-num.fr/View/CAT_000234. Thanks to Dan Edelstein for help understanding the political context.

CHAPTER 14: ANIMAL COLLABORATIONS

MAIN PRIMARY SOURCES:

E. Geoffroy Saint-Hilaire, "Observations," 471 (collaborative swallows, hooded crow), 473–76 (shark/pilot fish); E. Geoffroy Saint-Hilaire, "Description des crocodiles," 185–86 ("ready-made ideas"), 191 (intelligence, division of labor, property), 198 (good hearing), 201–5 (plover partnership), 205 (modern authors' refusal of animal intelligence), 223 (crocodiles and lions); E. Geoffroy Saint-Hilaire, "Histoire naturelle des poissons du Nil"; **I. Geoffroy Saint-Hilaire**, "Histoire naturelle des poissons du Mer Rouge et de la Méditerranée"; I. Geoffroy Saint-Hilaire, *Vie, travaux et doctrine*, 76 (galvanic experiments), 77 (near drowning), 80 (in charge of natural sciences), 81 (French civilization in Africa), 85–86, 89–90 (fish of the Nile, fish of the Red Sea, mummified animals, plover and crocodile, English caricatures), 93–94 (fish of the Mediterranean), 96 (anatomy of crocodile), 101–3 (torpedo fish, distraction from siege), 107 (threat to burn the collections), 112–13 (despite mummies, species vary); I. Geoffroy Saint-Hilaire, "Suite de l'histoire naturelle des poisons du Nil," 298–99 (upside-down catfish); **Lacépède**, "Rapport des professeurs," 235–37 (Geoffroy's plundered mummies, Lacépède's equivocal report).

SECONDARY SOURCES:

Bell, *Napoleon*, 35–41; **Englund**, *Napoleon*, 126–41; **Roberts**, *Napoleon*, 161–84; **Scurr**, *Napoleon*, 61–93; **Smith**, "The Ibis and the Crocodile"; **Bourdier**, "Esquisse," 33.

1. Hermanville, "Notice biographique," 470–80. Landrieu (*Lamarck*, 79) says Lamarck would soon have to resell the house and cites Hermanville, but Hermanville does not seem to say this. Moreover, Lamarck refers to having observed a hurricane from the house in July 1808: *Annuaire pour l'année 1809*, 184.
2. On the mutualist relationship of pilot fish and sharks—the fish eat parasites and food fragments from the shark's teeth and gills—see Morrissey and Sumich, *Introduction*, 274.
3. See, for instance, Décembre, "France," 1086, 1090.

CHAPTER 15: FATHERS AND SONS

MAIN PRIMARY SOURCES:

A. Lamarck to Cuvier, Feb. 20, 1830, 14–16 (Lamarck's marriages, André's adventures); **Latreille**, *Histoire naturelle*, 9:157 (rescue); **Jardine**, "Memoir of Latreille," 18–19 (parentage), 21 ("adopted father," relations with Lamarck), 26–28 (rescue).

SECONDARY SOURCES:

Bourdier, "Esquisse," 34; **Hamy**, "Premières relation de Latreille avec Lamarck," in *Débuts*, 176–81; **Dupuis**, "Pierre André Latreille"; **Nussac**, "Les débuts," 104–8 (Latreille's parentage), 121–29 (rescue, "clever Naturalist" on 126), 161 (childhood entomology), 192–95 (Latreille and Lamarck), 218 ("study of insects delighted"); **Damkaer**, *Copepodologist's Cabinet*, 124–30.

1. J.-B. Lamarck to Guillaumot, 10 vendémiaire an 9 (Oct. 22, 1800), BMHN 1998 no 191.
2. Latreille, "Discours," 1941.

CHAPTER 16: THE CREATION OF BIOLOGY

MAIN PRIMARY SOURCES:

Lamarck, *Discours d'ouverture*, 12–14 ("time and circumstances," waterbirds), 15 ("I will . . . prove"); Lamarck, *Recherches sur l'organisation*, 50 (habits make organs), 134 (bimanous, human evolution); Lamarck, "Discours d'ouverture du cours," 148 ("certain authors"); **BVMFB, Bissy** Notes, 1 (first day), 32–33 (22 prairial), 72–73 (*Grapsus* crabs), 135 ("limit of animality," "end").

"BIOLOGIE":

Lamarck, *Hydrogéologie*, 8, 188; Lamarck, *Recherches sur l'organisation*, vi, 186, 202; Lamarck, *Philosophie zoologique*, 1:xviii; Lamarck, *Histoire naturelle des animaux*, 1:49.

SECONDARY SOURCES:

Bange and Corsi, "Les auditeurs" (Lamarck's students); Bange and Corsi, "Students" (enrollment numbers); Bange and Corsi, "Chronologie"; **Bourdier**, "Esquisse," 35–38; **Simmons**, *Vital Minimum*, 61 (Villermé); **Corsi**, *Lamarck*, chap. 5, 192–225; **Robert and Cougny**, "Bissy," in *Dictionnaire*, 1:330.

NAPOLEON:

Bell, *Napoleon*; **Englund**, *Napoleon*, 180–85 (Concordat), 212–13 (Jacobins), 243–46 (self-coronation), 389–92 (relations with the church), 454 (scorn for atheism); **Roberts**, *Napoleon*, 255, 273, 349 (Légion d'honneur and associated lycées); **Scurr**, *Napoleon*, 121–24 (Concordat); **Woloch**, *Napoleon*, 66–74 (Jacobins), 45, 182–84 (Laplace), 101, 163, 165 (Fourcroy).

1. For example, Villermé, "Mémoire sur la taille."
2. Joseph Barboza from Brazil, Benjamin Franklin Harris from the United States.
3. BMHN MS 742/1/element 151. The manuscript, a plan for an ultimately unwritten book, is undated. Lamarck wasn't the only one: Several authors arrived at "biology" independently around the same time, including Thomas Beddoes, Karl Friedrich Burdach, and Gottfried Reinhold Treviranus. See Corsi, "Biologie."
4. Buffon, *Histoire naturelle*, 14:18, 21.
5. Cuvier, "Cours du Lycée. Géologie. Plan général," IF GC 3111, fossils "do not themselves form the layers that contain them," 57; organized matter "has not always existed," 58; "fossils have often been deposited in calm waters, and have not at all been transported,"

Notes

63; "the causes still currently active as such cannot have caused the changes of which we see traces," 64; G. Marzari Pencati, "Note prese al Corso di Cuvier. Corso di Geologia all'Ateneo nel 1805," MS, Biblioteca Bertoliana, Vicenza, S.C. 28 (7), 1–20, cited in Corsi, *Lamarck*, chap. 5, 74–86, esp. 84.

6. Thibaudeau, *Mémoires*, 83 (baubles). Lamarck's daughter at school: Bonneville de Marsangy, *Mme. Campan à Écouen*, 104.

INTERLUDE: THE WEATHER

MAIN PRIMARY SOURCES:

Cuvier, "Éloge," v (cloud-watching); [Cotte], "Mémoire" (Lamarck's 1777 presentation to the Academy of Sciences, of which nothing remains except Cotte's description); Lamarck, "Sur la forme," 149 (not random), 153 (types of clouds); Lamarck, "Nouvelle définition," 113 (addition in 1805 of seven more types, *brumeux* [misty], *terminés* [terminated], *en lambeaux* [ragged], *boursouflés* [puffy], *en barres* [bar shaped], *coureurs* [running], and *de tonnerre* [thunder]); Lamarck, *Annuaire pour l'an VIII*, 85–88 (atmospheric constitutions, complicating factors), 93–95 ("invitation to amateurs," annotation instructions, annotated yearbooks request), 96–113 (metric system), 97 ("two seconds"); Lamarck, *Annuaire pour l'an X*, 24–25 (calendar columns example), 50–51 ("reason to expect"), 77–89; Lamarck, *Annuaire pour l'an 1807*, 171–72 (funnel cloud); Lamarck, *Annuaire pour l'an 1808*, 1 ("friends of nature"), 200 ("most imposing and beautiful"); Lamarck, *Annuaire pour l'an 1809*, 184–87 (hurricane), 164 (bodily influence of atmosphere); Lamarck, *Annuaire de l'an XIII*, 97–98 ("spectacle of sky," "immense laboratory," "intimate relations"); Lamarck, "Mémoire sur le mode de rédiger"; Lamarck, *Météorologie*, 451–53, 474 (chagrin at colleagues' reaction), 475–76 ("what a strange thing"); Lamarck "De l'influence de la lune," 430 ("indivisible and determinable instants"), 431 (twenty years of hesitation); Lamarck, "Extrait d'un mémoire," 148 ("not an opinion"); Howard, *Essay*, 6 ("learned"); Laplace, *Philosophical Essay*, 3–4 (infinitely intelligent being); Laplace, "On the Notion of Power" and "On Causality," in Hahn, *Pierre-Simon Laplace*, 224–32 (contingency a figment); Garnerin, *Gazette national, ou, Le moniteur universel*, Oct. 8, 1805, 59–60 ("modern Icaruses"); Haeghens et al., *Annuaire météorologique*, ii–iii ("three friends of Meteorology"); Newton, "General Scholium," 389–90; Martins, "Introduction biographique," lxxxiv (posthumous glory); Chaptal, *Mes souvenirs*, 106–7 and n1 (Napoleon/Bourgoin episode); A. Lamarck to Cuvier, Feb. 8, 1830, IF GC MS 3252 / f. 358–59, 3–4 (Auguste's bitter reflection).

INFLUENCE OF THE MOON ON THE ATMOSPHERE, ATMOSPHERIC CONSTITUTIONS:

Lamarck, *Annuaire pour l'an VIII*, 77–89; Lamarck, *Annuaire pour l'an IX*, 109–11; Lamarck, "Extrait d'un mémoire"; Lamarck, "De l'influence de la lune"; Lamarck, "Météorologie"; Toaldo, *Essai*, 89–94.

GOVERNMENT WEATHER AGENCY:

Lamarck, *Annuaire pour l'an XI*, 85–86, 154; Lamarck, *Annuaire pour 1807*, 203; Lamarck, *Annuaire pour 1810*, 150; Lamarck, "Météorologie," 469–73, 476; Lalande, "Histoire de l'astronomie pour l'année 1801," 26.

"FRENCH NEWTON":

Le moniteur universel, Nov. 4, 1811; *Journal de l'Empire*, May 18, 1812, 4; Pierre-Joseph-Spiridion Dufey, *Nouveau dictionnaire historique des environs de Paris* (Paris: Charles Perrotin, 1825), 11.

SECONDARY SOURCES:

Packard, *Lamarck*, 79–82; **Landrieu**, *Lamarck*, 133–49; **Bourdier**, "Esquisse," 34–35; **Bange and Corsi**, "Chronologie"; **Daston**, "Cloud Physiognomy"; **Alder**, *Measure of All Things* (resentment of the metric system); **Delange**, "Phénomènes," 134 (negative response to Lamarck's meteorology); **Schiavon and Rollet**, *Pour une histoire du Bureau des longitudes*.

WOMAN PARACHUTISTS:

Noyes, *Lady Icarus*, 101; **Kotar and Gessler**, *Ballooning*, 81–82; **Marck**, *Elles ont conquis le ciel* (women's organs).

1. See Hagan, Forbes, and Richmond, "Atmospheric Tides," 164.
2. See, for example, *Journal de Paris*, Jan. 18, 1800, 534; *Le Courrier des spectacles*, March 16, 1800, 2; *Gazette national, ou, Le Moniteur universel*, Aug. 5, 1801, 1301; *Mercure de France*, Nov. 22, 1801, 398–99; *La Clef du cabinet des souverains*, Oct. 4, 1802, 8; *Gazette national, ou, Le Moniteur universel*, Aug. 31, 1804, 1504; *Gazette national, ou, Le Moniteur universel*, Oct. 4, 1804, 39; *Journal de Paris*, July 8, 1808, 1356; *Journal de Paris*, Oct. 17, 1808, 2157; *Journal de Paris*, Dec. 22, 1808, 2584; *Journal de Paris*, Dec. 31, 1808, 2648–49; *Journal de Paris*, Dec. 14, 1809, 2510–11. Lalande criticized Lamarck's meteorological theory: *Mercure de France*, March 7, 1807, 472. René Tourlet, a doctor and science journalist, defended Lamarck: *Gazette national, ou, Le Moniteur universel*, Feb. 4, 1808, 139–40. Many thanks to Pietro Corsi for help tracking the news of the yearbooks.
3. *Gazette national, ou, Le Moniteur universel*, Oct. 24, 1797, 131; *Journal de Paris*, Oct. 24, 1797, 135; Priscilla Lamure, "Le premier saut en parachute," *Écho de presse*, Retronews, March 13, 2018. On Garnerin, see Perrin, *La Vie rocambolesque d'André Garnerin*.
4. *La Chronique universel*, June 15, 1798, 2. Later, in London, Garnerin tried another feat: "I launched a cat with a parachute, in miniature, which encompassed a column of air 38 inches and a half in its basis. The descent was gradual, and the cat fell, with his little vehicle, in the garden of a man who insists on receiving three guineas for indemnification of the trespass committed by poor puss, or at least his picture with the parachute." "M. Garnerin's Account," *Derby Mercury*, Aug. 12, 1802, 3.
5. Burkhardt (*Spirit of System*, 10, 225n20) identifies the memoir in question as "Mémoire sur les variations de l'état du ciel," later published in the *Journal de physique* 56 (1802): 114–38. The original source is a letter from Étienne Geoffroy Saint-Hilaire to Georges Cuvier, Institut de France, Fonds Cuvier, MS 3225 (12).
6. Only in the twentieth century, apparently, did writers start referring to this hypothetical being as "Laplace's Demon." See, for example, *The Monist* 41 (1931): 3.
7. "Clarke's Second Reply," in Leibniz and Clarke, *Correspondence*, 12, 13 (see also 33); and Newton, *Opticks*, query 31, 403. Newton's friend and translator Samuel Clarke represented him in an epistolary debate with G. W. Leibniz in 1715–16. Clarke expressed Newton's view that God reigned over the laws of mechanics: They held true only as a

result of his immediate presence, and he could suspend them at any moment. Leibniz objected that it represented a poor view of God's power to think he couldn't create a mechanical system that would run on its own without his continual intervention. See "Leibniz's Second Letter," in Leibniz and Clarke, *Correspondence*, 9.

8. Laplace (*Exposition*, 479) agreed with Leibniz that the world, once created, should run on its own without divine intervention, though he dismissed Leibniz's idea of a "preestablished harmony" coordinating the movements and changes undergone by all substances in the world in accordance with the laws of mechanics. For evidence that he wasn't an atheist, see Laplace's letter to his son, *Œuvres*, 1:v–vi. Roger Hahn (*Pierre-Simon Laplace*, 67, 172–73) is among those who describe Laplace as an atheist. His main evidence is a draft of a letter to Laplace from the geologist Jean-Étienne Guettard in which Guettard accuses Laplace of having denied God's existence. It's impossible to be certain about Laplace's inner state of belief, but in his public and private writing, although he rejected Christian doctrine and denied the existence of miracles, he retained faith in a divine intelligence behind the workings of nature. Laplace's conversation with Napoleon: Antommarchi, *Derniers momens*, 1:282. The story found its way into Napoleon's "memoirs," which, though written in the first person, were in fact compiled from various sources by Étienne-Léon de Lamothe-Langon; see Lamothe-Langon, *Mémoires*, 1:262. The story's popularity: see, for example, Dawkins, *God Delusion*, 46.

9. Lubbock, *Herschel Chronicle*, 310; Laplace, *Exposition*, 479. According to a fellow astronomer, François Arago, Laplace was unhappy with Antommarchi's implication that he was an atheist and asked Arago to persuade the publisher to delete the passage: Faye, *Sur l'origine du monde*, 110–11.

10. Despite his support for a government meteorological service, Chaptal declined to endorse Lamarck's theory. See Petit-Perrin, "La météorologie," 97–98, and Chaptal-Lamarck, lettre du 23 vendémiaire an XI (16 octobre 1802), AN Serie F/20/1.

11. Victor Hugo, *Napoléon le petit* (1852); Karl Marx, *The Eighteenth Brumaire of Louis Bonaparte* (1852).

12. "Historique," Météo et Climat; "L'Office national météorologique (1920–1945)," Météo-France.

CHAPTER 17: THE ZOOLOGICAL PHILOSOPHY
....

MAIN PRIMARY SOURCES:

Arago, *Histoire*, 93–94 (Napoleon story); **PVIF**, séance du 11 novembre 1809 (Tuileries reception Nov. 16); ***Journal de Paris***, Nov. 17, 1809, weather for Nov. 16: 0–3°C. **Lamarck**, *Philosophie zoologique*, 1:i, ii ("singular gradation"), iii ("extremely remarkable"), 7 ("little by little"), 1 (interesting study of animals), 10 (invertebrates as foundation of zoology), 56 ("supreme author"), 361 ("only concerned"), 362 (astronomy much easier), 2:187–88 ("baseless," "purely imaginary"), 464 ("naturally prone," clock thought experiment); **Newton**, *Chronology*; Lucretius, *De rerum natura* I.551, II.78–79; **De Maillet**, *Telliamed*, 181; **Diderot**, *Rêve de d'Alembert*, 85, 91; **Buffon**, *Des époques*, 67–68; **Hutton**, "Theory of the Earth" (first presentation of deep time); Hutton, *Theory of the Earth*, vol. 1, part 1, chap. 1, §I (machine, "Divine wisdom"), §II (nature not quiescent), §IV ("no vestige of a beginning"), chap. 2 ("materials . . . necessarily removed," "benevolent intention"), chap. 3 ("almighty power," "land . . . continually wasting"); vol. 2, part 2, chap. 14 ("living world," "endless diversity").

SECONDARY SOURCES:

Englund, *Napoleon*, 212–13, 454 (Napoleon and atheism); **Jordanova**, *Lamarck*, 71 (transformism), 85 (religious faith); **Brin**, *Concept of Time* (time in the Bible); **Buchwald and Feingold**, *Newton and the Origin of Civilization*, 171, 186–88, 223–25, 404 (early modern controversy over age of world, biblical chronology, Newton). **Cf. Dal Prete**, *On the Edge of Eternity* (argument: the Enlightenment invented the idea of a conflict between Christian faith and an ancient earth). **Gould**, *Time's Arrow, Time's Cycle* (origins of deep time).

1. In his *Investigation into the Principles of Knowledge*, 2:500–501, Hutton suggested that organisms less well adapted to their environments would die off and better adapted ones thrive, whereby the divine machinery maintained all forms of life in their most perfect states.

CHAPTER 18: THE POWER OF LIFE AND THE INNER FEELING OF EXISTENCE

MAIN PRIMARY SOURCES:

Lamarck, *Philosophie zoologique*, 1:xi–xii (incapacitated animal experiments useless), xiv (hint of interior power), 4 (physical/moral, see also 218 and 2:290–92), 8 (changing environment, new actions, habits, penchants), 65 (nature begins again daily), 83–85 ("irritability"), 266 (contesting view of virtually everyone), 364 (Cabanis, see also 2:292–93, 431), 378 ("individuality"), 385 (reduction a mistake), 387 ("repeated sudden movements"), 456 (water the "true cradle"); 2:61 (passages life-death), 67 (spontaneous generation simplest organisms), 144, 230 (radiates, fistulides), 148 ("special organ of feeling"), 175 (animal feelings topic of zoology), 187–88 (animals have feelings), 226 ("inner emotions"), 274 (sensations, thoughts not miraculous), 306 (force transported into animals, see also 338–39), 310–11 ("purely passive machines"), 322–23 (walking dog), 330–45, esp. 333 (the will), 337–39 (acts of will in complex animals), 346 (matter causes ideas), 427 (dreaming dog), 456 (power of habits); Lamarck, *Hydrogéologie*, 188 (incalculable series of centuries).

"POWER OF LIFE":

Lamarck, *Recherches sur les causes*, 2:289; Lamarck, *Mémoires*, 249, 333, 395; Lamarck, *Hydrogéologie*, 117; Lamarck, *Philosophie zoologique*, 1:375–76, 2:127; Lamarck, *Histoire naturelle des animaux*, 1:134, 160–61, 183. Lamarck (*Philosophie zoologique*, 1:412) uses the word "orgasm" to denote the basic state of organic matter, endowed with the power of life. In animals, this state renders them irritable, whereas in plants it is too weak.

INNER FEELING OF EXISTENCE:

Lamarck, *Philosophie zoologique*, 1:xiii–xv, 84, 2:220, 255, 279–301, 279 (obscure, powerful), 280 ("active motor"), 281 ("intimate and continuous," "this *me*"), 285–86 (being startled, sore left eye story, inner feelings from theater, art, music), 286–87 (deaf pianist), 290 (moral and physical inner emotions), 339.

HYPOCEPHALON:

Lamarck, Philosophie *zoologique*, 2: 224–31 (general summary), 224 (medullary mass), 227 ("simple and fugitive"), 228 ("think, reason, invent"), 337 (faculty of will), 348 ("wrinkled, pulpy hemispheres"), 367, 375 ("organ of intelligence").

SPONTANEOUS GENERATION:

See chapter 12.

SECONDARY SOURCE:

Heilbron, *Electricity* (electrical science in the seventeenth and eighteenth centuries).

1. Newton, *Principia*, bk. 3, Propositions 1–9.
2. Nollet, *Essai*, frontispiece, 36 (suspended boy); Nollet, "Observations," 18; *Histoire de l'Académie royale des sciences pour l'année 1746*, 7–8 (240 people at Versailles"); Priestley, *History*, 125 (Carthusian monks, royal guard at Versailles, whose number he reports as 180); Sigaud de la Fond, *Précis historique*, 293 (Collège de Navarre).
3. Franklin, *Experiments and Observations*, letter 2; Coulomb, "Second mémoire."
4. On Newton: Lamarck, *Hydrogéologie*, 253; on magnetism: Lamarck, *Réfutation*, 234; on Nollet: Lamarck, *Recherches sur les causes*, 1:90; on Franklin: Lamarck, *Recherches sur les causes*, 1:133, 202–3; AN, *Inventaire dissolutif*, 11.
5. Legallois, *Expériences*, 28–32.
6. Cabanis, *Rapports*, 1:134.

CHAPTER 19: THE GIRAFFE AND COMPANY

MAIN PRIMARY SOURCES:

Lamarck, *Philosophie zoologique*, 1:241–42 (diminished use, moles), 243–44 (sound, ears), 245–46 (snakes' legs), 249–51 (swimming animals versus perching birds, shorebirds), 251 (water-birds, anteaters, hummingbirds, lizards), 251–52 (fish eyes), 253 (snakes' snout apertures), 253–55 (herbivorous mammals), 257–58 (lions, tigers), 262–63 (sloth), 235 (two new laws of nature), 2:451–53 (seal), 454–55 ("prolonged leap"), 463 (chart of origins of animals); Lamarck, *Histoire naturelle des animaux*, 1:181 (four laws), 200–201 (explanation of fourth law); **Darwin**, *Origin*, 6th ed., 186–88.

LAMARCK'S INVOLVEMENT WITH THE MENAGERIE:

PVAP, 7 germinal an X (March 28, 1802), AJ/15/103, p. 31, no. 45/213 (young bull); PVAP, 27 germinal an X (April 17, 1802), AJ/15/103, p. 41, no. 55/103 (goats); PVAP 17 floréal an X (May 7, 1802), p. 52, no. 66/213 (honey for the elephant); PVAP, 27 prairial an X (May 17, 1802), AN AJ/15/103, pp. 56ff., no. 70/213 (elephant cured, lionesses); PVAP, 7 thermidor an X (July 26, 1802), p. 104, no. 118/213 (lion cubs); Lamarck's initial attempts to get a kangaroo: PVAP, 7 floréal an X (April 17, 1802), AJ/15/103, p. 47, no. 61/213; PVAP, 27 prairial an X (May 17, 1802), AN AJ/15/103, pp. 56ff., no. 70/213.

LAMARCK'S GIRAFFE PASSAGES:

Lamarck, *Philosophie zoologique*, 1:256; Lamarck, *Recherches sur l'organisation*, 208.

SECONDARY SOURCES:

Landrieu, *Lamarck*, 59; **Packard**, *Lamarck*, 53; **Burkhardt**, "Lamarck, Evolution, and the Inheritance of Acquired Characters," 793; Burkhardt, *Leopard*, chaps. 1.12 and 2.7 (arrival of kangaroos just after Geoffroy had taken over from Lamarck as overseer of the menagerie).

1. For recent science on vestigial organs: Louchart and Viriot, "From Snout to Beak"; Thewissen et al., "Evolutionary Aspects."
2. For recent science on vertebrate ears: Pfaff, Schultz, and Schellhorn, "Vertebrate Middle and Inner Ear."

CHAPTER 20: A VERY BRANCHY TREE

MAIN PRIMARY SOURCES:

Lamarck, *Philosophie zoologique*, 1:58–59 (branching series), 26–27, 145–46 (platypus and echidna), 75–81 (against catastrophes), 86 (perfect/imperfect), 108 ("higher" animals), 165 ("lower" animals), 141–42, 220 (circumstances, irregular gradations), 144–45 (whales and snakes don't need legs), 154 (snakes), 2:457 (*Gordius*); Lamarck, *Histoire naturelle des animaux*, 1:133 (continuum as primary pattern); Lamarck, *Système analytique*, 131–32 (very branchy), 142 (power of circumstances), 149–50 (human superiority).

CHARTS OF ANIMALS:

Lamarck, *Recherches sur l'organisation*, "Tableau" between 36 and 37; Lamarck, *Système des animaux*, 8, 35; Lamarck, *Philosophie zoologique*, 1:277–80, 2:463; Lamarck *Histoire naturelle des animaux*, 1:457.

SECONDARY SOURCES:

Ritvo, *The Platypus and the Mermaid*, 6–20 (first European encounters with monotremes); **Day**, *Claiming a Continent* ("New Holland" and the colonization of Australia); **Gould**, "Tree Grows in Paris" (Lamarck's embrace of branchiness).

1. "Dinosaurs Among Us," American Museum of Natural History; "The Dinosauria," University of California Museum of Comparative Zoology; Pääbo, *Neanderthal Man*; Svante Pääbo, interview, *Guardian*, Jan. 12, 2023.
2. Lamarck mentions dictating this work, but doesn't say to whom, in *Système analytique*, 4. Cuvier ("Éloge," xxx) says Lamarck dictated to his "fille ainée." Latreille ("Discours") says Lamarck's eldest daughter was his particular helper. Bourguin ("Lamarck," 100) says Lamarck's eldest daughter helped him with his last work. Hermanville says the same

("Notice," 461). Packard (*Lamarck*, 55) says Lamarck dictated to Cornélie, but also that Cornélie's is the only name he knows of any of Lamarck's children (51n). Landrieu (*Lamarck*, 100) says Lamarck dictated to Rosalie. Bange and Corsi ("Chronologie") say he dictated to Cornélie. Bourdier ("Esquisse," 39) says he dictated but doesn't say to whom.

CHAPTER 21: THE NATURAL HISTORY OF ANIMALS WITHOUT VERTEBRAE

MAIN PRIMARY SOURCES:

Lamarck, *Histoire naturelle des animaux*, 1:iii–iv ("only general theory"), 277–303 ("man"), 451 (Savigny, Lesueur, Desmarest), 451–52 ("ascidians," "far from simple"), 458 ("we will never succeed"), 461 (two separate tables), 454 ("order of formation"), 2:182 ("extremely pretty varieties"), 209 ("fringed tentacles"), 121, 138, 180, and 474 (Lamouroux), 470–72 (Suriray), 465 (Venus girdle), 544 ("Ophiures," "brittle stars"), 3:295 (insect puberty), 7:600ff. (Spirula); **Prévost**, "Funérailles," 7 (Blainville difficult, unsociable); **Flourens**, *"Éloge,"* 8 (Cuvier on Blainville); **Blainville**, "Palaéontologie," liv–lv; **Cuvier**, "Extrait," 159–61 (citing Péron).

SAVIGNY:

Lamarck, *Histoire naturelle des animaux*, 1:451, 2:339, 392, 403–17, 3:83–85, 93–122, 303 ("extraordinary sagacity," lepidoptera mouths), 5:42, 115, 280–89, 302–51 (Savigny's studies of the mouths of lice and crustaceans, heads of annelids).

SECONDARY SOURCES:

Bourdier, "Esquisse," 38; **Landrieu**, *Lamarck*, 79, 243–44 (Lamarck as artist); **Goulven**, *Lamarck*, 93; **Bange and Corsi**, "Les auditeurs": Robert Maskell Patterson, Carlos de Gimbernat (Spanish geologist), Omalius d'Halloy, Constant Prévost, Cyprien-Prosper Brard, Ducrotay de Blainville; **American Philosophical Society** Papers of Robert M. Patterson; **Ashworth**, "D'Omalius d'Halloy"; **Buffetaut**, "Blainville"; **Jouannet**, "Notice historique de . . . Brard"; **Damkaer**, *Copepodologist's Cabinet*, 195–98 (Suriray), 116–22 (Lesueur, Péron, and the Baudin expedition); **Nicard**, *Étude*, 105–7 (Blainville's machinations for Lamarck's professorship); **Pizzetta**, "Ducrotay de Blainville," 276–80; **Lauzun**, *Les Lamouroux*, 75–128; **Bell**, *Napoleon*, 94–102; **Englund**, *Napoleon*, 405–46; **Roberts**, *Napoleon*, 710–81; **Scurr**, *Napoleon*, 219–74; **Laband**, *Land Wars*, 145–50 (Janssens); **Lever**, *Sade*, 512–31, 518 ("libertine dementia"); **D'Estienne**, "La Maison de Charenton," n12; **Ducoulombier**, "L'aigle et le pou," 356 (typhus epidemic of 1814); **Sebastiani**, "'Monster with a Human Visage'" and "Enlightenment Humanization and Dehumanization" (role of orangutan in discussion of human races); **Gould**, "Tree Grows in Paris," 127 (Lamarck's sketches of *Spirula*). **Tree of Life web project**: "*Spirula spirula*"; **Gillispie**, "Scientific Aspects" (Savigny, tools: 455, dissections: 460–61); **Pallary**, "Marie Jules-César Savigny" (sorrel story: 10, apparently from M. Michelin, see 6n, who presumably knew Savigny personally).

1. Sade died from "prostrating gangrenous fever," which could be associated with typhus.
2. Catalogue of Life: Lamarck, Linnaeus, Cuvier.
3. Suriray's given names were Jacques Simon Amand, but he styled himself just "Suriray."

4. Agasse, *Tableau encyclopédique* (1827), vol. 3, plate 465, nos. 5a and 5b; Landrieu, *Lamarck*, 243-44. Lamarck was not a great artist but a competent one and often made quick sketches; he also worked closely with commissioned artists, for instance for the plates of the *Encyclopédie méthodique*, making many corrections on the proofs.

CHAPTER 22: THE SELF-MADE PRIMATE

MAIN PRIMARY SOURCES:

Darwin, *Descent of Man,* 1:51 (ancient primate became less arboreal), 203 (cites Lamarck); **Delamétherie,** *Vues physiologiques,* xii ("bipedal by habit"); **Buffon,** *Histoire naturelle,* 14:18, 21 (quadrumanes). **Lamarck,** *Recherches sur l'organisation,* 134 ("What a subject of meditation"), 135 (occipital hole); Lamarck, "Homme" (universal human inclinations, 270); Lamarck, *Histoire naturelle des animaux,* 1:299–301 (human extremes of good and evil); Lamarck, *Philosophie zoologique,* 1:262 (if people weren't separated), 348–52 (human racial categories, quadrumanes coming down from trees, strain of standing upright, "wretched, anxious life," new wants and abilities), 2:361–64 (Gall, intellectual faculties not all innate, "dangerous opinion"); Lamarck, *Système analytique,* 150–51 (Caucasians most perfected), 153 (language, new wants, powers, faculties, "no *mother language*"), 158–61 (from equality to oppression, see also 281–84), 306–7 (exercise makes the difference in intelligence); Lamarck, *Histoire naturelle des animaux,* 1:280 (oppression in civilized Europe), 281 (humans constantly change); **Blainville,** "Sur une femme"; **G. Cuvier,** "Extrait"; G. Cuvier, "Femme de race Boschismanne" (dissection report); G. Cuvier, *La règne animal,* 1:95 ("caucasian race"); G. Cuvier, *Discours sur la théorie de la terre,* civ (degraded negro race); **Daubenton,** "Memoire sur les differences" (occipital hole); **Flourens,** "Notice"; **Geoffroy Saint-Hilaire and F. Cuvier,** *Histoire naturelle des mammifères,* vol. 1, illustrations of Baartman between introduction and first section, Cuvier's report; **Camper,** *Dissertation physique,* 42, 48 (racial ranking); **Geoffroy Saint-Hilaire and G. Cuvier,** "Histoire naturelle des Orangs-Outans," 457–59 (skull protrusion, intelligence); Buffon, *Histoire naturelle,* 4:388 ("the Black and the White" produce fertile offspring, so "all men come from the same stock and are of the same family"); **Gall and Spurzheim,** *Anatomie et physiologie du système nerveux*; **Spurzheim,** *Observations,* 1818; **Virey,** "Homme," 12, 152–53 ("negroes"), 83–84 (Baartman).

SECONDARY SOURCES:

Bourdier, "Esquisse," 38; **Corsi,** "Idola Tribus" (Delamétherie and Lamarck); Corsi, "Julien-Joseph Virey"; **Badou,** *L'énigme*; **Honour,** *Image,* chap. 2; **Holmes,** *Hottentot Venus,* 41–46, 91–109, 134ff.; **Qureshi,** "Displaying Sara Baartman"; **Crais and Scully,** *Sarah Baartman*; **Blanckaert,** *La Vénus hottentote*; Blanckaert, "Les vicissitudes de l'angle facial," 417–18, 430, 438, 440, 445; Blanckaert, "Les conditions d'émergence"; Blanckaert, "La mesure de l'intelligence"; **Gould,** *Mismeasure*; **Staum,** *Labeling People*; **Bénichou and Blanckaert,** *Julien-Joseph Virey*; **Duvernay-Bolens,** "L'homme zoologique," 16, 23–24; **Panese,** "La fabrique du 'Nègre'"; **Reynaud-Paligot,** "Anthropologie raciale," 3; **Stocking,** *Race, Culture, and Evolution,* 39; **Smith,** *Nature,* 92–113 ("polygenism," belief in multiple human species), 119–21 (Buffon on human kinds), 136–38 (Linnaeus on human classification).

1. "Sara" and "Sarah" are anglicized forms of "Saartjie"; the name appears in all these forms. Baartman's degree of consent seems unknowable. She clearly had little choice about the

conditions of her life, but during a trial following a suit brought by the abolitionist African Institution in 1810, she testified that she was fully consenting and had signed a contract. Saidiya Hartman (*Scenes of Subjection*) has called attention to the potential violence and voyeurism involved in recounting stories like Baartman's. I take her point to heart, but I think it's important to present these events because they are fundamental to understanding the thinking of Cuvier, Blainville, and the others, and because these men— especially Cuvier—have gotten a free pass from generations of historians. I've tried to relate the details respectfully and to avoid re-subjecting Baartman's memory to the humiliation she suffered in her lifetime.

2. There is considerable confusion in the literature regarding who was present with Baartman in the garden. Badou, Honour, and Holmes all say, for instance, that Léon de Wailly was among the artists. The de Wailly who signed the portrait is listed in the archives of the Muséum d'histoire naturelle both as Pierre-François de Wailly and as his son Léon de Wailly. But it must have been the father, who was a painter at the Muséum. The son was eleven years old and later became a writer, not an artist. Nicolas Huet's portrait is in the BMHN archives, Portefeuille 69, folio 2. It is very similar to the portrait signed by Charles Philibert du Saillant, comte de Lasteyrie, in the first volume of *Histoire naturelle des mammifères*. Jean-Baptiste Berré also made several depictions, which are published in Flourens, "Notice," 144.

3. On the superiority of French: Condillac, *Essai sur l'origine*, 264ff.; Diderot, "Lettre sur les sourds et muets"; d'Alembert, *Essai*, 253; Rivarol, *Discours sur l'universalité*. The so-called "quarrel of inversions" began with arguments over whether French or Latin had a more natural and intuitive syntax. See Mah, "Epistemology of the Sentence."

4. Déterville, *Nouveau dictionnaire*. Thanks to Pietro Corsi for help interpreting Déterville's practice.

CHAPTER 23: THE BIRDS AND THE BEES

MAIN PRIMARY SOURCES:

Lamarck, *Histoire naturelle des animaux*, 1:187 (body's response to sights, ideas), 4:55 (male/female bumblebees), 223 (dragonflies), 3:276 ("little ones" find food), 275 (sexing an insect, "press its belly," see also 318), 335 (fleas), 347, 357–58 (egg laying), 459–60 (scale insects); 4:73 (*Andrena* bee), 271 (beetles mating), 6:279–80 (mollusks, snails, slugs sex, Lamarck cites Geoffroy, *Traité*, see 18–19); Lamarck, *Philosophie zoologique*, 1:258–59 (kangaroos), 302 (Hymenoptera, for example, wasps), 2:149–50 (sex/feeling coextensive); Lamarck, *Système des animaux*, 220 (beetles), 224 (glowworms), 263–73 (sawflies); **Lacépède**, *Histoire naturelle de l'homme*, 122–23 ("uterine fury"); **Moreau de Sarthe**, *Histoire naturelle de la femme*, 1:162–63 (voices), 198 (natural state of marriage), 2:672 ("habits and inclinations"), 674–75 (kangaroos), 675–76 (Buffon as model), 679 ("upheaval and storm," mental and emotional capacities), 680 ("uterine influence"), 681 (elephant's trunk); **Buffon**, "Discours sur la nature des oiseaux," 3, 49–50; **de Gouges**, "Réflexions," first page (nature "had no part"); de Gouges, "Déclaration," 5 ("bloated with science"); **Desmarest**, "Kanguroo ou Kangurou," 356.

SECONDARY SOURCES:

Blanc, *Marie-Olympe de Gouges*; **Faucheux**, *Olympe de Gouges*; **Mousset**, *Olympe de Gouges*; **Barlier**, *Société des amis des noirs*.

1. The first European to depict a kangaroo might have been Sydney Parkinson, the botanical and natural historical draftsman aboard Captain James Cook's ship during his first voyage of 1768–71. The ship sailed to Australia and, before dying of dysentery, Parkinson sketched a kangaroo. On the basis of this sketch, George Stubbs made his painting *The Kongouro from New Holland* (1772). See Ashby, "Earliest Strange Creatures."
2. See J. Laskey, "The Kangaroo Described," *Gentleman's Magazine* 79 (1796): 467–69; Smith, *Wonders of Nature*, 126.

CHAPTER 24: DARKNESS

MAIN PRIMARY SOURCES:

PVIF, 6:350 (Aug. 3 and 10, 1818, Desfontaines reports), 474 (Aug. 9, 1819, Lamarck presents sixth volume), 8:290 (counting Lamarck present); **Audouin**, *Histoire des insectes nuisibles*, 13, 110; **Bourdon**, "Lamarck," 343–44 (ink splatter memory); Bourdon, *Principes*, 252–53 ("infinite skill"); Bourdon, *Mémoire sur le vomissement*; Bourdon, *Notions d'hygiène pratique*, 2; **Buchez**, *Traité*, 1:76 (race results from habits), 89 (justification of slavery), 91 (race not invariable); **Lamarck**, *Système analytique*, 42–43 (mistakenly ascribing intentions), 290 ("organic phenomenon"); Lamarck, *Philosophie zoologique*, 2:415–17 (imagination); Lamarck, notes on "physique terrestre," BMHN MS 756, 1v (like being an architect), 3r–v (always interposed hypotheses), 6r (like using a magnifying glass); **PVAP**, AN AJ/15/127 no. 231, July 11, 1828 (last Museum meeting attendance); **PVAS** 6: 474, 9 August 1819 (last Academy of Sciences attendance); **Geoffroy Saint-Hilaire**, "Discours," 210 ("vast emptiness"), 217–18 (Lamarck's children), 219 (government benefits); Geoffroy Saint-Hilaire, *Fragments*, 81 ("I loved and venerated him"); **Bourguin**, "Lamarck," 202 (Latreille's bread remark), 220 (*Waverley* novels); **Cuvier**, "Éloge," xx ("poet"), xxix ("credulous"); **A. Lamarck** to Eugène de Lamarck, June 11, 1865, 378 (criticism of father); **Latreille**, "Discours" (emotional eulogy); **Blainville and Maupied**, *Histoire des sciences*, 3:342 (on Cuvier's eulogy).

ROSALIE'S AND CORNÉLIE'S DEVOTION:

Lamarck, *Système analytique*, 4; **Latreille**, "Discours"; **Cuvier**, "Éloge," xxx; **Bourdon**, "Lamarck," 353–54; **Bourguin**, "Lamarck," 220; **Panckoucke**, "Lamarck," 487.

SECONDARY SOURCES:

Packard, *Lamarck*, 51–57; **Landrieu**, *Lamarck*, 68 (André's death), 98 (Latreille replacement), 100 (Rosalie's help), 101 (Lamarck's burial, daughters' plight), 104–5 (herbarium, finances, estate), 110–11 (Lamarck's chair succession); **Bourdier**, "Esquisse," 38–41 (broken femur: 40); **Duval**, "Le transformiste," 341 (eye strain); **Bange and Corsi**, "Chronologie"; Bange and Corsi, "Les auditeurs": "Jean-Victor Audouin," "Philippe Buchez," "Jean Baptiste Isidore Bourdon"; **Mc-William**, *Dreams of Happiness*, 123ff. (Buchez's Christian socialism); **Simmons**, *Vital Minimum*, 61 (Lamarck and Buchez; "Lamarck's influence showed in the work of progressive social economists"); **Hermanville**, "Notice," 461; **Lewalski**, *Life of John Milton*, 408–12 (Milton and daughters); **Outram**, *Georges Cuvier*, 57, 144, 172, 188 (daughter and stepdaughter); **Rook-maaker**, "Mauled by a Rhinoceros" (Duvaucel); **Pelseneer**, "Premiers temps," 75 (Cuvier's eulogy); **Appel**, *Cuvier-Geoffroy Debate*, 132.

1. *Moniteur universel*, Sept. 9, 1817, 996. Auguste said his brother died on July 18, 1818, on St. Pierre Island off the coast of Newfoundland. A. Lamarck to Cuvier, Feb. 20, 1830, 16. The earlier report must be the more reliable, having been published *before* Auguste said André had died.
2. For example, Henry Fuseli, *Milton Dictating to His Daughter* (1794), and Richard Westall, *Milton Dictating to His Daughters* (1802).
3. Auguste's son, Eugène, became an officer in the marines and died in Saigon at age forty-one, leaving behind a wife and two daughters. His letters, which his widow published in a posthumous volume titled *Letters from a Sailor*, contain a trove of information about the early diplomatic contacts between France and Japan. Auguste's daughter, Louise, would marry Pierre-Jules Callon, a mining engineer and professor at another of the *grandes écoles*, the School of Mines, and an officer of the Legion of Honor. They in turn would have two sons: Eugène, who became a judge at the Court des comptes (the administrative court that conducts audits of public and many private entities); and Georges, another engineer and officer of the Legion of Honor. Eugène gave rise to the family that, in 1957, modified their name to "Callon de Lamarck," rendering visible their evolutionist lineage. E. de Monet de Lamarck, *Lettres*; Geneanet: Guillaume Emmanuel Auguste Monet de Lamarck, Louise Marie Monet de Lamarck; Lamarck, *Lettres*; Geneanet: Guillaume Emmanuel Auguste Monet de Lamarck, Louise Marie Monet de Lamarck; *Journal officiel*, Nov. 7, 1957, 10483.
4. Landrieu (*Lamarck*, 63) notes that in 1817, Lamarck earned 6,200 francs, including 1,200 as a member of the institute, Haüy and Geoffroy Saint-Hilaire 10,200 francs (including 4,000 as professors at the Sorbonne), and Cuvier 41,200 francs (including as vice president of the Council of State). For the average carpenter's wages, see Sauvigny, *Bourbon Restoration*, 251. Bange and Corsi ("Chronologie") report that the estate (books and furniture) was assessed the following June at 12,000 francs.
5. A. Lamarck to Cuvier, Feb. 8, 1830, IF GC MS 3252 / f. 358–59, 1 (Cuvier's refusal to see Auguste), Feb. 20, 1830.
6. Martins, "Introduction biographique," xix.

CHAPTER 25: THE BATTLE OF THE MOLLUSKS

MAIN PRIMARY SOURCES:

Eckermann, *Conversations*, 479–81 (Goethe-Soret dialogue); **Cuvier**, *Recherches* (1824), lvii–lxiv (extinct species not varieties of living ones), 139–41; Cuvier, *Le règne animal*, 1:xxi, 57–60 (embranchments); Cuvier, "Mémoire sur les céphalopodes," in *Mémoires*, 43 ("vain systems"); Cuvier, "Considérations," 2–3 ("expressed my sentiment"), 4 (labyrinth), 9 (since Aristotle), 18 ("enormous differences"); **Geoffroy Saint-Hilaire**, "Recherches," 136–37 (Teleosaurus), 150–51 ("precisely in keeping," recommendation to young people); Geoffroy Saint-Hilaire, "Mémoire où l'on se propose," 210 ("work of six days"), 217 ("learned and venerable colleague"); Geoffroy Saint-Hilaire, *Principes*, 20 (calling a halt), 22–23 ("Animality"), 24 (argument became theological, heated retort), 35–52 (report on Laurencet/Meyranx), 66–67 (crutches), 98–101 (hyoid),

107 ("increasingly expand"), 166 (not facts, interpretation), 184 (Lamarck's strength of mind), 190 ("power and elevation"), quoting Cuvier: 24 (pantheism), 25 (reduce nature to slavery), 65–66 (role animal must play), 142 (jellyfish/elephant), 145 ("crowd of animals"), 188 ("positive facts"); Geoffroy Saint-Hilaire, "Mémoire sur le dégré d'influence," 63 ("ambient world"), 81 ("profound physiologist"); **Bory de Saint Vincent**, *Le moniteur universel*, July 28, 1830, 4 (review of Geoffroy); Bory de Saint-Vincent, "Création," 44 (dodo); **Goethe**, "Dernières pages" (reproducing last two articles, originally in *Revue encyclopédique* and *Annales des sciences naturelles*), part 1, 568–69 (Cuvier/Geoffroy, facts/ideas), part 2, 66–67 (Arago's revolutionary move); **Quinet**, "Discours sur Geoffroy Saint-Hilaire," 340; **Balzac**, *Le père Goriot*, 1 (dedication to Geoffroy); Balzac, "Avant-propos," 18–19 (Humanity/Animality, one animal); **Balzac**, *Louis Lambert*, 142–43 (museum lecture); **Sand**, *Histoire de ma vie*, 9: 204–7 (Sainte-Beuve); **Karénine**, *George Sand*, 1:411 ("presumptuous carelessness"); **Flourens**, *Ontologie naturelle*, 23–26 (species fixity).

SECONDARY SOURCES:

Appel, *Cuvier-Geoffroy Debate*, 1 (Goethe/Soret), 143–74, esp. 147 ("completely mistaken," "cannot find words"), 197 (Raspail quotation); **Pilbeam**, *Constitutional Monarchy* (July Revolution), chaps. 4–5; **Pelseneer**, "Les premiers temps"; **Corsi**, *Age of Lamarck*, 218–29 (Bory); Corsi, "Revolutions of Evolution" (shifting political significance of evolution, Geoffroy); **Richards**, "Foundations of Archetype Theory" (Goethe's transformism); **Jack**, *George Sand*, 159–72.

1. Appel (*Cuvier-Geoffroy Debate*, 131–32), citing Bourdier ("Lamarck et Geoffroy Saint-Hilaire"), writes that Geoffroy rejected both the idea of an inherent tendency toward complexity and the inherited effects of use and disuse. Instead, he believed that new forms were created by the direct action of the environment upon the fetus, causing changes in the size and branching of arteries. This is at odds with what Geoffroy says in "Recherches sur l'organisation des gavials," 150–51, namely that transformation happens "precisely" in accordance with Lamarck's two laws, which he then spells out in his footnote: first, that systematic usage of an organ develops it, and second, that these developments are inherited. Geoffroy also says he thinks certain anatomical variations that Lamarck describes are more due to "changes in the distribution of arteries," and he refers readers to his own work on human monstrosities (Geoffroy, *Philosophie anatomiques*), where he attributes monstrous variations to the direct action of the environment on the fetus. In sum, Geoffroy was equivocal on the sources of variation. Later ("Mémoire sur le dégré d'influence," 81) he developed the idea that the environment causes small changes in the developing fetus that, added together over generations, produce new forms, and he criticized Lamarck's choice of particular proofs showing that animals' actions lead to modifications in the long term: again, clear as mud. It appears that Geoffroy wanted neither to accept nor quite to reject the most notorious of Lamarck's ideas: that animals could transform themselves by acts of will and pass these changes to their offspring.

2. PVAS, 9:414; *Gazette médicale de Paris* 1 (1830), 110; AS, *Pochette de séance* for March 8, 1830.

3. *Journal des débats*, March 2, 1830 ("poetic idea"); March 30, 1830 ("philosophical principle").

Notes

CHAPTER 26: THE BATTLE OF THE BOILED BROTH

MAIN PRIMARY SOURCES:

Pouchet, *Théorie positive de l'ovulation*; Pouchet, *Hétérogénie*, 61–64 (Lamarck); Pouchet, "Note" (microorganisms in hay infusions); **Pasteur**, *Oeuvres*, 7:326 ("spiritualist doctrine"); Pasteur, "Générations," 35–36 ("question of spontaneous generation"), 187–89 (corpuscles in dust), 190–91 ("entire months"), 223 (prize competition), 328–46 (scientific soirée), 328 ("all minds awake," "accessible to experiment," "to combat"), 332 ("What a conquest"), 333 (designer God), 334 ("question of fact," "enough poetry"), 337 ("completely illusory," "no one among you"), 343–45 (Mer de Glace), 342 ("Never will the doctrine"), 346 (blood/urine), 385 ("swan-necked"); Pasteur to Pouchet, Feb. 28, 1859, in "Générations," 629 ("unwittingly," devoid of proof); **Owen**, *On the Anatomy*, 3:814 (Cuvier/Pasteur comparison); **Flourens**, *Examen*, 45, 170.

SECONDARY SOURCES:

Geison, *Private Science*, 11–52, 112–37; **Latour**, "Pasteur et Pouchet"; **Price**, *French Second Empire*; Price, *French Second Republic*; **Séguin**, *Louis Napoléon*, 123–78 (Napoléon III election, coup), 145–46 (Loi Falloux allowing religious instruction); **Englund**, *Napoleon*, 445 (Napoleon II); **Jack**, *George Sand*, 294–309; **Schopp**, *Alexandre Dumas*, 441–70.

1. Royer, "Préface du traducteur," in Darwin, *L'origine des espèces*, v–lxiv. Darwin's friend and correspondent the American botanist Asa Gray published a book of essays making the case for the reconcilability of evolution and religious faith: Gray, *Darwiniana*.
2. Pouchet produced the oxygen artificially to preclude contamination; he doesn't specify how, but it was likely by heating mercury oxide.
3. Sillerud, *Abiogenesis*; Kauffman, *At Home*, chap. 2.

CHAPTER 27: DARWIN'S GRANDFATHER PROBLEMS

MAIN PRIMARY SOURCES:

E. Darwin, *Botanic Garden*, part 1, 92–93 (French Revolution), 107–12 ("Vegetable Glandulation"), part 2, 40–41 (Vallisneria); E. Darwin, *Temple of Nature*, 25–26 ("wondering eyes"), additional notes: 3 ("animalcules"), 44 ("finer links"), 159 ("births and deaths"); E. Darwin, *Zoonomia*, 1:505; **C. Darwin**, *Origin*, 1st ed., 489–90 ("entangled bank"), 3rd ed., xiii ("justly-celebrated"), xiv ("my grandfather"), 135 ("need hardly say"), 6th ed., xiv ("use and disuse"), 3 ("complex and little known"), 8 ("changed habits"), 98 ("not as yet"); Bastian, Beginnings; C. Darwin, "Preliminary notice," 92 ("old men"), 95 ("present generation"); C. Darwin, *Movements and Habits*, 206 ("vaguely asserted," "tendrils ready"); C. Darwin, *Power of Movement*, 1 ("circumnutation"), 573 ("like the brain"); C. Darwin, Notebook B, 1 ("Zoonomia"), 18 ("each species changes"), 78 ("Man is derived"), 216–17 ("Lamarck's 'willing' doctrine"); C. Darwin, *Autobiography*, 77 (Lyell's *Principles*), 43–44 (poetry), 138–39 (novels); C. Darwin, *Variation*, 2nd ed., vol. 2, chap. 27 ("pangenesis"), 399 (little universe); C. Darwin, *Descent*, 613 ("remarkable conclusion"); **DCP**: Darwin to Alfred Russel Wallace, Aug. 28, 1872, vol. 20, no. 8488 (Archebiosis), Oct. 21, 1859, vols. 17, 24, no. 6951 ("transformist"); Darwin to Joseph Hooker, Jan. 11, 1844, vol. 3, no. 729 ("Heaven forfend"); Darwin to Lyell, Oct. 11, 1859, vol. 7, no. 2503

("Creation of Monads"), March 12–13, 1863, vol. 11, no. 4038 ("wretched book"); Lyell to Darwin, Oct. 3, 1859, vol. 7, no. 2501 (Geoffroy and Lamarck); Oct. 4, 1859, vol. 13, no. 3132 ("Lamarck's monads"), March 15, 1863, vol. 11, no. 4041 ("change of name"); March 15, 1863, vol. 11, no. 4041 ("whole orang"); **DP**: Darwin Library and CUL-DAR119, "Books to Be Read/Books Read" (Lamarck's *Natural History*);

SEXUAL SELECTION:

C. Darwin, *Origin*, 6th ed., 69–70; C. Darwin, *Descent*, 162, 236, 316, 337, 353, 371; **Lamarck**, *Système des animaux sans vertèbres*, 16 ("monade," "animate point"); Lamarck, *Histoire naturelle des animaux*, vol. 1: "Supplément," 453–54, 458 (no simple line); **Lyell**, *Principles of Geology*, 2:8 ("internal sentiment," "fictions"), 9 ("unpardonable," "staggering"), 21 ("defective," "fallacious"), 23 ("Author of Nature"), 27 ("semblance"), 30 (ancient Egyptian animals); Lyell, *Geological Evidences*, 388–89 ("thirty years"), 390 ("element of time"), 392 ("thirty or forty centuries"); [**Chambers**], *Vestiges*, 1st ed., 195 ("*os coccyges*"), 204–5 ("*chemico-electric*," simplest forms), 214 ("law of organic development"), 222 ("*type next above it*"), 230–31 (Lamarck).

SECONDARY SOURCES:

Secord, *Victorian Sensation*, 365–79; **Levine**, *Darwin the Writer* (importance of writing to Darwin's thinking and his science); Levine, *Darwin Loves You* (Darwin's "alternative form of enchantment").

1. Wallace, a naturalist and explorer, was a sort of parallel author of the theory of natural selection, having arrived independently at an essentially similar theory of transmutation of species: Wallace, "Tendency of Varieties."
2. In France, Darwin's big book, once published, had a similar, though limited, effect on Lamarck's reputation. The zoologist and anthropologist Jean-Louis-Armand de Quatrefages de Bréau, a correspondent and friendly critic of Darwin's, published *Darwin et ses précurseurs français* in 1870, in which he associated Lamarck closely and respectfully with Darwin, although he ultimately rejected both theories (see, for example, 13, 42–59, 362). A new edition of *Philosophie zoologique* appeared in 1873, edited by Charles Martins, who in the "Introduction biographique" represented Lamarck as unappreciated and maligned. In 1914, the Belgian malacologist Paul Pelseneer referred to an actual resurgence of Lamarckism after 1870, citing both Quatrefages and Martins ("Les premiers temps," 83), but the evidence is sparse.
3. The word "evolution" did not denote Darwin's theory of descent with modification until the sixth edition of the *Origin of Species*, at which point Herbert Spencer had attached it to Darwin's theory in his own writings. Riskin, *Restless Clock*, 221.

CHAPTER 28: HE CUT OFF THEIR TAILS

MAIN PRIMARY SOURCES:

Weismann, *Essays*, 69 ("whole principle ... collapses"), 95–96 (Bernoullis, Gausses, Handels), 98 ("in my opinion"), 423 ("completely abandon"), 425 ("duty"), 447 ("contempt," "no one can be prevented"), 432–33 (mouse experiment), 434 (dogs, cats, sheep); Weismann, *Germ-Plasm* (body cells/germ cells); Weismann, "On the Mechanical Conception," 708 ("directive power"), 710–12 ("Universal Cause," *Final Cause*"), 712 ("behind," absolutely opposed"); **Brown-Séquard**, "Faits nouveaux"; **Poulton**, "Dr. Romanes' Article"; Poulton, "Lamarckism Versus Darwinism"; Lankester, "Functionless Organs" ("pure Darwinism"); **Romanes**, "Lamarckism Versus Darwinism" ("Neo-Darwinian"); **Mayr**, *Growth of Biological Thought*, 535 ("school of selectionism"); **Morgan**, *Critique*, 33 ("common sense").

DCP: Darwin to Charles Lyell, March 25, 1865, vol. 13, no. 4794 ("detestable"), Feb. 25, 1860, vol. 8, no. 2714 ("rubbish"); Darwin to J. D. Hooker, Nov. 3, [1864], vol. 12, no. 4650 (clever/empty), June 23, 1863, vol. 11, no. 4218 ("words & generalities"), Dec. 24, 1866, vol. 14, no. 5321 ("unintelligible"); Hooker to Darwin, Dec. 14, 1866, vol. 14, no. 5305 ("thinking pump").

"INHERITANCE OF ACQUIRED MODIFICATIONS":

Spencer, *Principles*, 1:247 ("inheritance of an acquired peculiarity"), 249 ("inheritance of acquired modifications"); **F. Cuvier**, "Essais," 217 ("que les facultés acquises se propagent par la génération et deviennent héréditaire").

"SURVIVAL OF THE FITTEST":

Spencer, *Principles*, 1:444, 453–54, 457, 468, 474; **Darwin**, *Origin*, 5th ed., 72.

"EVOLUTION":

Spencer, *Essays*, 1:1, 9; Spencer, *Principles*, vol. 1; **Darwin**, *Origin*, 6th ed., 189, 201, 202, 215, 282, 424; Darwin, *Origin*, 1st ed., 302 ("descent with modification").

SECONDARY SOURCES:

Churchill, *August Weismann*, 52–56 (friendship with Haeckel), 325–26; **Celestin**, *Charles-Édouard Brown-Séquard*, 143, 163, 197–207, 214; **Maienschein**, "Epigenesis and Preformationism" (embryological meaning of "evolution"); **Richards**, *Tragic Sense of Life* (Haeckel's Lamarckian Darwinism).

GERMAN RESEARCH UNIVERSITIES:

Clark, *Academic Charisma*; **Howard**, *Protestant Theology*, 4–5; Ringer, *German Mandarins*, 102–27; **Lenoir**, *Instituting Science*; **Turner**, "Humboldt in North America," esp. 292–93; **Riskin**, *Restless Clock*, 253–57.

1. See, for example, Mukherjee, *Gene*, 72; Zimmer, *She Has Her Mother's Laugh*, 57.
2. The first to use the term "Weismann barrier" might have been Arthur Koestler in *The Case of the Midwife Toad*, 126, 128, 131.
3. "Evolution" appears once in Darwin, *Variation* (1868), 2:35. Previously, "evolution" had designated a theory of embryology in which organisms "evolved" from preexisting homunculi in the egg or sperm.
4. Ward (*Neo-Darwinism and Neo-Lamarckism*, 51) refers to this as the first use of "neo-Darwinian" or any of its forms. Romanes collaborated with Darwin on a paper titled "On New Varieties of the Sugar-Cane Produced by Planting in Apposition," which was read at the Linnaean Society a fortnight after Darwin's death. See Darwin to Romanes, Jan. 6, 1882, DCP, vol. 30, no. 13600, n1.

Notes

CHAPTER 29: INSIDE MAMMOTH CAVE

....

MAIN PRIMARY SOURCES:

Packard, *Lamarck*, 4 (childhood house visit), 42 (house photograph), 396 ("neo-Lamarckism), 400–401 (American environments); Packard, "Century's Progress," 597 ("distinctively American," triumvirate with Hyatt, Cope); Packard, "introduction to *Standard Natural History*," 1:lii (Cuvier, Agassiz, "popular prejudice"), lxvi (Lamarckian school), liv ("neo-Lamarckianism," "In other words"); **Packard and Putnam**, *The Mammoth Cave*, 6–8 (entry into cave, first thirteen hours, Mammoth Dome), 9–10 (Snowball room), 11 (worms somewhere), 12–13 (cave camel crickets), 16–17 (rapid myriopod, blind crawfish), 19 (blind crustaceans and *Campodea*), 26–28 (cave animals modified forms of outside animals), 26 (natural selection too slow); **Randolph**, *Mammoth Cave*, 19 (Putnam's changed direction); **Wright**, *Guide Manual*, 7–8 (Bransfords); **Anon.**, "Introductory," *American Naturalist* 1, no. 1 (1867): 3 ("followers of Lamarck"); **Cope**, "Progress," 218 ("first presented," "adverse influence," "resuscitated"); Cope, *Origin of the Fittest*, 147 (higher and lower races), 161 (women childlike), 168 (perfect humanity), 171 ("Divine Spirit"); Cope, "Two Perils"; **Ward**, *"Neo-Darwinism and Neo-Lamarckism,"* 12, 53, 65 ("Neo-Darwinian song"); Ward, *Pure Sociology*, 132 (races), 392 (romantic love), 449 ("gynaecocratic"); Ward, *Lester Frank Ward and the Welfare State* (progressive politics); **Weismann**, *Essays*, 448n296 ("distinguished American biologists"); **Galton**, *Inquiries*, 24–25 ("cultivation of race," "eugenics"), 332 ("preponderating"); Galton, *Hereditary Genius*, 14 ("no patience," "pretensions"), 32 ("mental capacity"); **Baldwin**, "New Factor in Evolution," 443 ("young children," "conscious agency"), 444 ("reasoning"), 450 ("intelligent"); Baldwin, "New Factor in Evolution, Continued," 552 ("direct substitute"), 536–37 ("social heredity"), 547 ("Evolution . . . not more biological than psychological"); Baldwin, "Autobiography," §2 ¶1 ("problems of genesis," "focus"); Baldwin, *Mental Development*, 344–47 (Helen and Elizabeth); Baldwin, *Darwin and the Humanities*, 42 ("By the use of the brain, the organism becomes the instrument of mind"); **Elliot**, "introduction to *Zoological Philosophy*," lxxxviii ("final state of oblivion").

NATURAL SELECTION, LAMARCKISM AND SOURCE OF VARIATIONS:

Packard, *Lamarck*, 339, 350–51, 382–83, 391–92, 398–99; **Cope**, *Origin of the Fittest*, 405–6, 422–23.

SECONDARY SOURCES:

Numbers, *Darwinism*, 34 (American neo-Lamarckism); **Pfeifer**, "Genesis," 161 (Packard coined "Neo-Lamarckism"); **Cockerell**, "Biographical Memoir of Alpheus Spring Packard," 197 (Agassiz), 201 (Hyatt, Cope); **Mead**, "Alpheus Spring Packard," 44–47 (Agassiz); **Tozzer**, "Frederic Ward Putnam," 125–26, 128, 130 (Agassiz); **Winterer**, *How the New World Became Old* (the development of geologic deep time in America); **Lyons**, *Making Their Mark*, 21 (Bottomless Pit); **Gould**, *Mismeasure*, 106–9; Porter, *Rise of Statistical Thinking*, 128–46; **Kevles**, *In the Name of Eugenics*; **Rafferty**, *Apostle of Human Progress* (Ward); **Lanham**, *Bone Hunters* (Cope); **Weber and Depew**, "preface to" *Evolution and Learning* (Baldwin effect); **Radick**, "Animal Agency"; Radick, "Baldwin Effect" (Baldwin and interpretations of Darwin); **Cahan**, "Genetic Psychologies" (Baldwin and Piaget); **Kingsley**, "Edward Sylvester Morse (1838–1925)," 549, and **Benson**, "Why American Marine Stations?," 10 (Packard, Hyatt, Morse, and Putnam meeting).

1. [Packard], "introduction "to *Standard Natural History*, 1:liv; Ward (*Neo-Darwinism and Neo-Lamarckism*, 53) refers to this introduction as written by Packard and to this use of "neo-Lamarckianism" as the first.
2. In addition to being a creationist, Agassiz was a polygenist, meaning that he believed the races of human to be distinct species: "Sketch," lxxiv–lxxv.
3. The initial name was the Society of Naturalists of the Eastern United States; the society adopted its current name in 1886.
4. Cockerell, "Biographical Memoir," 201; see also Kellogg, *Darwinism Today*, 134.

CHAPTER 30: THE GOSPEL ACCORDING TO SAINT WILLIAM

MAIN PRIMARY SOURCES:

Weismann, *Germ-Plasm*, 40–42 ("biophors"), 69 (vital particles), 75–77 ("idants," "ids"), 57 ("determinants"), 413 ("inexhaustible"), 415–17 (external influence, nutrition); Weismann, *Essays*, 375–76 ("chromatin"); **De Vries**, *Intracellular Pangenesis*, 6–7 ("intracellular pangenesis"), 49 ("pangen," English translation of "pangene"), 19, 225 (mosaic); **Darwin**, *Variation*, 2nd ed., 2:23 (mosaic); **Mendel**, *Experiments in Plant Hybridizations*, 313 ("hybrid"), 318 (garden beds, pots), 320 ("dominant," "recessive"), 325 (mathematical ratios), 332 ("factors"); **De Vries**, "Sur la loi de disjonction," 846 (rounding numbers), 847 (equal chances); **Correns**, "G. Mendel's Law," 42–44 (numerical ratios), 46 ("chance"); **W. Bateson**, "Problems of Heredity" (published lecture), 56–57 (Galton), 60; W. Bateson, *Mendel's Principles of Heredity*, 15 (random), v–vi (preaching the gospel); W. Bateson, "Progress of Genetic Research," 91 ("genetics"); W. Bateson, "Common-Sense," 371 (genetics, eugenics, nature of offspring), 372 (defiance of common sense), 373 (feebleminded), 380–81 (education), 383 (racial discrimination), 384 (born unequal), 387 (equality of political power); **B. Bateson**, *William Bateson*, vi, 203 (scientific Calvinism), 73 (discovery of Mendel on train); **Galton**, "Average Contribution"; Galton, "Diagram of Heredity"; Acts 19:24–40 (Demetrius); **Weldon**, "On the Ambiguity" 2 (reservations); **Johannsen**, "Genotype Conception," 130, 132, 133, 159 ("gene"); **Wheeler**, "On Instincts," 303 ("mortal sin"); **Stresemann**, *Ornithology*, 330–31; **Mayr**, *Toward a New Philosophy*, 7 (not testable, philosophy), 27 (Lamarck and Darwin wrong, Weismann right), 30–31 ("programmed," "bird that starts," "no more nor less"), 40 ("subhuman"); Mayr, *Growth*, 342, 352, 359 (Lamarck's courage); Mayr, *Animal Species*, 602, 604–6, 634 (importance of behavior), 108, 404, 495 (genetic basis of behavior), 594 (genetic basis not deterministic), 636 (humans and higher mammals capable of learning); Mayr, "Emergence of Evolutionary Novelties," 352 ("new habits and patterns of behavior"), 371–72; Mayr, *What Makes Biology Unique*, 57 (mistaken), 61 (goal coded); *American Biology Teacher* 54, no. 1-8 (1992): 413 (Mayr: "I was a Lamarckian").

SECONDARY SOURCES:

Churchill, *August Weismann*, 346, 512, 644n53 (meiosis); **Riskin**, *Restless Clock*, 266, 445n71 ("chromosome"); **Gayon**, "From Mendel to Epigenetics," §2, esp. 226; **Keller**, *Century of the Gene*, 1–3, 112–15, 136; **Radick**, *Disputed Inheritance*, 1 ("Scientific Calvinism"), 2, 126–34, 127 (de Vries, Correns), 141–42 (Bateson discovery of Mendel); **Kampourakis**, *Making Sense of Genes*, 1–10, 109–10, 128; **Holmes**, *Graft Hybrid*; **Charlton**, *Understanding Gregory Bateson*, 13 (Bible-reading atheist, son's name); **Provine**, *Origins of Theoretical Population Genetics*, 51–52, 70–89; **Haffer**, *Ornithology* (Mayr).

1. "Meiosis" was the 1905 coinage of a botanist, John Bretland Farmer, and a biologist, J. E. S. Moore. In Weismann's terminology, "germ plasm" is made of nuclear-rod "idants," which are each made of several "ids," which are in turn made of hundreds or thousands of "determinants," which are made of the actual bearers of heredity, characters, and qualities: "biophors." The biophors are of various kinds, and each corresponds to a different part of the cell.

2. Johannsen had first proposed the term in *Elemente der exakten Erblichkeitslehre* (Jena: Gustav Fischer, 1909), iv.

3. See, for example, Carter and Eckerman, "Symbolic Matching"; Scott, *Animal Behavior*, 193; Kamil and Balda, "Cache Recovery and Spatial Memory"; Jones and Kamil, "Tool-Making and Tool-Using"; Burton, Burton, and Taylor, *Bird Behavior*, 58; Millikan and Bowman, "Observations"; Kroodsma, *Acoustic Communication*.

CHAPTER 31: THE SCANDAL OF THE MIDWIFE TOAD

MAIN PRIMARY SOURCES:

Kammerer, *Inheritance of Acquired Characteristics*, 14 (geneticians, "rigid dogmatism"), 16–17 (Einstein, the "Great War," racial boundaries), 51–55 (midwife toad experiments), 88–103 (salamanders), 99 ("follows the father"), 261–63 (socialism), 265 (Darwinism and humaneness), 266–67, 271 (race mixing), 269 (craniology), 281 (goodwill), 304 (prevent births in prison), 343 (environment and inheritance), 353–57 (productive eugenics), 359 (one human race, not a single aggressive race); "Paul Kammerer's Letter to the Moscow Academy of Sciences"; **Darwin** to Romanes, Jan. 1, 1882, DCP, vol. 30, no. 13592 (Brazilian sugarcane), March 29, 1882, DCP, vol. 30, no. 13748F (grafting sugarcane; bears "a note in Romanes's hand, 'My last letter from Darwin'"); **W. Bateson**, "Dr. Kammerer's Testimony," 344–45 ("incontrovertible"); W. Bateson, *Problems of Genetics*, 188–90 (almost universal), 211 (mistakes of observation), 234 (skepticism); **Shaw**, *Back to Methusaleh*, xxi ("Neo-Lamarckian"), liii ("Lamarck-Shavian invective"); **Einstein** to Heinrich Zangger, Feb. 27, 1920, in *Collected Papers*, 9:277–78; Noble, "Kammerer's Alytes."

GRAFT HYBRIDS:

Darwin, *Variation*, 1:413–27, 2nd ed., 2:360 (Romanes's idea of "graft hybrids as evidence for pangenesis); **Romanes** to Darwin, Jan. 14, 1875, DCP, vol. 23, no. 9816, July 12, [1875], DCP, vol. 23, no. 10059; **Kammerer**, *Inheritance of Acquired Characteristics*, 137–49.

SECONDARY SOURCES:

Koestler, *Case of the Midwife Toad*, 16–17, 21, 74 (Kammerer's military service), 81 (trip to England), 114, 124 (Kammerer was framed), 139, 142 (Einstein), 117–22 (suicide), 122 (Przibram's loyalty and fate); **Taschwer**, *Case of Paul Kammerer*, 526, 535; **Gliboff**, "Case of Paul Kammerer," 527 (*Dummkopf*), 528 (pad wasn't fabricated), 529 (Kammerer's style), 530–31 (origins of Vivarium, Kammerer's position), 539 (socialism), 542–43 (nuptial pad atavism possibility), 545–46 (antisemitism, "sticky Jew"), 546–49 (Bateson, evidence issues, Baur), 551 (World War I at Vivarium), 552–56 (England, Moscow), 554–55 (Kammerer might have injected the ink long before); **Holmes**, *Graft Hybrid*, 6, 80–88, 189, 191; **Vargas**, "Did Paul Kammerer Discover Epigenetic Inheritance?" 668–69, 675; **James A. Secord**, "Introduction," in *Correspondence of Charles Darwin*, 30:xvii–xxxi, on xviii (experiments with Romanes); **Logan**, *Hormones*,

Heredity, and Race (political association of Lamarckism with Jewish science in interwar Vienna); **Brauer,** "'L'art eugenique'" (Lamarckian eugenics late nineteenth, early twentieth centuries, environment and experience); **Gould,** "Zealous Advocates," 625 (latent nuptial pads possible recurrence).

1. A resounding silence echoes through Darwin's letter to Romanes, written six years before the abolition of slavery in Brazil: the absence of any reference to slavery. Darwin had witnessed Brazilian slavery firsthand in 1836, near the end of the *Beagle* voyage. In his *Journal* (499–500), he'd recorded his revulsion at the "heart-sickening atrocities" he saw there: "On the 19th of August we finally left the shores of Brazil. I thank God, I shall never again visit a slave-country." Nevertheless, when it came to his exchange with Romanes regarding the experiments on sugarcane, if Darwin gave any thought to the people harvesting it, he kept it to himself.

CHAPTER 32: THE BIOLOGIST WHO JUST COULDN'T SEEM TO COME IN FROM THE COLD

MAIN PRIMARY SOURCES:

Lysenko, *Science of Biology Today,* 32, 60 ("wrest" favors, quoting the biologist Ivan Michurin); **Zhou and Liu,** "Hybridization by Grafting" (return of graft hybridization); Osborn, *Impressions,* 79 ("Darwin's bulldog"); **T. H. Huxley,** "Reception," 551 ("How extremely stupid"); T. H. Huxley, "On the Hypothesis"; T. H. Huxley to Darwin, Nov. 23, 1859, in DCP, vol. 7, no. 2544 (stake, claws, beak); **Darwin** to John Collier, Feb. 16, 1882, in DCP, vol. 30, no. 13689 (challenge to a duel); **J. Huxley,** *Evolution,* 18 (Weismann), 22 (eclipse, random mutations, see also 54, random "with regard to evolution"), 23 (Bateson, particulate), 24 (large mutations), 26 (unification of biology around Mendelianism, rebirth of Darwinism), 28 (phoenix), 47 (particulate), 458 (repudiation of Lamarckism, literary men), 465; J. Huxley, *New Bottles,* 13 ("managing director," "transhumanism"), 17 ("transcend"), 64–65 (Weismann's germ plasm versus soma remains essential), 306 (urgency of eugenic policy); J. Huxley, *Essays,* 1–2 (blind force), 8 (machinery of heredity), 28 (facts indifferent to desires); J. Huxley, *If I Were Dictator,* 15; J. Huxley, "Case for Eugenics," 287 ("rough and ready"); J. Huxley, "Humanist Frame," 24 (free but unequal); J. Huxley, *UNESCO,* 21 ("biological inequality"), 38 ("truly scientific eugenics"); J. Huxley, *Soviet Genetics and World Science,* 185–87 (Communism and Lamarckianism); **J. Huxley and Haddon,** *We Europeans,* 216 ("pseudo-scientific"), 236 ("dangerous myth"); **Bateson,** "Heredity and Variation," 222 (resisting chromosome theory); Bateson, "Presidential Address to the Zoological Section," 238 (mutations); **Dawkins,** *Extended Phenotype,* 26 ("dogmatic and hysterical"), 172 ("devastate"); Dawkins, *Selfish Gene,* xxi ("vehicles"); **Dennett,** *Darwin's Dangerous Idea,* 320; **Fisher,** *Genetical Theory,* 20–21 ("particulate"), 227 ("inverted birth rate"), 261–63 (family allowances); **Dobzhansky,** *Mankind,* 13–14 ("sound core" of eugenics); **Mayr,** "Where Are We?," 6 ("improvident moron"); Mayr, "Biology of Race," 94 ("underlying"); **Wilson,** *On Human Nature,* xv–xvi ("necktie-free," "Marxist ideology"), 79, 219 (Lamarck wrong); Wilson, *Sociobiology,* 562 ("biologicized"), 564 (ethical genetic divergence), 575 ("genetic basis"); **MacBride,** *Study of Heredity,* 243–50; **Haldane,** *Heredity and Politics,* 92 (eugenic policies assume anti-Darwinian understanding of "fitness"), 155 (contradictory to believe in fixity/degeneration), 175–76 (some aspects of "eugenic programme" vehicles for racial and class prejudices); Haldane, "Mathematical

Notes

Darwinism" (doubts practicability of eugenic policies); Haldane, *Everything Has a History*, 286 (questions equation of economic success with biological value); Haldane, *Inequality of Man*, 15–16 ("problem of the American Negro"), 46–47 ("innate human diversity"); **Zirkle**, *Evolution* (ideological association of Lamarckism with Marxism); **Gould**, "Pleasures of Pluralism" (against modern synthesis); **Lewontin**, *Triple Helix*, 42 (Lamarck wrong), 48, 126 (against selectionism, genetic determinism); **Schrödinger**, *What Is Life?*, 69 ("characteristic feature"); Schrödinger, "Mind and Matter," 113 (*"used organ,"* "as if Lamarck were right").

SECONDARY SOURCES:

DeJong-Lambert, *Cold War Politics*, chaps. 3 and 4; **Gordin**, "Lysenkoism"; **Joravsky**, *Lysenko Affair*, chaps. 7 and 8; **Graham**, *Lysenko's Ghost*, 17–19 (Lamarck), 70–79 (Vavilov), 96 (graft hybridization); **Levins and Lewontin**, "Problem of Lysenkoism"; **Gissis and Jablonka**, *Transformations of Lamarckism*, 84–86 (return of graft hybridization); **Holmes**, *Graft-Hybrid*, 18–23; **Smocovitis**, *Unifying Biology* (the modern synthesis); **Bashford**, *Huxleys*, 1, 310–65 (Julian Huxley, eugenics); **Yates and Mather**, "Ronald Aylmer Fisher"; **Weindling**, "Julian Huxley and the Continuity of Eugenics"; **Adams**, *Evolution of Theodosius Dobzhansky*, 223 (American Eugenics Society); **Borrello and Sepkoski**, "Ideology as Biology"; **Bowler**, "E. W. MacBride's Lamarckian Eugenics"; **Radick**, "Animal Agency," 43–44 (MacBride); **Schneider**, "Toward the Improvement" (French Lamarckian eugenics); **Lecourt**, *Proletarian Science* (politics of Lysenko in the West).

1. UNESCO, *Race Concept*, 10, 15–16 (statement), 27, 56 (Fisher's objection). The 1951 statement was a revision of an earlier one from the previous year; the book presents both the revised statement and the criticisms of the original one.
2. Bashford (*Huxleys*, 1, 321) writes that Huxley was an "early driver of Unesco's anti-racist 'Statements on Race,'" which were released after he left, but the exact nature of his involvement seems unclear.
3. "Human racial variation in behavior" is from referee comments on a paper by J. Philippe Rushton for the journal *Ethology and Sociobiology*: Borrello and Sepkoski, "Ideology as Biology."
4. Haldane was a Communist and admirer of the Soviet Union until after World War II, when he left the party. See Gordin, "Lysenkoism"; and Subramanian, *Dominant Character*. He also promoted the Lamarckian idea, though without mentioning Lamarck, that life originated in spontaneous generation in the primordial soup: "Origin of Life."
5. Dobzhansky (*Mankind*, 16, see also 4, 21, 139) also warned that Lamarckism, though it should have been "a dead issue," had been revived as Lysenkoism.

EPILOGUE: THE LIFE-MADE WORLD

MAIN PRIMARY SOURCES:

Lala [formerly Laland] et al., *Evolution Evolving*, 3 (Mojave woodrats), 4–8 (whales), 5–6 (methylation, mice), 8 (tit egg swap), 53 (histone modifications), 148 (four categories), 151 (RNA effects, epigenetic inheritance summary), 157 (microbiome), 161–62 (animal cultures); **Lala et al.**, "Extended Evolutionary Synthesis"; **Pigliucci and Müller**, *Evolution*; **Slagsvold and Wiebe**, "Social Learning in Birds," 366, 969–77; **Carey**, *Epigenetics Revolution*, 305 (caterpillar/

butterfly); **Darwin**, *Journal*, 380 ("perfect gradation"); Darwin, *Formation of Vegetable Mould*, 172 (moles, larvae, insects), 312–13 ("degree of intelligence," "When we behold"); **McNew et al.**, "Epigenetic Variation" (epigenetics of Darwin's finches); **Dias and Ressler**, "Parental Olfactory Experience" (mouse olfactory experiments); **Yehuda**, "How Parents' Trauma" (intergenerational trauma); **Veilleux et al.**, "Molecular Processes" (damselfish); **Liew et al.**, "Intergenerational" (coral); **Weyrich et al.**, "Environmental Change-Dependent Inherited Epigenetic Response" (guinea pigs); **Lea et al.**, "Resource Base" (baboons); **Waddington**, "Canalization," 563 ("statistically minded"); Waddington, *Strategy*, 29–30 ("epigenetic landscape"), 78 ("Naturally," "acquired characters"), 151 ("extremist"); Waddington, "Epigenotype," 10; Waddington, *Behind Appearance* (science and painting); Waddington, *Scientific Attitude*, 33 ("full of passion"); Waddington, *Science and Ethics*, 92–93 ("cannot shelve . . . ethics"); **Johannsen**, "Genotype Conception," 132–34; **Cowles and Bogert**, "Preliminary Study"; **Bogert**, "Thermoregulation"; **Muñoz**, *Bogert Effect*, 49 ("major architects"), 61 ("mainstream"); **Pennisi**, "Natural Inspiration" (interview with Muñoz: "nimble," not "passive vessels"); **Huey, Herz, and Sinervo**, "Behavioral Drive"; **Grant and Grant**, *40 Years*, 291 ("interbreed," "behavioral barrier" "culturally inherited"), 299 ("eco-behavioral-genetic"); **West-Eberhard**, *Developmental Plasticity*, 20, see also 11; **Whitehead and Rendell**, *Cultural Lives of Whales and Dolphins*; **Odling-Smee**, "Niche Constructing Phenotypes," 76–77 (woodpecker finch); Odling-Smee, *Niche Construction*, 289 ("eco-evo-devo"), 374 ("tunneling, exuding"); **Lewontin**, "Organism and Environment," 160–61; Lewontin, *Triple Helix*, 48, 68 (niche construction); "**Coast Redwood (Sequoia sempervirens)**," accessed Dec. 14, 2024, ucanr.edu/sites/forestry/Ecology/Identification/Coast_Redwood_Sequoia_sempervirens_198; **Lamarck**, *Philosophie zoologique*, 2:101 ("powerful cause"), 127 ("gradually composed").

SECONDARY SOURCES:

Ward, *Lamarck's Revenge* (epigenetic inheritance); **Carey**, *Epigenetics Revolution*, 19–20; **Keller**, *Century of the Gene*, 77–80, 117–20.

1. Lamarck, *Système analytique*, 154–55. Thanks to Pietro Corsi for drawing my attention to this passage, which also appears in Lamarck, "Homme," 270–71. See also Burkhardt, *Spirit of System*, 214; Jordanova, *Lamarck*, 58–70.

INDEX

Note: Italicized page numbers indicate material in photographs or illustrations.

Index